The Business of Water

A Concise Overview of Challenges and
Opportunities in the Water Market

(A Compilation of Recent Articles from *Journal AWWA*)

THE BUSINESS OF WATER
A Concise Overview of Challenges and Opportunities in the Water Market
(A Compilation of Recent Articles from *Journal AWWA*)
Edited and with an Introduction by Steve Maxwell

Project Manager and Senior Technical Editor: Melissa Valentine
Production: Melanie Schiff

Library of Congress Cataloging-in-Publication Data has been applied for.

Printed in the United States of America
American Water Works Association
6666 West Quincy Avenue
Denver, CO 80235

ISBN 1-58321-556-5
978-1-58321-556-2

Table of Contents

This page intentionally blank.

A Chinese philosopher once said, "Water is good; it benefits all things and does not compete with them." Hundreds of years later, another philosopher, Mark Twain, stated, "Whiskey's for drinking, water's for fighting." So, who's right? Actually both gurus are correct—this seeming incongruence is an indication of the complexity of water issues.

Water is an essential element of life. Throughout history the ebb and flow of cultures can be traced to the availability of water. The quality of life in any area is, in large part, determined by water resources and how they are managed. Water, particularly the lack of it, has resulted in major legal battles as well as physical conflicts.

Although much has been written about regional and global water issues, there really hasn't been a thoughtful look at the actual business of water. The commercial side of water is going to play an increasingly important role in how we ultimately manage water. That is why I am excited about this book. It provides the first comprehensive look at the current status of the water industry, what some future scenarios might be, and what we need to do to be effective stewards of our critical water resources.

I met Steve Maxwell at an AWWA meeting and was immediately impressed by his knowledge of what was going on with the various aspects of the water business and implications for the future. Because of his excellent background in helping organizations devise business strategies specifically to address water issues, I convinced him to write regular articles on this topic for the *Journal AWWA* so that our readers could benefit from his insights. This book encompasses Steve's insights, as well as those of various other contributors to *JAWWA*, in one place.

This collection of articles reflect the opinions of the individual authors. But, I believe they highlight the critical issues that we need to be discussing to insure the best future for water.

I know you will enjoy this book.

Jack W. Hoffbuhr
Executive Director
AWWA
January 2008

This page intentionally blank.

Introductory Overview

BY STEVE MAXWELL

It is time to develop a new mind-set with respect to the impending world water crisis. There simply is no substance more critical to life than water—we cannot survive without it for more than a few days. Ultrapure, processed water has advanced industrial economy and increased the standards of living for much of the world's people. Modern irrigation techniques have provided food for much of the world's expanding population and transformed deserts into productive agricultural oases or major metropolitan areas. Yet we continue to deplete and pollute our limited water resources at an alarming rate. Simultaneously, we are allowing our water treatment and distribution infrastructure to crumble. The twin challenges of water quantity and water quality represent an inexorable planetary crisis—and unfortunately, one that is still not evident to much of the general public and many policy makers.

At the same time, the commercial water industry continues to undergo rapid and tumultuous change. Given the dramatic growth in regulation, extensive ownership changes, increasing public concerns, and inevitable increases in water prices, change is likely to be the one constant in this business into the foreseeable future. In this compilation of important recent articles from the *Journal AWWA,* the key trends and drivers (summarized in the Table 1-1) behind the estimated $400 billion world market will be summarized, and the critical issues and challenges that shape the future of the water business will be concisely discussed.

Regretfully, much of our population still seems to believe that water should be essentially free, forgetting that it costs money—and lots of it—to collect, clean, store, and distribute water. Many of our treatment plants and distribution pipelines were built 50 to a 100 years ago and are rapidly decaying, with leakage rates as high as 50 percent in some older cities. More ominously, many of our underground and surface water sources are irreversibly contaminated or are drying up from decades of overuse. Yet, public officials often seem to be rewarded for keeping spending down, not for insuring that their communities will have vital water resources in the future. City councils strongly prefer to avoid raising rates, even though large *percentage* increases in water rates would amount to no more than $10 or $15 per month for the majority of the population. **(See Article 1.)**

The primary reason for the inattention to this growing crisis is that water remains very cheap relative to its real value. **(See Articles 2 and 3.)** Americans currently pay an average of a quarter of a penny per gallon for the clean drinking water that magically comes from our taps—about $20 to $30 per month for the typical family. There is not another product whose real value so far exceeds its price—or for that matter, one whose price is often so unrelated to its real cost of delivery. Eventually, we must all pay to correct the water pollution problems that we have created, and we must rebuild the infrastructure that we have allowed to decay—huge replacement costs that current water prices do not properly reflect. We are facing serious water problems in the near future, and we need to take more dramatic steps now to begin aggressively attacking these problems.

Fortunately, as each year passes, more countries, institutions, and individuals are realizing that this is not just a water problem but an impending water crisis. As the articles in this collection will demonstrate, the challenges are perhaps becoming clearer and better understood, but the solutions are still elusive—and in too many cases, we still aren't doing enough to really address or begin to correct the problems. And as we increasingly understand the phenomenon of global warming, it is becoming clear that this phenomenon may affect rainfall patterns and water availability in complex ways not yet truly understood—but which will certainly exacerbate an already dire water situation.

Viewed from a more positive perspective, however, as water problems grow, so do opportunities for the broader commercial water industry to provide better and faster solutions. These opportunities and challenges, and the forces that are shaping the commercial water industry, are the focus of the articles in this compilation.

THE GLOBAL WATER PROBLEM: HOW BAD IS IT?

Over the course of human development, and particularly during the last several hundred years, man has severely altered and damaged the earth's natural hydrologic cycle. The water dilemma in which we find ourselves today has resulted from centuries of unfettered industrial *expansion*, exploding population *growth*, and population *shifts* to more arid regions. Perhaps worst of all is our nonchalant, uninformed,

TABLE 1-1   The water market "in a nutshell"

Key Drivers Behind the Market

- water quality and water scarcity problems are truly reaching crisis proportions, worldwide
- more significantly, public awareness and understanding of water problems is increasing
- regulation and enforcement are increasing, and new policies and approaches are emerging
- huge economic (and human) capital investments are required—and much more focus is needed

Resultant Trends and Developments

- outsourcing or privatization continues but remains controversial, particularly in the US
- "musical chairs" in the industry—ownership rearrangement and consolidation continues
- a strong surge of public and private equity investment interest in the industry
- greater efficiency—more focus on *reducing demand to increase supply*
- increased focus on water recycling and reuse—technologies and attitudes
- incremental but continuing technological advance will help address some of the problems
- consumers and residential users increasingly vote according to their "pocketbooks"
- new paradoxes and new questions continue to emerge and confuse the situation

Inescapable Conclusions

- water prices bear little relationship to the actual cost of delivery or the true value of water
- interest in water may be high—but mechanisms still lacking to connect dollars with needs
- delivered water prices must—and will—rise to higher and higher levels
- new ways must be found to manage water as an economic commodity, while providing it for all

Source: Tech*KNOWLEDGE*y Strategic Group.

and unfortunately continuing belief that our standard of living can continue to increase forever, while all the accompanying environmental impacts will somehow take care of themselves.

The impact on water resources comes from all corners of our modern society. Our industrial economy has caused vast chemical pollution of most of the world's natural waterways; tens of thousands of dams have changed the course of natural waterways for flood control and irrigation purposes; extensive irrigation programs have turned arid regions into arable farmland; and double-cropping practices to maximize agricultural yield have depleted natural aquifers at accelerated rates. Perhaps most significantly, the increasing combustion of fossil fuels is inexorably leading to the warming of the planet—and we are only beginning to understand the long-term effects of climate change on the hydrologic cycle. **(See Article 4.)**

The past decade has seen a plethora of reports and "doomsday" forecasts for future water availability pour forth from the federal government, trade associations, and international think tanks. And the prognosis continues to worsen. For example, a recent World Bank report suggested that India will essentially run out of sufficient water within two decades. In 2005, the USEPA produced perhaps the most widely cited study in terms of the

US situation—the *Drinking Water Infrastructure Needs Survey and Assessment*—that called for the investment of $277 billion over the next twenty years, simply to maintain our domestic drinking water infrastructure at acceptable levels. Stories about water quality and quantity problems appear increasingly on the front pages of the mainstream media as well.

To illustrate the depth and magnitude of the world's water problems, consider the following alarming facts:

■   Many of the world's cities still dump their untreated sewage directly into the natural waterways or the ocean. A visit to coastal mega-cities such as Lagos, Nigeria, or Sao Paolo, Brazil, makes one truly marvel at the natural treatment capacity of our oceans; given the amount of raw sewage that is discharged directly into the oceans, it is remarkable that they are still relatively clean in many areas. Coastal mega-cities pollute beaches and waterways destroying local flora and fauna, not to mention economic opportunities and recreational activities.

■   The United Nations estimates that more than 10 million people per year die from consuming unsanitary water. This is a difficult statistic to accurately measure, because so many diseases are transmitted through unsanitary water conditions. However, there can be no doubt that millions continue to die as a direct result of contaminated water.

■ Two and a half billion people worldwide—almost 40 percent of the world's population—have no access to basic sanitation.

■ 45,000 dams around the world are estimated to have displaced approximately 80 million people. The vast, recent dam constructions in China are only the most visible of such projects. The ecological or social impacts of these large dams are long lasting, while their economically useful lives may be relatively short.

■ There are approximately 79,000 dams in the U.S. that are categorized as "large"—many of which are no longer functional or safe. It has become apparent that it will cost less to remove them rather than to try to fix them. However, dam decommissioning and removal is rarely performed. Only about 500, mostly much smaller, dams have actually been removed.

■ Natural wetlands—ecologically designed to regulate and clean our surface waterways—are being destroyed at an alarming rate.

■ Aquifers worldwide are being depleted at a much higher rate than natural processes can replenish them—and the long-term impact for agriculture, world food supply and international trade balances is potentially catastrophic. We "eat" much more water than we drink. The water content of different types of food may indicate the need for radically different agricultural and international trade patterns in the future. (Indeed, this concept of *virtual water* and comparative advantage in agriculture—growing more water-intensive crops in wetter regions and more water-efficient crops in drier regions—is a macroeconomic concept of growing interest, but it would require massive shifts in international trade.)

■ There are also numerous secondary effects of growing water shortages. For example, as aquifers run dry, not only do inhabitants above run out of water, but the ground they live on may also begin to collapse. This is becoming a significant problem in certain cities around the world.

■ As a grim summary of the various points previously listed, it is now widely predicted that about half of the world's population will live with chronic water shortages by the year 2050.

In short, we are rapidly creating a situation of severe *water stress* in many parts of the world. And as a result, water stress will inevitably lead to political stress. As opposed to other broad environmental problems like air pollution and global warming, water problems—and solutions—tend to be regional in nature, and in many cases are specific to individual watershed basins. There are 260 major river basins in the world that collectively cross 145 national boundaries—and about 60 percent of the world's population lives within those 260 basins. This fact alone indicates that we are facing serious political problems in the water arena in the future.

Political conflicts over water have been going on for thousands of years, and there are a number of current "hot spots" around the world that could explode. For example, India has strained relations with both its eastern and western neighbors—Bangladesh and Pakistan—over the two key watersheds that flow into those countries, the Ganges and the Indus. The Jordan River flows along the Syrian-Lebanese border and is used by Jordan, Israel, and the Palestinian territories. Nine often contentious countries depend on the Nile for most of their water.

So, are we finally facing the real "revenge of Malthus"? The infamous Club of Rome report *The Limits to Growth* in the early 1970s focused the world's attention on the issues of depleting natural resources and caused a huge international furor—yet it barely mentioned water. The quantity of water on the earth cannot be increased or decreased. However, we *can contaminate* some water so much that it is no longer usable, and *we can* clean some sources of water, like seawater, such that they become more usable. In many parts of the world, by decreasing water quality, we are effectively decreasing the quantity.

Many of these impending problems will inevitably erupt in the future—perhaps in the very near future—and they should be top policy priorities for governments worldwide. The articles in this compilation were written by various experts from the water industry who comment on future trends and possible advances in policy analysis, technology, regulation, and resource management. These articles will suggest new ways of thinking and new approaches to old problems—approaches that will hopefully help us to avert even greater future problems. **(See Article 5.)**

THE WATER INDUSTRY: HOW BIG IS IT?

Although we casually refer to the *water industry*, there is, strictly speaking, no such thing. There is instead a sprawling conglomeration of many, fundamentally quite different businesses that all have something to do with the delivery of clean water—but which cannot be accurately classified under any single heading. As it is loosely defined, the water industry includes a very broad array of sectors extending well beyond the nation's water and wastewater utilities—such as concrete pipe manufacturers, specialty chemical producers, measurement and monitoring firms, tank manufacturers, treatment equipment manufacturers, new technology developers, manufacturer's representatives, engineers and consultants of all types, testing laboratories, contract operators of water plants, and many others. The only real similarity between these entities is that they are somehow involved in some aspect of providing clean water.

A diffuse and fragmented industry like the water and wastewater treatment business is very difficult to classify. Because it is challenging to define this industry, it is more difficult—if not impossible—to accurately estimate its size, growth characteristics, and other economic attributes. Nonetheless, there *are* some conventional wisdoms and some rough market statistics for the overall water-related business. The size of the domestic US water and wastewater industry is generally estimated at around $110 billion per year, as summarized from one source in Table 1-2. **(See Article 6.)**

An analysis of these summary figures, and the more detailed data behind them, is revealing. The fees that individuals and businesses pay to utilities for primary water and sewage services comprise almost two-thirds of total annual spending on water in the United States. The vast majority of these revenues pass through the municipal and public agencies—approximately 55,000 water utilities and about 16,000 wastewater utilities—not private companies.

The right column of Table 1-2 serves to underline this disparity. Projected long-term growth rates for different sectors of the overall water-related industries range from 2 to 10 percent per year. And certain specific "subniche" areas that are not broken out here—such as the membrane filtration or ultraviolet radiation treatment technologies—may be growing at much higher rates, possibly in excess of 15 percent per year.

The growth rates of individual sectors are experiencing slow but perceptible change. For example, even though water treatment chemicals comprise a very large market, the average growth rate of this sector is widely assumed to be slowly declining. Likewise, it appears that the explosive recent growth of the contract operations business will slow somewhat as that sector

Table 1-2   The US Water Industry
(Revenues in Millions)

Business Sector	2006 Revenue	'07–'09 Growth %
Water Treatment Equipment	$9,790	6%–8%
Delivery Equipment	$12,170	3%–5%
Chemicals	$4,150	2%–4%
Contract Operations	$2,490	6%–8%
Consulting/Engineering	$7,810	4%–6%
Maintenance Services	$1,850	3%–5%
Instruments and Testing	$1,470	3%–5%
Wastewater Utilities	$37,490	4%–6%
Drinking Water Utilities	$36,610	3%–5%
Total U.S. Water Industry	**$113,830**	**4%–6%**

Adapted from the *Environmental Business Journal*, 2007.

matures, particularly if public opposition to privatization increases.

On the other hand, growth rates of traditionally less glamorous infrastructure sectors like pipe rehabilitation are likely to increase in the future, as more and more capital is inevitably poured into maintaining and upgrading the infrastructure. A high percentage of future spending will target items such as steel and concrete pipe, pumps, valves, and storage tanks. This may not be the most glamorous side of the business, and the companies in this sector have not attracted much attention from the investment community—but this is where many of the dollars will be spent. **(See Article 7.)** A recent study suggested that trenchless pipe renovation will likely grow to $5 billion annually within the next several years—a growth of 500 percent from current levels.

Although market data for the water industry in the US market is sparse, similar information for most of the world is truly speculative. However, while the US is clearly the world's largest individual market, it is increasingly very clear that opportunities abound for water companies in the rest of the world. Most knowledgeable observers believe that the world market is about three to four times the size of the US market. Several reputable parties have determined that the level of world business is around this level. The British publication *Global Water Intelligence* recently presented a set of assumptions that estimated a global water market to be $420 billion per year.

So, how big *is* the water market? The truth of the matter is that, even when individual sectors are precisely defined and the figures carefully totaled, we do not have a very accurate estimate of how big this business really is. Certain conventional wisdoms are held within the industry, but it is typically difficult to document these estimates with solid market data. However, the world water market is huge, and—perhaps most critically—many of the key geographic markets, such as China, are at an earlier and much more rapid stage of growth than is the United States.

Good market data and growth trends are critical for effective strategic planning, both in the private and public sectors. However, with a total market that is agreed to be somewhere in the range of hundreds of billions of dollars per year, and given the truly critical underlying needs and factors that are driving this market, individual firms don't really need to worry whether the world market is $300 billion or $400 billion a year. The growth of the overall business will probably continue to range around 6 to 8 percent a year—a little in excess of GNP and population growth rates—and this will vary quite widely depending on individual end market sector and geographic region. Hence, for analytical and strategic planning purposes, it is more meaningful to talk about the growth and profitability characteristics of individual market sectors. **(See Articles 8 and 9.)**

THE WATER BUSINESS—KEY DRIVERS AND TRENDS

Several key factors—economic forces, social demands, and political realities—are driving a rapid evolution, and a rapid expansion, of the broader water industry. These drivers, in turn, give rise to various trends and effects that will likely be key attributes of this business for years to come. Critical among these trends are the following:

■ *Water quality and scarcity problems are reaching crisis proportions worldwide.* It is best to start at the beginning. Water quality and scarcity problems are the main concerns and drivers behind the challenges, all of the regulations, all of the commercial business opportunities, and ultimately, the projected growth for the water business over the coming decades.

Clearly, the world has both significant water quantity and water quality problems, but the quantity aspect of this equation—the absolute scarcity of water—is beginning to become a greater urgency in many regions. **(See Article 10.)** Dozens of reports and studies have pinpointed the fundamental lack of clean water as one of the most serious long-term threats facing mankind; indeed, one can scarcely pick up a news magazine or the daily newspaper without seeing an article about water problems. The United Nations Millennium Development Goals identified access to clean drinking water as one of the primary international objectives of this century, asserting that the world should "halve by 2015 the proportion of people without sustainable access to safe drinking water"—a goal that few expect to be achieved. Although it has become commonplace, and even somewhat trite, to describe water (in terms of economic and political significance) as "the oil of the 21st century," it could well turn out to be true.

■ *Public awareness and understanding of water problems is increasing.* As water scarcity and quality problems have become more serious, the public has gradually become better informed and increasingly concerned about the water problems that may be passed on to future generations. Again, one needs only to look at the front pages of the popular media to see how broad and widespread this recognition is becoming. And as the general populace becomes more aware and concerned about the myriad of future water challenges, public perceptions and demands will become more critical drivers for future water utility management. **(See Articles 11 and 12.)** However, even though water issues *are* gradually getting more attention, massive public education programs are still needed around the world, to inform the public of the real nature of the water quality and quantity problems.

■ *Regulation and enforcement are increasing, and new policies and approaches are emerging.* As public awareness and concerns about water grow, greater government review and legislation have occurred, ultimately resulting in a vast network of regulatory control and enforcement. As opposed to the arena of broader environmental legislation—where public interest, regulation, and enforcement have waxed and waned during the past three decades—when it comes to drinking water, the American public seems insistent on ever-stronger and broader regulatory protection.

A huge superstructure of regulatory controls has been promulgated by federal and state agencies over the past three decades. Public agencies and private companies are struggling not only to understand but how to comply with this rapidly expanding and ever more complex regulatory regime. It has become a major cost to the public water and wastewater utilities. **(See Articles 13 and 14.)** Regulatory controls are becoming stronger in almost every other region of the globe as well.

Nonetheless, even in the economically most advanced and most highly regulated countries such as the United States, water pollution problems continue to grow, and new water contamination issues continue to be discovered. One recent study—the Environmental Working Group's *National Assessment of Tap Water Quality*—found that "tap water in 42 states is contaminated with more than 140 unregulated chemicals that lack safety standards," and suggested that the USEPA should be doing a far more thorough job of regulating drinking water – above and beyond the vast regulatory structure that is already in place. For example, the new arena of *xenobiotic* contaminants—man-made health, beauty, and medical compounds that are now found in natural waterways—have only recently been detectable, and their impact on human health remains uncertain.

Overlying all of these traditionally regulated areas is growing concern about potentially "introduced" compounds in water distribution systems—fear of terrorist activities and security concerns regarding contamination of primary drinking water supplies. Despite all of the advances in providing clean water that have been accomplished over the last 50 years, there is still essentially no monitoring of the drinking water distribution system—that is, monitoring the quality of water once it leaves the treatment plant and is discharged into almost a million miles of underground distribution system piping. This area will be the subject of new regulation, and it represents a commercial opportunity that will likely explode over the next several years.

■ *Huge economic (and human) capital investments are required.* For municipalities and industry to comply with all these expanding regulations and for the United States to maintain and expand necessary water and wastewater infrastructure, huge capital expenditures will be required, into the long-term future. **(See Article 15.)** Numerous and extensive studies have predicted how large this collective expenditure must be, and it is certainly in the range of hundreds of billions of dollars over the next two decades. The so-called "spending gap"—the difference between the current amount of US infrastructure spending and the levels that may be required to maintain that infrastructure into the longer-term future—continues to increase. Where all the money will come from to address these issues represents an increasing conundrum. And it is not just a matter of dollars—the utility industry is also facing a human resource shortage and is increasingly desperate for new and younger talent to join the industry. **(See Article 16.)**

The critical question of *how*—not *if*—we should fund these required capital expenditures is likely to turn into a huge and contentious political debate in coming years. For example, the large metropolitan centers that often have the oldest and most run-down infrastructure with the greatest capital needs are often the very same areas with shrinking city-center populations and a declining tax base. It seems likely that the federal government will eventually have to intervene, although that may not happen until major public health crises occur. These required billions and billions of investment dollars represent a huge and unresolved future crisis for the United States—but they also constitute a huge opportunity for firms serving the water and wastewater treatment industry.

INDUSTRY TRENDS AND DEVELOPMENTS

These four interlocking industry drivers produce various trends and developments in the water business—with respect to supply, demand, market, and competitive conditions, and in terms of how utilities should manage and distribute increasingly precious water resources. These key trends are highlighted below:

■ *Outsourcing or privatization continues—but remains controversial.* Privatization has been replaced by *public-private partnership* not only in the vernacular, but also in terms of the operating philosophy that must be practiced in order for such ventures to be successful. Regardless of what they are called, the outsourcing of water and wastewater operations constitutes one of the more controversial aspects of the water industry today.

In many parts of the world, private operation of drinking water assets is taken for granted. The French—and more recently the British—are the world's major participants in private water management and operation. Privatization and contract water operations are significant and growing in many other parts of the world. *Global Water Intelligence* reports that about 10 percent of the world's population is currently served by private operators—a figure that is expected to grow to 16 percent by 2015. More than 45 percent of the population

in Western Europe is now served by private operators, with rapid growth occurring in the Mediterranean and North African regions. In the United States, however, only 10 percent of the population has its water provided by private companies.

The arguments *against* privatization revolve around concerns about the motives of private, for-profit firms, and the sense that access to clean water, as a basic human right, should be equitably allocated and priced. Many people simply believe that water resources are a part of our natural heritage that should not be entrusted to private companies to own, manage, and disburse. Given the isolationist attitudes that have prevailed since the events of 9/11, the fact that many of the contract operators are foreign-owned companies has led to further concern and suspicion among the US public.

On the other hand, the basic economic drivers *supporting* privatization remain strong and growing. Few municipalities enjoy overflowing coffers, and few public officials who wish to be reelected want big tax increases on their watch. Most public works managers find themselves in a quandary—technical requirements, regulatory complexities, and overall costs continue to increase; at the same time, the general public remains resistant to increasing taxes and users fees. The perception that water is a free commodity persists, and it may take some severe water shortages or public health disasters before the public really understands these complicated issues. In some cases, one solution to this intensifying dilemma may be to turn to private companies to finance, build, and operate their water or wastewater systems.

■ *"Musical chairs" in the industry—ownership rearrangement and consolidation continues.* One major impact of the exploding investment interest in the water industry has been a trend towards consolidation on the vendor side of the business and an ongoing rearrangement of ownership of key industry assets. This has almost resembled a game of "musical chairs" during the last several years—as many major international industrial companies have sought to exploit the opportunities offered by this growing business.

The first widespread consolidation came in the early 1990s, with the widely heralded "foreign invasion" of the US water industry—British, French, and German companies buying many of the large US businesses. However, this phenomenon began to reverse direction in the early part of this decade; these early and largely foreign consolidators were gradually displaced by major US industrial corporations, which began to make dramatic moves to acquire assets in the water treatment and purification business. Veolia, Suez, RWE, and the other European water companies have largely exited all but their infrastructure management businesses in the United States. While industrial firms such as General Electric, Danaher, ITT, Pentair, and Siemens are emerging as the new diversified water service and equipment companies. (For a summary of consolidation activity over the last several years. **(See Articles 17 and 18.)**

With so many major industry assets changing hands so quickly—even multiple times—the competitive situation in the water treatment equipment industry is very "fluid." The situation is gradually becoming clearer, but a key question remains—which of these companies will ultimately be the major players in the next generation of this industry? Most observers are predicting they will be the various diversified international industrial companies—ITT Industries, GE Water, Pentair, Siemens, and perhaps several others who have not yet made a strategic move. This activity also gives rise to several key concerns. How can companies like RWE that were such committed buyers a few years ago become such eager sellers? What was wrong with their strategies? Will the new owners of these assets have sounder strategies? And most importantly, what will be the ultimate impact of this large-scale ownership rearrangement on employees, shareholders, and customers?

When people discuss industry consolidation, it is usually within the context of private companies—the "commercial" sector of the business. However, there is growing recognition and discussion of the fact that, in many areas, consolidation in the municipal sector of the industry would make good economic sense, too. Water and wastewater utilities are both very capital-intensive businesses, and there is no doubt that scale conveys a distinct operating, technical, and financial advantage. Yet, the municipal sector of the business continues to be made up primarily of very small local players— almost 85 percent of all municipal systems are categorized as *small*. Many tiny utilities are struggling to maintain operations in an increasingly complex technological and regulatory environment. As the efficiencies and economies of scale of larger operations continue to increase, it seems increasingly possible—indeed necessary—that we will begin to see broader consolidation within the utility sector as well. **(See Article 19.)** However, it is clearly politically and financially far more difficult to merge municipally or governmentally owned systems—even though such mergers may make good economic sense.

There are other larger concerns regarding the coalescence of smaller and local utilities into larger and larger "super-regional" utilities. How would a gradual consolidation of the public water utility sector affect the delivery of water and provision of sewerage services? Would such super-regional utilities privatize themselves and or even consider floating public stock? Would we perhaps see more water and hydropower, based utilities merging to more effectively utilize their common resource and interest—water? **(See Article 20.)**

■ *A strong surge of public and private equity investment interest in the industry.* As widespread recognition of water problems has increased and as the need for innovative new services and products has become clear, hundreds of other strategic and financial buyers have also swarmed into the water industry to try to establish a foothold. And, as in many other sectors of the worldwide economy, the private equity (PE) community is rapidly making its presence and power felt in the water business. **(See Article 21.)**

Private equity firms raised $400 billion of new capital in 2006 and conducted deals worth over $570 billion in just the first three quarters of the year—representing some 25 percent of all merger and acquisition transactions around the world. The total amount of capital sitting in private equity funds is at an all-time high, and the availability to stretch that capital through additional borrowing has rarely been as attractive as it has been the last few years. As a result, typical PE fund managers find themselves with huge amounts of capital that they have to (relatively quickly) invest somewhere. With this kind of explosive activity, it is little wonder that private equity firms are also carefully searching the water industry for attractive acquisitions.

The water industry is attractive to PE investors for several reasons. First, it is perceived to represent strong and very consistent growth over the long-term future, and certain sectors offer the allure of high profitability at the same time. The stability and predictability of water utility businesses is a key factor behind the currently high level of interest in infrastructure-related water businesses. The water business offers another key characteristic that PE firms look for—it is a relatively fragmented industry ripe for consolidation. This offers PE firms the opportunity to consolidate businesses to build larger, stronger, and more valuable companies. The only drawback from the PE firm's perspective is the huge premiums currently being paid for companies in this sector—something that financial buyers often cannot afford.

Among the experience that private equity ownership brings to water companies—beyond cash alone—is the ability to recognize, acquire, and enhance undervalued assets, and to run them in a strictly economic manner— attributes which some less aware water companies could benefit from. In addition, private equity firms may bring extensive business operations experience, distribution and marketing contacts, and experience with financial management and control systems that can help companies become more profitable. For many firms, the private equity investor may also bring a more stable operating environment, in which both the potential conflicts with a larger corporate owner as well as the pressures of public ownership, are effectively removed. Finally, from the perspective of the seller, careers are much more secure than they might be in a sale to a strategic buyer.

In fact, the PE firm will typically only consummate an investment if it is confident in the capabilities and commitment of existing management.

Heightened investment interest in the water industry is not just a private equity phenomenon. There is a much greater interest in water investments by the broader public as well. Publicly traded water stocks, though few in number in the United States, continue to be highly sought after by individual investors, as evidenced by continuing high valuations. The marketplace is also starting to see more activity in the area of initial public offerings of stock in the broader water and related industries. The return of American Water Works Corporation to the public market, after its brief ownership by Thames/RWE is being eagerly anticipated by stock analysts and investors, because it will once again be the largest publicly traded water utility in the United States. Mueller Water and Basin Water are two other recent entrants to the public market. On the engineering side of the business because of the strength of water design and construction capabilities, there have also been several new issues, including Kellogg Brown and Root, SAIC, and ICF Consultants.

■ *Greater efficiency—more focus on "reducing demand to increase supply."*
Smarter, more efficient, and more sustainable use of our existing water resources is increasingly considered as a new "source" of water—as well as the least expensive and best opportunity we have to extend our overall water resource availability. However, despite improvement and increasing public attention during the last several years, there are still considerably more options available in terms of more efficient conservation, use, and reuse of scarce water resources.

For example, many water distribution systems currently incur leakage and loss of as much as 20 to 30 percent of total treated drinking water volumes. Loss rates in the main distribution systems (referred to as *nonrevenue* water) are even higher in many parts of England and France where water mains are often well over a hundred years old. Distribution system losses are clearly one of the first areas that should be addressed— it is clearly much easier and less expensive, and more environmentally sound, to fix existing water mains than to build new dams, reservoirs, or seek other sources of supply. This consideration again underlines the expansive growth that the infrastructure sector of the marketplace—pipes, meters, pumps, and tanks—is likely to experience in the coming years.

From a broader conservation perspective, the last several years of significant droughts across much of the western United States have dramatically illustrated how much water we waste—and how much water we can conserve once we are really forced to confront the issue. Most of us could without much hardship be far more efficient and miserly in our use of clean water,

but to date, water availability and water prices have not really forced us to confront this issue. However, greater public awareness of conservation issues and local conservation programs has progressed in many areas of the country. **(See Articles 22 and 23.)**

An interesting but frustrating paradox is beginning to emerge in this regard. As consumers work harder and smarter to use less water at the urging of their local utility, the revenues paid to the utility or water agency in some regions have fallen significantly. This has led to an entirely new and generally unanticipated problem—declining revenues to the utility. This leaves the utility in the politically unenviable position of having to ask consumers to pay more even though they are consuming less.

Perhaps the area in most need of efficiency improvement is agricultural irrigation. Almost 70 percent of our total water usage worldwide goes to agricultural irrigation, and 40 percent of our food supply comes from artificially irrigated lands. However, irrigation can also be a hugely wasteful process as it is typically practiced— not only wasting water, but also contributing to salt buildup and gradually less fertile soils. Some sources estimate that 30 percent of all artificially irrigated land has already been rendered infertile because of careless irrigation practices. New and more efficient technologies for the use of water in irrigation, through such practices as drip irrigation—"more crop per drop"— offer great promise. More efficient water usage, better drainage systems, and increasing use of certain types of wastewater for agricultural irrigation should all be important policy objectives that can collectively constitute another important new "source" of freshwater.

■ *Increased focus on water recycling and reuse—technologies and attitudes.* More efficient water usage is perhaps best reflected by the growing interest in water reuse and recycling—a sector that many pundits believe to be one of the most exciting growth opportunities in the entire water business. As the boundary between water and wastewater continues to fade—as wastewater increasingly comes to be viewed as another source of primary water—there are increasingly strong economic reasons to recycle and reuse wastewater for a broad array of applications. Water recycling initiatives, from the individual residence to the large municipality or major industrial installation, are growing rapidly and are a key driver behind technological advance in the industry. However, these terms are often used rather vaguely or interchangeably, and hence some definitions and clarifications are in order. **(See Article 24.)**

Most wastewaters can easily be recycled and cleaned to levels where they can be reused for primary drinking water—and this can occur in both a direct and an indirect manner. *Indirect* reuse of wastewater—that is, after treated wastewater has been discharged into and then later withdrawn from a river, or after it has been discharged into and then withdrawn from an underground aquifer, and put through a primary treatment process—has clearly been practiced on some basis since the dawn of man. Such indirect reuse is virtually dictated by the earth's natural hydrologic cycle.

For example, it is estimated that on some of the major river systems in the United States, water is used and reused up to 20 times as it travels to the sea. The discharge waters from one wastewater treatment plant essentially comprise the raw water intake at a primary drinking water plant located downstream. In fact, as a result of 30 years of steady progress under the Clean Water Act, the discharged waters from wastewater treatment plants are sometimes cleaner than the supposedly "natural" rivers and streams into which they flow.

What is usually meant by *water reuse,* however, is a slightly more *direct* type of reuse. It implies more of a "man-made" process without the long-term intervention of nature and the hydrologic cycle—that is, more immediate treatment and recycling of wastewater for primary use purposes. Although direct reuse of wastewater for domestic purposes has been technologically feasible for years, widespread direct reuse for drinking water purposes seems to be far in the future. Indeed, today such direct reuse for drinking water purposes is only commercially practiced in a few very arid locations around the world. Putting a "black box" treatment system on the outside of a home to treat sewage and recycle it directly back into the tap—often referred to as *toilet to tap* in the popular media—is unacceptable to most people. Scare stories in the press tend to reinforce this reticence, even though from a technological perspective, it is fairly straightforward to recycle wastewater to drinking water standards. Wider acceptance of wastewater reuse represents a major public education challenge, but eventually more and more direct reuse seems certain to happen, especially in the more arid regions of the world.

There is one very critical statistic to consider when evaluating the potential impact of reuse as a means of extending our water resources—a fact that often goes relatively unnoticed, but one that should eventually make direct reuse much more feasible on a wide scale. *Only a tiny percentage of our primary water supply is actually used for drinking.* Compared to the roughly 130 gallons of water per capita per day that the water utilities of the United States currently treat to drinking water standards, most individuals drink less than a gallon a day. Even if we also consider the portion of our total per capita water consumption that is used to cook and clean—which we might also want treated to drinking water standards—this still represents a small percentage of total water consumption. In other words, the vast majority of the 130 gallons of water used per person per day could be recovered and treated for a variety of other uses without anyone having to drink directly recycled wastewater. Hence, if only small incremental gains could be made in terms of percentage wastewater reuse for nonpotable purposes, overall water availability concerns could be very substantially impacted.

More and more cities around the world are beginning to seriously look at recycled wastewaters for drinking water supplies. This is especially true for cities in arid regions or remote internal areas far from the oceans. Arid *and* coastal cities, like many of the major metropolitan centers in the Middle Eastern countries, may also use seawater desalination as a new source of drinking water. They are among the world leaders in large-scale desalination plants. **(See Article 25.)** However, it is evident that carefully planned and efficient reuse systems are often going to be a much less expensive source of additional water than are large, energy-intensive desalination plants or lengthy pipelines to reach new supply sources.

One thing is certain——as water prices continue to rise, there will be ever-increasing incentives for more careful conservation, recycling, and reuse. With greater economic incentives, individuals and households will begin to use and reuse water more carefully, and industrial companies will rethink their approaches and retool their manufacturing systems, to use less water and to better recycle their wastewater streams.

■   *Incremental but continuing technological advancement will help address some of the problems.* Few observers believe that there are any truly revolutionary future technological breakthroughs that will transform the treatment and use of water. However, incremental technological advances are ubiquitous, and thousands of technology developers are actively working on developing and commercializing better systems across all sectors of the industry. **(See Articles 26, 27, and 28.)** Perhaps one of the most well-known examples of technological advance in water treatment is the improvement in efficiency and unit cost reductions that have been recognized in reverse osmosis—one of the primary technologies involved in seawater desalination, as well as water treatment and recycling. It is the rapidly declining cost and improving efficiency of reverse osmosis that has made membrane treatment of raw water and desalination of seawater economically feasible.

Other technologies used for the treatment and preservation of drinking water are showing incremental advances and continual cost reduction. A review of the agenda for any technology conference in this industry quickly suggests the breadth of technological approaches that are being applied to water treatment. Beyond the more widely known techniques, such as membrane filtration, UV radiation, chlorination, demineralization, and ion exchange, there is a bewildering array of newer and developing technologies. Briefly, this includes such technologies as electro-coagulation, sonication, cavitation, ozonation, electro-deionization, biocidal disinfection, electrodialysis reversal, multi-stage bubble aeration, and various alternative chemical treatments. There is also a new and emerging focus on systems that mimic or enhance natural water treatment methods, such as natural attenuation and constructed wetlands development.

The proper application of these existing and improving technologies—either separately or in the right combination with each other—can help to solve many of the world's water problems. The key challenge in many areas, however—particularly in the less economically developed parts of the world—is money, not technology. **(See Article 29.)** Indeed, many observers of the global water situation believe that the simpler and "lower tech" approaches, such as sand filtration and enhanced natural wetland treatment, rather than reverse osmosis and the like, will be easier to implement and less expensive. Ultimately, they may play more significant roles in helping to solve the vast majority of the world's water quality problems. In summary, despite the lack of potential technological "silver bullets," existing technologies appropriately funded and applied can go a long ways towards solving the world's water problems.

■ *Consumers and residential users increasingly vote with their pocketbooks.* As the general public has become more aware and concerned about water, individual consumer preferences and demands are becoming a more important driver behind the commercial water industry. A critical consideration is the growing concern among consumers about the safety of their tap water. Remarkably, the Metropolitan Water District of Southern California reported several years ago that almost *two-thirds* of its customers were concerned about the quality of the water coming out of their taps. Because of the fact that water utilities have not effectively marketed the true value of their product and partly because *real* water quality problems have occasionally occurred, consumers are increasingly either buying bottled water or considering ways to further treat the tap water coming into their homes.

This is an area of controversy in the water utility industry. Approximately 55,000 agencies that provide drinking water, most of which are municipally or government-owned and operated, are at odds with the small but growing residential treatment and POE/POU (point-of-entry/point-of-use) equipment manufacturers and bottled water distributors. The utility industry maintains that public tap water is not only safe but indeed is truly one of the great economic bargains of all time. The latter group cautions that the only way you can really be sure of the quality of your water is to treat it again within the confines of your own home—or drink it out of a prepackaged bottle.

The explosive growth of the bottled water industry over the past few years is a spectacular example of how retail customer perceptions—rightly or wrongly—can create and drive new markets. The extent of this phenomenon is staggering. It is now the second-largest beverage category in the United States. It is estimated that Americans consumed over 7 billion gallons in 2005, an 11 percent increase over 2004—or 26 gallons per person—more than $10 billion worth in total. According to the Beverage Marketing Corporation, the reasons are clear. "Bottled water is a healthy, safe, ready-to-drink commercial beverage, which is becoming increasingly affordable—a great beverage alternative. Bulk and single-serve packaging options facilitate a variety of uses." Providers continue to market bottled water by promoting it as something completely different than tap water. Typically, it is not much different, but apparently the concept sells to a large portion of the American public. The rapidly growing business has been quickly consolidated into the hands of the major international food and beverage companies, such as Nestlé, Coca-Cola, Pepsi, and a few others.

Although (a) there is little evidence that bottled water is necessarily safer than tap water, (b) the bottled water industry is currently only lightly regulated, and (c) the "transportability" of water can easily and obviously be accomplished by other means, the bottled water business seems to be a triumph of marketing—living proof that "perception is reality." The Pacific Institute has calculated that if all the money spent on bottled water ($50 billion over the past decade and still growing at a rate of 10 percent per year) was redirected, we could build all of the treatment systems and infrastructure necessary to deliver good clean drinking water to every person worldwide.

A more substantive and serious question that arises from the tap water safety issue, and that is increasingly debated within the industry, has to do with the social and economy-wide efficiency of centralized water treatment versus distributed or localized treatment of water. This issue refers to the fact previously mentioned above—very little of our centrally treated drinking water is actually used for drinking. If we are only actually drinking 1 percent of the water that is treated to our very stringent regulatory standards, does it make any sense from a broader social and political perspective to treat all of our water to these exacting standards? Would it make more sense, from an overall economic perspective, to treat municipal water to lower standards at much less expensive and simpler central plants, and gradually build an infrastructure where local areas or individuals treated that water to their own requirements for their specific purposes at the point of use? In other words, would it make more sense—from a broad economic perspective—to save money in the construction of such sophisticated central facilities, and have each home and business treat that very small amount of water that they drink or otherwise use via POE/POU devices at the point where it is actually consumed?

This would obviously require a massive rebuilding of the entire water infrastructure system in the country, and there would be difficult questions about how public health could be protected and insured under such a distributed system. Nonetheless, over time, more and more consumers may treat or retreat water to their own specifications at the point of use. Some water utilities have already experimented with providing POE treatment devices to their customers. However, a widespread move toward distributed treatment is not likely to happen any time in the near future.

■ *New paradoxes and new questions continue to emerge.* Particularly in the United States, the water business seems increasingly confused by a series of emerging ironies or paradoxes. For example, although the pressing need for vast water infrastructure expenditures is becoming progressively clear, it is less and less obvious where the funding for these massive investments will come from. Local agencies and municipalities continue to suffer from tighter fiscal constraints, and the federal government, at least at this point, shows no inclination of becoming involved. Projections about the limitless future of the water business are beginning to be tempered by the reality of national fiscal constraints. There are huge needs, but will we ever be able to fund the solutions? As previously mentioned above, it may take some sort of public health calamity resulting from the weakened water infrastructure to force federal attention toward this entire problem.

The water industry has never been hotter from an investment perspective. Investors of all kinds have been rushing to investigate the water industry— corporate and strategic investors, such as General Electric and Siemens, hundreds of private equity groups such as CD&R and the Carlyle Group, and literally millions of private individual investors. The stock prices of water companies have been driven to high levels, M&A transactions are occurring at stratospheric valuations, and various types of new water-specific investment fund vehicles are seemingly being issued every day. However, it often seems that there is a multitude of investors chasing relatively few real and attractive investment vehicles. Investors lament the paucity of pure-play water stocks in the United States, and are increasingly turning to foreign stock markets to invest in attractive water companies. The fact that the water infrastructure is largely owned and operated by municipal agencies makes it difficult to invest in.

At the same time as this frenzy of investment interest has intensified, there has been constant drumbeat about the need for new investment, and the truly vast capital expenditures that will be required to maintain and expand the drinking water and wastewater infrastructure. Every month seems to bring forth a new study of future water needs and the hundreds and hundreds of billions of dollars that will be required for the country's utilities to continue to provide clean water. The most recent USEPA estimate of capital requirements, for just the drinking water portion of the U.S. infrastructure system, is estimated to be $277 billion over the next two decades. The controversial "spending gap"—what the experts estimate that we *should* be spending versus what we actually are spending to correct these problems—continues to increase and seems to portend potential disaster in the future. Similar estimates of investment needs, with even larger magnitudes, apply for most of the rest of the world as well. Water distribution, treatment, and wastewater collection infrastructure is crumbling in many parts of the economically developed world and does not exist in many parts of the less developed world. In short, the needs in this industry are truly vast and the situation is demanding new investment dollars. **(See Article 30.)**

This situation represents another major paradox and dilemma for the water industry. On the one hand, there is a huge amount of investment money available, and on the other hand, huge infrastructure and capacity needs require huge investments. On one side, there are investors eager for good investment opportunities, and on the other, the public is insisting on the rebuilding and expansion of a dilapidated water system. Given the urgency of the world's water problems, this is a situation that demands new and revolutionary approaches— for more creative financial mechanisms that will allow private investors to invest their money for the public good, and concurrently to earn a competitive rate of return on those monies. There is a huge interest in water investments, and there are clearly huge needs. It is imperative to implement more effective ways to connect this supply of "water dollars" with the demand for water dollars. **(See Article 31.)**

As alluded to earlier, and in another somewhat counterintuitive trend, water conservation measures continue to improve in many arid parts of the United States. The resulting lower water usage has translated into lower revenues to the municipality and hence, fewer funds to invest in badly needed infrastructure. By "doing the right thing" and conserving water, consumers are putting a greater, rather than less financial burden on the water utilities. In short, responsible customers must be told to both use fewer gallons and pay more for the gallons they *do* use—not a very palatable outcome to either the water supplier or the customer.

In the future, the water industry will probably encounter more of these types of conundrums and paradoxes. They will serve to underline the drastic need in this industry, not only for innovative new technologies and massive capital investment, but also human resources—the technical, policy, and management talent to steer this complex business into the future.

■ *Water utilities must advance to the next level.* The job of the water utility is to provide safe, reliable, secure, and affordable water to the US public. Utilities are perhaps the last "natural" monopoly; and to the extent that all natural monopolies are regulated to protect the public interest, many water utilities have found themselves in an unfavorable position over the last several years. Staying current with technological alternatives and trying to comply with complex and often overwhelming regulatory requirements is becoming increasing difficult—particularly for the smaller and undermanned utility. Consumers are becoming more knowledgeable, they are becoming more demanding, and they are increasingly questioning the quality of the product. Attracting new talent and maintaining the human resource base, both in terms of management and operations, is a constant challenge for most utilities.

However, water utilities will continue to play a more critical role in the community and in industry. Even if there is a move toward more localized or more "distributed" treatment in certain areas, and even though the POE/POU and bottle water markets are booming, the vast amount of all primary water will be delivered by the water utility industry. Hence, utilities must continue to change, adapt, and improve their capabilities, operations, internal systems, and their levels of customer support or service. Utilities need to embrace new technological advances, more refined asset management plans are needed to optimize existing operations, and more dollars must be invested now to insure the dependability of the infrastructure over the longer-term future. Utility leaders, along with their municipal bosses, must take responsibility and inform the public about the inevitability of price increases, so that the necessary funds are in place to maintain infrastructure and avoid massive problems in the future.

Finally, utilities need to improve the marketing of their product. Utilities—both water and power—have never really focused on marketing, because they have a captive customer base. However, that is beginning to change at least marginally, as customers do in fact have other ways of obtaining and treating their water. In the future, utilities will have to promote the true economic value of primary drinking water and the great economic bargain that it represents for most Americans. If utilities take a greater leadership role in the future and help the public to better understand both the challenges and the benefits of the public drinking water system, they can improve their own economic situation while making an invaluable contribution to the solution of the world's water problems. **(See Articles 32 and 33.)**

CONCLUSIONS

A clear and inescapable conclusion of any review of the water business is the inevitability of continuously rising water prices in the future—indeed, the urgent need for increasing water prices in many parts of the globe. Water resources are poorly managed in many parts of the world, and water remains absurdly inexpensive. A recent authoritative special report from *The Economist* concisely concluded that water is "ill-governed and colossally under-priced" around the world. Discouragingly, it also reported that the United States is the most wasteful nation worldwide in terms of water usage.

The true *cost* of delivering clean water—as well as the average *price* of water—continues to creep slowly upward in most localities, but prices are not rising at the kind of rates that will be necessary to upgrade and maintain the infrastructure on a truly sustainable basis. Most of us do not really recognize the true value of water, and few of us pay anywhere near what it is really worth. Any review and analysis of the water situation, particularly in the United States, must come to the same conclusion—we simply must pay more for water, both to encourage more efficient patterns of use, as well as to finance a sustainable treatment and distribution infrastructure.

In a recent review of worldwide water pricing, the English publication *Global Water Intelligence* asserted that "there is no correlation between...water scarcity and water price." The report also observed that there is no other product whose price to the consumer is so totally unrelated to its cost—an observation that seems to often go unnoticed or ignored in most economic and political debates. Indeed, to paraphrase Benjamin Franklin's observations from over 200 years ago—it seems likely that we will only recognize the true value of water once our "well runs dry." This is beginning to happen.

As prices continue to increase, decisions about water usage will inevitably begin to take on greater significance in the overall economy, and in many ways, this will become a self-reinforcing cycle. And as prices rise, many of the trends previously discussed will intensify—greater reliance on reuse and recovery, more emphasis on conservation, a continuing trend towards more private-public partnerships, more rapid advancement in technology, and an ongoing consolidation in the industry. This is already starting to happen—water prices are rising faster than inflation, but in many areas prices need to increase at a much greater rate.

The outlook for the future is not entirely gloomy. As Table 1-3 shows, there is both good news and bad news, as far as the worldwide water industry is concerned. **(See Article 34.)**

Over time, water will be viewed more as a true economic commodity—one that can be bought, sold, moved around like other commodities, and of course, hoarded. Government subsidies and major federally funded dams, and water distribution and irrigation programs over the past hundred years have seriously distorted the operation of a free market in the United

Table 1-3   Water: Good news and bad news

Good News	Bad News
There is a lot of freshwater in the world.	It's not always where man needs it.
Water is free from nature.	Infrastructure needed to deliver water is expensive.
In many areas, water is readily accessible at a low cost.	People assume it will always be available and take it for granted.
Nature is constantly recycling and purifying water in rivers and lakes.	Man is polluting water faster than nature can recycle it.
There is a huge amount of water underground.	Man is using this water faster than nature can replace it.
5 billion people have reasonable access to freshwater.	Over 1 billion do not.
3.8 billion people have at least basic sanitation.	2.4 billion do not.
Millions are working their way out of poverty.	Affluent people use more water.
The pace of industrialization is increasing.	Industry will require more freshwater.
Industry is becoming more efficient in its water use.	Many industries are still using water unsustainably/inefficiently.
Awareness of water issues is increasing.	Translating awareness into action can be slow.

Source: "Facts and Trends" World Business Council on Sustainable Development.

States. However, market forces seem likely to increasingly exert themselves as clean water becomes more scarce. And it will increasing make more sense to evaluate, understand, and manage water resources on a watershed basis—independent of political boundaries.

At the same time and from the global perspective, the commoditization of water will have to be aggressively balanced by equity and fairness concerns—everyone needs water to live, and there will always be some who cannot afford to pay for it. Finding the right balance to this dilemma—*water as an economic commodity versus water as a human right*—will be one of the great social, economic, and political challenges of this century. **(See Article 35.)** And as previously mentioned, we must develop new and more creative financial mechanisms that will allow private investors to invest their money for the public good—better ways to connect growing investment interest with the huge capital requirements that are imminent.

Yes, water frequently falls from the sky. Yes, three-quarters of our planet is covered with water. And it is true, freshwater is still very abundant in many parts of the world. But it's not always clean, it's not always where we need it, and it costs hundreds of billions of dollars a year to collect, clean, and distribute. The world's population has increased four-fold over the last hundred years, but we still have the same amount of water that we had a million years ago. We cannot create new water, but we also cannot destroy it. And, unlike any other commodity, there is truly no substitute for water.

In closing, we need to remember and remind our friends and neighbors that the amount of freshwater is essentially fixed. We need to become much smarter and much more efficient in our treatment and usage of this increasingly scarce resource. The facts are simple—water is an essential prerequisite for life and for sustaining and improving our standard of living and our modern industrial economy. *We are not going to find a substitute for water.* As the global water crisis intensifies, we face numerous and daunting political and economic challenges, but there will also be almost limitless opportunities for creative, innovative, and well-managed solutions.

I hope that the articles in this compilation will shed new light on this challenge and will serve as useful background to those technology developers, investors, business managers, and policy makers who will contribute to solving these problems and meeting these challenges in the future.

Steve Maxwell
Boulder, Colorado
October, 2007

13

"Big Picture" Water Challenges

and Quandaries

As clean water becomes increasingly scarce, new and unpredictable challenges will inevitably emerge, and there will likely be revolutionary changes in the way we think about and manage water. Some of the "big picture" water challenges and new quandaries may include the following:

■ *Water rights may become the source of outright conflict and war.* This may sound overly dramatic, but in fact, such conflicts have already been occurring for decades. If all other methods of acquiring, storing, and using water fail, there is always the option of stealing water from your neighbor. Many regions, such as the Middle East, are moving inexorably towards serious battles over water. The fact is, abundant people are often located in areas where there is not abundant water—almost insuring that conflicts over water rights and water availability will eventually develop. The broad demographic trends in the United States over the past 50 years—population shifting from the Northeast to the arid Southwest—have accentuated this imbalance and may not be sustainable in the future.

■ *Will OPEC be replaced by some sort of "OWEC"—a cartel of water exporters?* This may seem unlikely today; however, in the future, those countries with more abundant water will take stronger measures to protect that resource and maximize its value. To protect itself against the burgeoning thirst of the country to its south, a public watch-dog group called the Council for Canadians has been formed to monitor the protection of that country's vast water resources.

■ *Water is being transferred from agriculture to US cities at an alarming rate.* Farmers, particularly in the Southwest, are finding that the economics clearly suggest they stop farming and instead sell their underlying water rights to neighboring (or far away) cities. What are the implications of allowing our farmlands to dry up in order to slake the thirst of our ever-expanding cities? What will be the long-term impact of this trend on a country long considered the "breadbasket" of the world?

■ *Why do we as a society continue to treat all our water to drinking quality standards when less than one percent is actually used for drinking?* From a broad social perspective, this seems highly inefficient. Should we as a society really be incurring the capital costs of providing drinking-quality water for firefighting, mixing cement, washing cars, or watering yards? How long will it take us, and what will it cost, to put in place the complementary infrastructures to allow the efficient and widespread usage of different grades of water? Is it even a realistic goal? Are there alternatives?

■ *Desalination has been widely promoted as a truly "new" source of freshwater—an increasingly economic means of actually "increasing the size of the pie."* The cost of reverse osmosis technology has come down dramatically in recent years, and there are now hundreds of desalination plants planned worldwide. But will these plants operate as efficiently and inexpensively as currently foreseen, and will the environmental impacts—the disposal or subeconomic use of the by-product salts—become a serious social or economic cost?

■ *As sediments continue to accumulate in reservoirs that were constructed over the last 100 years, will we be able to continue to provide controlled water supplies, particularly to areas like the Southwest?* Alarming statistics about the shrinking effective volume of major reservoirs, combined with recent droughts, have crystallized concerns about the fragility of long-term water supply in many regions. We have become experts at building dams, but what are we going to do with these dams when the reservoirs fill up and no longer fulfill their original purpose? Can we tear them down again in an economically and environmentally viable way, or will they become silent mud-filled monuments to a short-term planning mentality in the management of water resources?

■ *As the debate about global warming continues, the question is what, if anything, can we do about it?* And if the planet's climate does change, there are numerous questions regarding what this may imply for water resource management, storage, and utilization over the longer-term future. The impacts of continued warming on coastal regions (where much of the world's population resides) could be substantial, and many water utilities are now actively accounting for such an eventuality in their long-range planning.

■ *What will be the future impact of China on world water markets—both in terms of the supply and demand?* Catastrophic industrial spills into two major Chinese rivers have brought this issue to the front pages in recent months. As China industrializes, its thirst for water will continue to grow, and it will become a more significant consumer of water and related products and services. It is hard to find a water company of any type that is not either already "in" China, or trying to figure out how to get in—either to sell its products and services to this nascent but booming market, or to take advantage of China's huge and low-cost manufacturing labor force.

■ *Steady advances in analytical technology are allowing us to detect contaminants in our waters at lower and lower levels, and are in fact outstripping our scientific ability to even understand what it is that we are finding.* However, what we are finding is not encouraging. In fact, a new word, *xenobiotics,* has been coined to describe a new range of man-made contaminants—medical drugs, pharmaceutical compounds, personal hygiene and beauty products, and a whole range of modern household chemicals. Modern analytical instruments can detect these heretofore unknown contaminants throughout our natural waterways. Although toxicologists do not yet fully understand their impact, evidence is gathering that many of these substances may be detrimental to the human endocrine and reproductive systems.

■ *Security concerns surrounding drinking water will continue to mount.* The United States has approximately 55,000 public drinking water systems, 16,000 public wastewater treatment systems, and about 80,000 large dams and reservoirs—all connected to about 1 million miles of water distribution lines. Almost all of this infrastructure is highly vulnerable to attack. The events of 9/11 served to crystallize in the public eye how potentially vulnerable we are. Systemwide, real-time monitoring and better rapid detection technologies are sorely needed.

Source: TechKNOWLEDGEy Strategic Group

Census Data Shed Light On US Water And Wastewater Costs

BY SCOTT J. RUBIN
(*JAWWA* April 2005)

The cost of water and wastewater service in the United States has drawn increased attention in recent years. Some stakeholders believe that the cost of this service is too high, placing an undue burden on low-income households (Saunders et al, 1998). Others believe that water and wastewater service is a bargain compared with its value and the relative cost of other utility services, such as energy and telecommunications (USEPA, 1998).

Some stakeholders discuss the affordability of water and wastewater as if they were separate issues (USEPA, 1998). In communities where both services are provided centrally, however, it is common for water and wastewater utilities to have agreements that provide for the disconnection of water service if the wastewater bill is not paid. In other words, in order to keep water service in the household, it is necessary to pay both the water and wastewater bills. Obviously, disconnection of water service also effectively removes wastewater service from the home as well. The two services, therefore, are inextricably linked; both must be present in order for either service to be available to the household.

DATA QUALITY DEPENDS ON THE SURVEY

Some surveys have limitations. Periodically, the US Environmental Protection Agency (USEPA) and other entities conduct surveys in an attempt to estimate the cost of water and wastewater service in the United States and in particular locations (US Bureau of Labor Statistics & US Census Bureau, 2003; *Raftelis Financial Consulting, 2002*; USEPA, 2002). These surveys tend to be limited in scope geographically and generally focus on larger utilities and bigger metropolitan areas. The surveys usually collect information from utilities rather than households. Although utility surveys are important and provide useful information, they cannot determine the actual burden placed on households by paying the bills for water and wastewater service. ("Burden" can be represented by the percentage of income that the household pays for water and wastewater service.) In addition, in many communities, the water provider and wastewater provider are different entities. Utility surveys in these communities, therefore, may not fully capture the total household burden for these services.

The federal government's ongoing Consumer Expenditure Survey provides information at the household level for water and wastewater expenditures (US Bureau of Labor Statistics, 2002). That survey, however, does not differentiate between households that pay directly for water and wastewater and those that do not pay directly for service. Thus, the averages and aggregate data reported in that survey include a large percentage of households that reported having no expenditure for water and wastewater service.

Census long form gathered detailed information. Many of these problems are solved by using data collected by the US Census Bureau as part of the decennial census (Rubin, 1998). In particular, the "long form" survey for the 2000 census collected detailed information about household incomes and expenditures from a stratified sample of approximately 1% of US households, or about 1.2 million households. In developing this survey, the US Census Bureau divided the country into 532 areas, known as Public Use Microdata Areas (PUMAs).

A PUMA never crosses state boundaries and generally contains about 400,000 people. For those states with relatively small populations, the entire state constitutes one PUMA. The Census Bureau has made available the data from this survey, known as the Public Use Microdata Sample (PUMS) (US Census Bureau, 2003).

The PUMS contains data on each household's annual income, as well as its water and wastewater expenditures (combined into a single entry). The PUMS also includes demographic information about the household, such as race or ethnicity, age and number of household residents, and home location (i.e., whether the PUMA was in a metropolitan or nonmetropolitan area). In addition, the PUMS collected information about the home (housing unit), including whether it was owned or rented (tenure), the number of rooms in the home, and the characteristics of the building in which it was located (e.g., whether it was a one-unit detached house, a mobile home, or part of a multiunit building) (US Census Bureau, 2003).

All economic data in the PUMS were for calendar year 1999, as reported by respondents when the census data were collected in April 2000. As with any survey, the accuracy of the economic data in the PUMS depends on how respondents completed the survey. Presumably, the responses represent a mixture of precise information and respondents' estimates of the amounts spent on certain items.

Study set out to estimate water and wastewater costs. The objective of this study was to analyze the entire PUMS data set to estimate the cost of water and wastewater service for households with various characteristics. The database was compiled and summarized at the national and state levels using standard database, statistical, spreadsheet, and mapping software.

All results described here are based on the weighted results of the survey responses collected by the US Census Bureau. The weightings were developed by the bureau and are included in each PUMS record to "allow users to produce estimates that closely approximate published data in other products" (US Census Bureau, 2003). In other words, the weighted results produce estimates within a reasonably small margin of error for any relevant geographic area. As an example, the Census Bureau reported that there were 1,737,080 households in Alabama in 2000. Performing a weighted calculation on the PUMS data derived an estimate of 1,736,817 households in Alabama in 2000, a difference of 263 households, or 0.015%.

In the interpretation and use of these data, unless otherwise noted, all results are reported only for households that paid directly for water and wastewater service and had an annual household income of at least $1,000. The analysis was restricted to households reporting incomes of at least $1,000 to eliminate anomalous results that would be obtained by including the approximately 0.5 million households (0.5% of all households) that reported little or no income but still had substantial expenditures for water and wastewater and other services. It could not be determined from the PUMS documentation whether households that reported very low incomes did, in fact, have such low incomes (and paid for services by increasing debt or reducing savings) or if they excluded certain types of income from their responses. Income, age, and other characteristics were tabulated only for households that paid directly for service. ("Pay directly" for service means that a bill was received for the service and that the cost was not included in a rental or maintenance fee.) However, many households that did not pay directly for water and wastewater service paid those costs indirectly through rents or maintenance fees. Therefore, care must be exercised when using these results to draw conclusions about the effect of increases in the cost of water and wastewater service. Increases in water and wastewater costs can be expected to be passed on to those who pay directly for the service, but they also may be passed on through increased rents and maintenance fees to many of those who do not pay directly for service.

RESULTS SPOTLIGHT FACTORS IN US WATER AND WASTEWATER COSTS

In 2000, there were approximately 103.4 million US households with incomes of at least $1,000 during 1999. Of these, approximately 64.1 million (62%) paid directly for water and wastewater service in 1999. Among the households that did not pay directly for the service, 18.4 million (18%) reported that the cost of water and wastewater was included in their rent or maintenance fee or in a similar charge. The remaining 20.9 million households (20%) did not pay for service. Presumably most of these households supplied their own water and wastewater needs through private wells and septic systems.

Among the households that paid a water and wastewater bill in 1999, the average bill was $476 per year. Of those paying directly, the average household spent 1.6% of its income for water and wastewater service. However, 50% of households spent less than 0.8% of their income for water and wastewater service, whereas 25% spent more than the average of 1.6%.

As explained subsequently, these figures do not represent the effect of water and wastewater costs on households with characteristics differing from the average or median household. The following sections break down these figures by six household characteristics: size of household, age of residents, income, tenure, type of housing unit, and size of housing unit.

Size of household affected percentage of income paid for water and wastewater service. Table 1 shows that about half (49%) of the one-person households in the United States paid directly for water and wastewater service. One-person households paying directly spent nearly 2.4% of their income for water and wastewater, a higher percentage of income for the service than almost any other group or subgroup in the study.

As the number of people in the household increased, the average water and wastewater bill also increased. Increasing the household size from one person to two people added $72 per year to the bill. Thereafter, each additional person increased the bill by $36 to $45 per year. After a household contained two people, however, there was little change in the prevalence of households paying directly for water and wastewater service. As shown in Table 1, as households increased from one person to two, the prevalence of paying directly for the service increased from 49 to 64%. Larger households, however, saw a relatively minor change in this percentage, peaking at 70% of four-person households that paid directly for water and wastewater.

TABLE 1   Relationship between household size and water and wastewater bill

Number of People in Household	Number of Households	Households Paying Directly for Water and Wastewater Service %	Average Bill of Paying Households $	Average Bill as Percentage of Income
1	26,149,268	49.01	378	2.39
2	33,948,697	64.36	450	1.30
3	17,164,204	65.86	495	1.27
4	14,869,096	69.62	539	1.20
5	6,969,684	69.40	575	1.34
6 or more	4,334,157	67.81	637	1.61

TABLE 2   Relationship between age and water and wastewater bill

Age of Household Residents	Number of Households	Households Paying Directly for Water and Wastewater Service %	Average Bill of Paying Households $	Average Bill as Percentage of Income
One or more age 65 or older	24,720,579	66.35	453	1.94
None age 65 or older	78,714,527	60.59	484	1.36
One or more under age 18	37,603,344	66.09	533	1.42
None under age 18	65,831,762	59.62	440	1.57

A combination of increases in household income, the size of the water and wastewater bill, and changes in the likelihood that a household paid directly is reflected in the calculation of the average bill as a percentage of income. For households that paid directly for service, the average bill declined as household size increased until the household had four people (1.2% of income). Then, the water and wastewater burden increased, such that households with six or more people paid 1.6% of their income for the service.

Age of residents did not play a major role in cost. As shown in Table 2, the age of residents in the household did not appear to significantly affect the cost of water and wastewater. For example, compared with households without an elderly person, households with at least one person age 65 or older paid only $31 per year less for water and wastewater service (but 0.6% more of their income).

Households with at least one child age 17 or younger paid about $93 more per year than did households without children. This is to be expected, however, because households with minor children are likely to have more residents, on average, than households without minor children. The cost of water and wastewater as a percentage of income, however, did not significantly dif-

fer. Households with minor children paid approximately 1.4% of their income for water and wastewater, whereas those without minor children paid 1.6% of their income for the service.

Further analysis shows no meaningful difference in the results for households with an elderly or young person and those for all households of the same size. For example, households consisting solely of one person age 65 or older had an average water and wastewater bill of $385 per year, compared with all one-person households, which had an average bill of $378 per year. Similarly four-person households with at least one minor child had an average water and wastewater bill of $535 per year, compared with all four-person households, which had an average bill of $539 per year.

The age of residents did not appear to affect whether the household paid directly for water and wastewater service. In each age category, between 60% and 66% of households paid directly. Overall, the age of household residents did not appear to be meaningful in determining a household's water and wastewater bill or the percentage of the household's income spent for the service.

Annual household income was a factor. Table 3 shows that as the level of household income increased, the

TABLE 3   Relationship between annual household income and water and wastewater bill

Annual Household Income $	Number of Households	Households Paying Directly for Water and Wastewater Service %	Average Bill of Paying Households $	Average Bill as Percentage of Income
Less than 10,000	8,085,340	41.61	399	8.40
10,000 to 19,999	13,285,453	49.90	413	2.87
20,000 to 29,999	13,645,244	54.26	425	1.74
30,000 to 39,999	13,006,945	58.48	442	1.28
40,000 to 49,999	11,228,285	62.86	457	1.03
50,000 to 74,999	20,285,059	68.44	480	0.79
75,000 to 99,999	10,498,576	74.28	512	0.60
100,000 or more	13,400,204	77.28	584	0.39

household was more likely to pay directly for water and wastewater, and the amount paid for service increased. Specifically, only 42% of households with incomes below $10,000 per year paid directly for water and wastewater service. This figure increased steadily as income increased, with more than 77% of households with annual incomes of $100,000 or more paying directly.

The data also show that the annual cost for water and wastewater increased as income increased. Because the cost increased at a much slower rate than the change in income, the percentage of household income spent for water and wastewater service declined precipitously as income increased.

Because higher-income households were more likely to pay directly for water and wastewater service than were lower-income households, the median income of households that paid directly was 17% higher than the median household income (MHI) of the country as a whole. The national MHI among households with incomes of $1,000 or more in 1999 was $42,700. (This was slightly higher than the national MHI for all households of $41,994, because households with incomes less than $1,000 were excluded from this analysis.) In contrast, the MHI for households paying directly for water and wastewater in 1999 was $49,900.

From this, it is apparent that households that did not pay directly for water and wastewater service had a lower MHI than households that paid directly. Specifically, households that reported the cost of water and wastewater was included in their rent or in another fee had an MHI of only $30,000. Those who reported that water and wastewater had no cost (predominantly households that supplied their own service) reported a MHI of $36,300.

Even though households paying directly had a higher MHI, there are still millions of low-income

households that paid directly for water and wastewater. Of the 21.4 million households with annual incomes between $1,000 and $20,000, 10 million households paid directly for water and wastewater service. This represents one sixth of all households paying directly.

Tenure data indicated homeowners paid directly. More than twice as many homeowners paid directly for water and wastewater than did renters (78% versus 29%). Table 4 shows that among those who paid directly for water and wastewater, the average bill for homeowners was only about 15% higher than that of renters ($486 versus $422).

As a percentage of income, however, there was a larger difference between owners and renters. The burden for renters was 60% higher than it was for homeowners (2.2% of income versus 1.4% of income).

Type of housing unit influenced who paid directly. As shown in Table 5, the type of housing unit had a significant effect on whether the household paid directly for water and wastewater service. More than 81% of detached, one-unit (single-family) homes paid directly. This figure declined steadily as the type of housing became more clustered, culminating in buildings with 10 or more units having fewer than 10% of households paying directly for service.

Among those households paying directly, the percentage of income spent for service increased for single- and two-family buildings as the housing became more clustered, going from 1.4% of income for one-unit detached homes to 2.2% of income in two-unit buildings. Larger buildings, however, then saw a decline in the water and wastewater burden.

Size of housing unit was not a significant factor. Table 6 shows the effect of the size of the housing unit (the number of rooms in the home) on the water and wastewater bill. The average bill was not very different for

TABLE 4   Relationship between housing ownership and water and wastewater bill

Housing Ownership	Number of Households	Households Paying Directly for Water and Wastewater Service %	Average Bill of Paying Households $	Average Bill as Percentage of Income
Owned	69,041,712	78.32	486	1.38
Rented	34,393,394	29.15	422	2.22

TABLE 5   Relationship between type of housing unit and water and wastewater bill

Type of Housing Unit	Number of Households	Households Paying Directly for Water and Wastewater Service %	Average Bill of Paying Households $	Average Bill as Percentage of Income
1 unit, detached	64,022,653	81.05	488	1.42
1 unit, attached	5,757,810	67.49	461	1.64
Mobile home	7,227,628	47.38	370	1.96
2 units in building	4,315,621	43.45	514	2.21
3–9 units in building	9,425,711	19.02	431	2.09
10 or more units in building	12,577,743	9.56	319	1.69

In fact, larger homes had a slightly lower bill than did smaller homes ($442 versus $448).

Furthermore, the bill as a percentage of income declined steadily as the size of the home increased. This is not surprising because the size of the home would be expected to be highly correlated with household income.

Statistical analysis determines correlation among various factors influencing water and wastewater cost. Charges for water and wastewater are determined in large part by local factors. These typically include the location and quality of source and receiving water, the distance over which water or wastewater must be transported, the energy and labor costs, the age of the facilities, the availability of grants or low-interest loans, the degree of cross-subsidization to or from other municipal services, and the taxes imposed on utilities. It is not expected, therefore, that a statistical analysis of national data would be able to explain differences in water and wastewater costs with a high degree of certainty. Such an analysis, however, should be able to highlight several factors that affect the magnitude and direction of these costs at the household level.

A multilinear regression analysis, using stepwise regression, was performed to determine if certain factors were correlated with the level of household water and wastewater costs. Stepwise regression is a form of multivariate regression analysis that examines the statistical

The output of this type of analysis is a "constant" or base value (expressed here as dollars per year spent for water and wastewater service) and the amount by which each variable would increase or decrease the base amount. The results of the analysis were instructive, not because they can be used to predict water and wastewater costs for a particular household, but because they showed that certain factors were highly correlated with the level of these costs.

In particular, the results of the analysis showed that five factors were highly correlated with the level of household costs for water and wastewater service. As expected, these factors did not "explain" why a household's water and wastewater bill was a particular amount. The overall model explained only 7.5% of the variation in these costs nationwide (i.e., the R^2 of the regression model was 0.075).

The results were statistically significant, however, meaning that it was unlikely that the variations observed were the result of random chance. For the model as a whole (as measured by the F statistic) and for each of the variables in the model (as measured by the t statistic for each variable), there was a less than a 1 in 1,000 chance ($p < 0.001$) that the observed trends were the result of random occurrences.

As shown in Table 7, five factors were highly correlated with a household's cost for water and wastewater service: (1) number of people in the household,

TABLE 6   Relationship between number of rooms and water and wastewater bill

Number of Rooms	Number of Households	Households Paying Directly for Water and Wastewater Service %	Average Bill of Paying Households $	Average Bill as Percentage of Income
1–3 rooms in home	16,460,543	25.59	448	2.34
4–6 rooms in home	56,437,067	62.40	442	1.69
7 or more rooms in home	30,537,496	80.79	530	1.12

TABLE 7   Results of multivariate regression analysis

Variable	Coefficient—$/year	t Statistic	Significance—p
Constant	295.29	2,516.06	<0.001
For each person in household, add:	39.26	1,264.41	<0.001
If value of home is			
$100,000 to $249,999, add:	29.52	277.06	<0.001
$250,000 to $499,999, add:	113.51	630.96	<0.001
$500,000 or more, add:	299.58	926.58	<0.001
If located in metropolitan area, add:	48.76	496.49	<0.001
For each dollar in annual household income, add:	0.0003	343.68	<0.001
If type of building is			
Two-unit building, add:	52.89	193.01	<0.001
Mobile home, subtract:	−54.61	−259.26	<0.001
Building with 10 units or more, subtract:	−110.16	−320.37	<0.001

R^2: 0.075

F statistic: 574,122 (p <0.001)

(2) value of the home, (3) home location (i.e., whether the home was located in a metropolitan area), (4) annual household income, and (5) type of housing unit (e.g., mobile home).

Interpreting the results of the regression analysis began with the constant, which was $295. This can be viewed as the hypothetical "base" bill, (i.e., the annual cost if none of the other factors were present, which was not possible, of course). The other factors then acted to increase or decrease this base amount.

Specifically, the factors had the following effects on the hypothetical base charge.

• Each additional person in the household increased the water and wastewater bill by $39 per year.

• If the home had a value between $100,000 and $250,000, the bill increased by $30 per year; for homes with a value between $250,000 and $500,000, the bill increased by $114 per year; and for homes with a value of $500,000 or more, the bill increased by $300 per year.

• If the home was located in a metropolitan area, the bill increased by $49 per year.

• The bill increased by about $3 per year for each $10,000 increase in the household's annual income.

• If the housing unit was part of a two-unit building, the cost increased by $53 per year. (This increase may be the result of two units being served through a single water and wastewater connection; thus, one unit reported a higher cost whereas the other reported that the cost was included in its rent or maintenance fee.) However, the cost decreased by $55 per year if the unit was a mobile home or $110 per year if the unit was part of a larger building (10 units or more).

The results of the regression analysis identified several factors that appear to be important in determining the differences in water and wastewater costs throughout the United States. These national trends, however interesting, did not capture the important differences among the states.

STATE RESULTS INDICATED WIDE DIFFERENCES

Direct payment statistics varied from state to state. The state-by-state analyses included the 50 states and the District of Columbia. Table 8 shows the wide range of variability among the states. For example, nationally 62% of households paid directly for water and wastewater service. On a state-by-state basis, however, this figure ranged from 35% of households in Vermont to 80% of households in Alabama.

The average water and wastewater bill also varied significantly, from a low of $334 per year in Nebraska to a high of $721 per year in Hawaii; within the contiguous 48 states, the highest average bill was in California, with an annual average cost of $591. Figure 1 shows the range of differences among the states in the annual cost of water and wastewater service. Figure 2 shows the average percentage of household income spent for this service in each state.

Income was related to direct payment for water and wastewater. Nationally, 21% of households reported annual incomes less than $20,000, whereas roughly the same percentage (23%) had incomes of $75,000 or more. Among households paying directly for water and wastewater service, however, only 16% had incomes less than $20,000, whereas 28% had incomes of $75,000 or more. This can be expressed as a ratio of low-income to high-income households; in this case, the ratio is 0.6, meaning that there were only 60% as many low-income households as there were high-income households paying directly for water and wastewater service. In other words, on a national level, households that pay directly were skewed toward those with higher incomes. This was reflected in the previous discussion indicating that the MHI

of households paying directly for water and wastewater was $49,900, or roughly 17% higher than the MHI of all US households.

This trend did not hold true in each state, however. Table 9 shows a state-by-state analysis of the income distribution of households that paid directly for service. The results in several states differed significantly from the national average.

As previously noted, nationally there were only 0.6 times as many low-income households paying directly for water and wastewater as there were high-income households. In contrast, West Virginia had 2.2 times as many low-income households paying directly for service as it had high-income households. Thus, of the 506,000 households that paid directly in West Virginia,

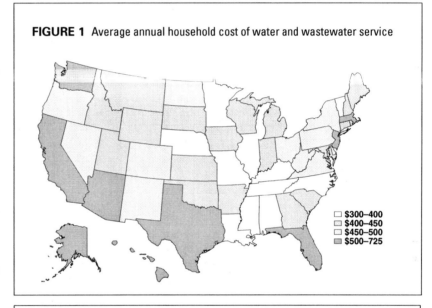

FIGURE 1 Average annual household cost of water and wastewater service

- $300–400
- $400–450
- $450–500
- $500–725

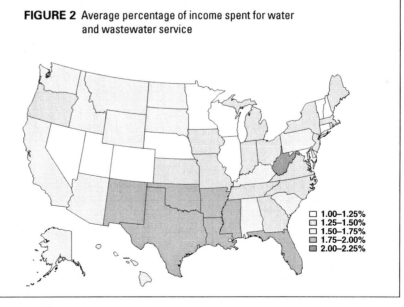

FIGURE 2 Average percentage of income spent for water and wastewater service

- 1.00–1.25%
- 1.25–1.50%
- 1.50–1.75%
- 1.75–2.00%
- 2.00–2.25%

TABLE 8   Summary of state results for all households

State	Number of Households	Households Paying Directly for Water and Wastewater Service %	Average Bill of Paying Households $	Average Bill as Percentage of Income
Alabama	1,684,863	79.96	371	1.64
Alaska	220,658	42.33	603	1.63
Arizona	1,865,750	69.66	552	1.65
Arkansas	1,018,569	77.26	418	1.85
California	11,285,634	60.56	591	1.52
Colorado	1,636,462	64.23	465	1.23
Connecticut	1,281,080	48.50	442	1.10
Delaware	295,250	58.65	456	1.31
District of Columbia	238,823	38.46	505	1.49
Florida	6,202,839	62.25	573	1.91
Georgia	2,941,022	68.91	426	1.43
Hawaii	394,954	53.87	721	1.62
Idaho	462,706	59.07	413	1.54
Illinois	4,508,577	62.86	426	1.22
Indiana	2,301,392	61.06	424	1.42
Iowa	1,135,366	70.17	421	1.37
Kansas	1,022,384	76.08	419	1.40
Kentucky	1,552,439	74.51	387	1.56
Louisiana	1,597,412	75.95	385	1.90
Maine	513,407	35.57	429	1.40
Maryland	1,949,916	58.19	432	1.01
Massachusetts	2,394,256	53.43	519	1.26
Michigan	3,721,234	55.58	420	1.34
Minnesota	1,876,421	56.41	390	1.01
Mississippi	1,010,191	77.55	372	1.89
Missouri	2,156,308	68.22	390	1.39
Montana	352,074	49.63	417	1.72
Nebraska	656,455	63.08	334	1.13
Nevada	736,094	60.42	472	1.22
New Hampshire	470,019	35.55	410	1.00
New Jersey	3,011,314	61.00	559	1.31
New Mexico	663,373	66.47	470	1.85
New York	6,869,263	41.60	450	1.25
North Carolina	3,071,642	57.33	378	1.37
North Dakota	253,107	60.63	451	1.54
Ohio	4,372,020	64.40	459	1.47
Oklahoma	1,312,418	73.56	443	1.87
Oregon	1,313,819	58.88	489	1.45
Pennsylvania	4,687,381	63.03	492	1.70
Rhode Island	400,788	53.11	420	1.29
South Carolina	1,496,062	65.82	419	1.65
South Dakota	285,788	67.32	418	1.50
Tennessee	2,185,992	74.60	380	1.45
Texas	7,222,150	72.53	555	1.92
Utah	694,329	72.48	439	1.20
Vermont	238,251	34.51	357	1.05
Virginia	2,663,683	58.75	465	1.32
Washington	2,239,923	60.50	569	1.58
West Virginia	717,626	70.47	465	2.22
Wisconsin	2,063,117	54.11	400	1.14
Wyoming	190,505	61.25	442	1.52

TABLE 9	Incomes of households paying directly for water and wastewater service			
State	**All Households Paying Directly for Service**	**Percentage With Incomes Less than $20,000 Per Year**	**Percentage With Incomes Greater than $75,000 Per Year**	**Ratio of Low Income to High Income**
Alabama	1,347,156	24.50	17.73	1.4
Alaska	93,408	8.16	40.94	0.2
Arizona	1,299,592	13.91	27.07	0.5
Arkansas	786,945	25.86	14.64	1.8
California	6,834,788	11.46	38.79	0.3
Colorado	1,051,159	10.96	33.10	0.3
Connecticut	621,355	8.97	40.71	0.2
Delaware	173,171	10.95	33.70	0.3
District of Columbia	91,853	11.88	44.27	0.3
Florida	3,861,267	16.78	24.35	0.7
Georgia	2,026,723	16.07	28.47	0.6
Hawaii	212,754	9.13	42.85	0.2
Idaho	273,341	16.39	19.95	0.8
Illinois	2,834,001	12.98	31.63	0.4
Indiana	1,405,117	15.79	21.86	0.7
Iowa	796,666	15.92	18.29	0.9
Kansas	777,787	10.90	22.47	0.8
Kentucky	1,156,694	22.94	19.16	1.2
Louisiana	1,213,237	26.37	18.05	1.5
Maine	182,637	15.87	22.05	0.7
Maryland	1,134,610	9.83	40.19	0.2
Massachusetts	1,279,369	10.00	41.70	0.2
Michigan	2,068,206	14.42	29.45	0.5
Minnesota	1,058,421	10.58	32.11	0.3
Mississippi	783,440	29.32	14.98	2.0
Missouri	1,471,012	17.90	22.04	0.8
Montana	174,745	21.46	15.18	1.4
Nebraska	414,087	15.13	21.44	0.7
Nevada	444,729	9.89	30.97	0.3
New Hampshire	167,101	9.49	33.10	0.3
New Jersey	1,836,784	9.22	44.64	0.2
New Mexico	440,923	22.36	18.51	1.2
New York	2,857,857	11.35	38.00	0.3
North Carolina	1,760,982	18.40	23.26	0.8
North Dakota	153,447	18.48	18.77	1.0
Ohio	2,815,702	15.92	24.70	0.6
Oklahoma	965,423	23.63	16.58	1.4
Oregon	773,564	12.64	26.12	0.5
Pennsylvania	2,954,611	17.75	24.44	0.7
Rhode Island	212,859	12.77	29.85	0.4
South Carolina	984,749	20.14	20.62	1.0
South Dakota	192,389	17.34	16.47	1.1
Tennessee	1,630,855	20.67	19.49	1.1
Texas	5,238,329	18.01	25.91	0.7
Utah	503,246	9.91	28.41	0.3
Vermont	82,210	12.94	23.20	0.6
Virginia	1,564,935	12.38	35.01	0.4
Washington	1,355,248	10.92	32.39	0.3
West Virginia	505,717	27.74	12.44	2.2
Wisconsin	1,116,360	12.31	25.98	0.5
Wyoming	116,683	18.64	18.56	1.0

28% had incomes less than $20,000 per year, but only 12% had incomes of $75,000 or more.

In total, there were 14 states with ratios of 1.0 or greater, i.e., states where there were more low-income (less than $20,000) households paying directly for service than there were high-income ($75,000 or above) households. The most extreme cases were in Arkansas, Mississippi, and West Virginia—all with ratios of 1.8 or more, meaning that their ratio of low-income to high-income households paying directly for service was more than three times the national average (ratio of 0.6).

In these states and others with high ratios, increased water and wastewater costs would affect low-income households at a significantly higher level than they would on a national basis. Conversely, states with very low ratios of low-income households paying directly for service would likely have fewer low-income households affected by a rate increase than would be expected from a national-level analysis. Such states include Alaska, Connecticut, Hawaii, Maryland, Massachusetts, and New Jersey—all of which had ratios of 0.2, meaning that high-income households paying directly for service outnumbered low-income households by about five to one.

Variability among states also seen in other factors. If similar comparisons were repeated for any other factor, the same type of variability would be found among the states. For example, the percentage of one-person households that paid directly for water and wastewater service varied from 23% in the District of Columbia to 72% in Mississippi (compared with the national average of 49%).

In summary, national averages provided the general relationships among factors (e.g., that the water and wastewater bill increased with income and household size or that it was little affected by the age of household members). An analysis conducted at the national level, however, cannot capture the substantial differences among the states. For every factor analyzed in this study, the range in variation among the states was substantial, forming wide ranges above and below the national average.

MAJOR DIFFERENCES SEEN IN LOCATION DATA

Classification of nonmetropolitan areas is problematic. Out of the 104 million households in the United States, approximately 83 million households were located in metropolitan areas, whereas only 21 million were located in nonmetropolitan areas (Rubin, 2003). Except in New England, metropolitan areas are determined at the county level, i.e., an entire county is either in a metropolitan area or it is classified as being nonmetropolitan. As discussed elsewhere, the economic and demographic characteristics of metropolitan areas differ significantly from those of nonmetropolitan areas (Rubin, 2003). In particular, the levels of income are substantially higher in metropolitan areas than in nonmetropolitan areas.

The PUMS data also showed substantial differences in the cost of water and wastewater service for metropolitan areas versus nonmetropolitan areas. However, any interpretation of the PUMS data must consider that the data set was able to classify 61.2 million households in metropolitan areas but only 4.0 million households in nonmetropolitan areas. The remaining households were in PUMAs that contained a mixture of metropolitan and nonmetropolitan areas. In other words, the PUMS data set contained data classifiable by metropolitan area for 74% of metropolitan households but only 19% of nonmetropolitan households. Thus, any conclusions drawn about differences between metropolitan areas and nonmetropolitan areas should recognize that the PUMS data did not make it possible to identify most nonmetropolitan households.

Costs were higher in metropolitan areas. With that caveat, Figure 3 shows that the average water and wastewater cost in metropolitan areas was 32% higher than in nonmetropolitan areas ($512 versus $386). However, because incomes in metropolitan areas tended to be at least 40% higher than they were in nonmetropolitan areas (Rubin, 2003), the average burden was lower in metropolitan areas (1.4% of income) than in nonmetropolitan areas (1.8% of income).

Further analysis showed that one-person households in nonmetropolitan areas were more likely to pay directly for service than one-person households in metropolitan areas (58% versus 45%). In addition, important differences between metropolitan and nonmetropolitan areas were found in the makeup of the type of housing (e.g., one-family detached, mobile home) and the direct payment for water and wastewater. Homeowners in metropolitan areas were more likely to pay directly than were homeowners in nonmetropolitan areas (84% versus 69%). In contrast, renters in metropolitan areas were much less likely to have to pay directly for water and wastewater than were renters in nonmetropolitan areas (24% versus 43%). In both instances, however, residents of metropolitan areas paid a lower percentage of their household's income for service (1.3% for owners and 2.0% for renters), compared with residents of nonmetropolitan areas (1.6% for owners and 2.7% for renters).

In summary, the cost of water and wastewater service in metropolitan areas tended to be significantly higher than in nonmetropolitan areas for any relevant characteristic (e.g., household size, income, and tenure). That cost difference, however, was more than offset by the higher income levels in metropolitan areas, such that households in nonmetropolitan areas paid a higher percentage of income for service than did comparable households in metropolitan areas (except, of course, when households with the same level of income were compared).

CONCLUSION

This analysis of US Census Bureau survey data yielded a number of significant findings.

• Nationally, about six of every ten households paid directly for water and wastewater service. The other households either had the cost included in a rental payment or maintenance fee (18%) or did not pay for service (20%).

• The average annual cost of water and wastewater for a household was $476 per year and ranged from a low of $334 in Nebraska to a high of $721 in Hawaii.

• The median household income of those paying directly for service was about $7,700 per year (17%) higher than the national median income.

• The average household spent 1.6% of its income for water and wastewater service, ranging from a low of 1.0% of income in New Hampshire to a high of 2.2% of income in West Virginia.

• The number of people in a household had a relatively small effect on the annual cost of water and wastewater service, increasing the bill by less than $40 per year for each additional person.

• Increased income had a small effect on the cost for the household. Each $10,000 increase in income could be expected to increase the bill by approximately $3 per year.

• Homeowners were almost three times as likely as renters to pay directly for service, but there was little difference in the average bill between these groups.

• If other factors were held constant, the number of rooms in the home showed no relationship to the cost of service.

• The value of the home was highly correlated with the water and wastewater bill, at least among higher-priced homes (those with a value of $100,000 or more).

• Some groups often thought to have little direct responsibility for paying for service did, in fact, pay bills directly. These groups included mobile home dwellers (47% paid directly for service), households with incomes less than $10,000 per year (42% paid directly), households with incomes between $10,000 and $20,000 per year (50% paid directly), and one-person households (49% paid directly).

This article highlighted some of the important differences in these factors throughout the United States. State-by-state variations in all of these factors and in the cost of water and wastewater service will lead to national-level analyses or conclusions having

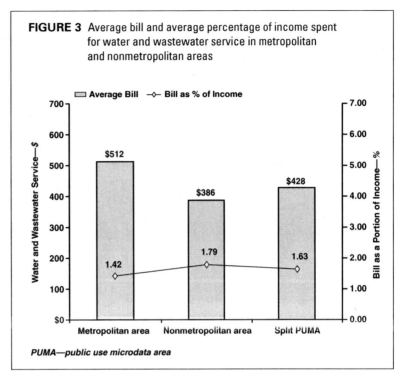

FIGURE 3 Average bill and average percentage of income spent for water and wastewater service in metropolitan and nonmetropolitan areas

PUMA—public use microdata area

very different effects around the country. For example, although as a group few renters (29%) paid directly for service, in nonmetropolitan areas the percentage of renters who paid directly was 43%.

In conclusion, the US Census Bureau's Public Use Microdata from the 2000 census provides an important source of information about the costs and burdens of water and wastewater service at the household level. This article represents an initial effort to summarize those data, characterize the costs of US water and wastewater service, and begin identifying and explaining some of the tremendous variation in costs in various parts of the United States. Other researchers are encouraged to use this rich data set to further refine this analysis and to continue to mine the data for additional insights into the costs and burdens associated with the provision of water and wastewater service.

ACKNOWLEDGMENT

The author would like to thank the reviewers for their helpful comments and the National Rural Water Association (NRWA) for funding the research summarized here. The full report of that research, *The Cost of Water and Wastewater Service in the United States,* is available on the NRWA website at www.nrwa.org and on the author's website, the Public Utility Home Page, at www.publicutilityhome.com.

REFERENCES

Raftelis Financial Consulting, 2002. *Raftelis Financial Consulting 2002 Water and Wastewater Rate Survey.* Charlotte, N.C.

Rubin, S.J., 1998. A Nationwide Look at the Affordability of Water Service. Proc. AWWA Ann. Conf., Dallas.

Rubin, S.J., 2003. Economic Characteristics of Small Systems. *Critical Issues in Setting Regulatory Standards* (2nd ed.). National Rural Water Assn., Duncan, Okla.

Saunders, M. et al., 1998. *Water Affordability Programs.* AwwaRF, Denver.

US Bureau of Labor Statistics, 2002. *Consumer Expenditure Survey 2000.* BLS, Washington.

US Bureau of Labor Statistics & US Census Bureau, 2003. *Current Population Survey.* http://www.bls.census.gov/cps/cpsmain.htm, accessed June 1, 2003.

US Census Bureau, 2003. *2000 Census of Population and Housing, Public Use Microdata Sample, United States: Technical Documentation.* US Census Bu., Washington.

USEPA, 1998. *Variance Technology Findings for Contaminants Regulated Before 1996.* Ofce. of Water, Washington.

USEPA (US Environmental Protection Agency), 2002. *2000 Community Water System Survey.* EPA 815-R-02-005A, Washington.

ABOUT THE AUTHOR

Scott J. Rubin is an attorney and consultant whose practice is limited to matters affecting the public utility industries. A member of AWWA and the American Bar Association Public Utility Law Section, he has researched water rates and the affordability of water service for almost 15 years and has consulted on affordability issues for AWWA and the National Drinking Water Advisory Council.

Water Is Cheap—Ridiculously Cheap!

MAKING SOME KEY COMPARISONS REGARDING WATER MIGHT HELP US REALIZE THAT THE DRINKING WATER WE TAKE FOR GRANTED IS WORTH FAR MORE THAN WHAT WE CURRENTLY PAY FOR IT.

BY STEVE MAXWELL
(*JAWWA* June 2005)

W ater quality and water availability are huge challenges that will face the United States—and the rest of the world—over the coming decades. Everywhere you look today, there are stories about water shortages, water pollution, political wrangling over water rights ownership, and the huge capital expenditures that will be required to maintain our water infrastructure. The ultimate conclusion to many of these issues is that in one way or another we will all inevitably pay more for water. As a society, we simply have to spend more of our dollars to correct the water pollution problems that we have created, to ensure that everyone has access to this basic human right as the planet's population continues to expand, and to finance the vast treatment and distribution infrastructure that we need.

At the same time, however, there is growing political resistance and social concern about water prices rising over the long term—rate shock. Although water is still very cheap relative to its real value (as outlined later), water rates and user fees are still political hot potatoes in many cities and towns throughout North America. Some city councils and mayors live in fear of taxpayer revolts and are often afraid to raise water rates, even though water bills are still a tiny part of an average family's monthly expenses. In other cities, battles are raging between municipal officials and private contractors who have tried to force rate increases, and controversies over water issues are increasingly resulting in the expulsion of elected officials. But are all of these political battles, financial concerns, and instances of public hand-wringing really justified? This article discusses the real value of water versus the average price of water relative to the other necessities and various luxuries of life. (See Article 3.)

Water is vital to each and every one of us, and without it, life cannot exist. Relative to its true value and significance, water isn't expensive—it is ridiculously cheap! In fact, public drinking water availability in the United States is one of the great economic bargains of all time. You'd have to look far and wide to find another product whose real value to the consumer is so high relative to its price—and to find a commodity whose price is typically so unrelated to its actual cost. A look at the facts and some interesting anecdotes quickly confirms this.

WHAT DO AMERICANS PAY FOR WATER?

First, let's look at what we actually pay for our drinking water in the United States. Clearly, the situation in terms of water supply and delivered costs varies significantly across the country, as might be expected given the range of climates, weather patterns, and conditions of water infrastructure in different regions of the United States. Although what we pay for water varies widely, the average price is about $2.50 per 1,000 gal (3,785 L) or approximately $20 per month for the average US family.

The *2004 Report on Water and Wastewater Rates,* authored by AWWA and Raftelis Financial Consultants Inc., found that the average cost of water to the US consumer was $19.11/1,000 cu ft (28.32 m³), or $0.0026/gallon. (To convert costs per gallon given in this article to cost per litre, divide the dollar amount by 3.785.) Rates vary widely, though they tend to be higher in the northeastern part of the country, and—contrary to intuition—lower in the South and West. According to the 2004 Water Pricing Survey published in the September 2004 issue of *Global Water Intelligence,* Boston, Mass., residents pay $0.004/gallon of water, whereas residents of Denver, Colo., pay $0.0018 cents/gallon—and residents of Las Vegas, Nev., that shining oasis in the desert, pay just $0.002/gallon. AWWA's detailed analyses of regional water rates have clearly demonstrated that water pricing is often political rather than economic.

WHAT DO OTHER COUNTRIES PAY FOR WATER?

How do water rates in the United States compare

with those of other countries? As shown in Table 1, a recent report summarized average water prices in various developed countries. This report estimated the average price in the United States to be $0.0023/gallon, which is fairly close to the figure cited previously. The evidence here is pretty clear—Americans generally pay less for their water than do most people in other developed countries.

Germans pay almost a penny a gallon, whereas only the Canadians—with their relatively small population and vast water resources—pay less than Americans do. Of course, this kind of analysis is only applicable to the more developed countries of the world—the vast majority of people in less-developed countries don't even have the option of buying clean drinking water.

THE RELATIVE COST OF WATER

It is instructive (and perhaps a bit amusing) to look at the price of water compared with the prices of other key products that many of us buy and use every day. A look at Table 2 helps us realize just how cheap tap water really is.

There is not much doubt as to which of these substances is the most critical to humans—we can't live for more than seven or eight days without water. But water remains hundreds or even thousands of times cheaper than the other liquid commodities or extravagances that we frequently buy.

Perhaps of particular interest here is the price of bottled water, which is between a thousand times and ten thousand times the cost of average tap water but is (in most cases) barely distinguishable from tap water. The US Food and Drug Administration loosely regulates the bottled water industry. However, in a publication titled *The World's Water 2004–2005: The Biennial Report on Freshwater Resources,* the Pacific Institute (an independent water think tank) said that bottled water "standards vary from place to place, testing is irregular and inconsistent, and contaminated source water may lead to contaminated products." Although most of the US population seems to be highly resistant to (or even shocked at) the notion of rising water rates, for most Americans, a 20% increase would be roughly equivalent to buying a couple of containers of bottled water a month!

In that same publication, the Pacific Institute estimated that on a worldwide basis the total annual consumer expenditures for bottled water approach $100 billion annually—"a vast sum that both indicates consumers are willing to pay for convenient and reliable drinking water and that *society has the resources to make comparable expenditures to provide far greater quantities of water for far less money by investing in reliable domestic supplies*" (italics added). In other words, if we were to spend this money on building public systems instead of buying bottled water, we could easily

TABLE 1	Average price of water by country
Country	Average Price US$/gal (US$/L)
Germany	0.0084 (0.0022)
Denmark	0.0083 (0.0022)
United Kingdom	0.0057 (0.0015)
Holland	0.0054 (0.0014)
France	0.0053 (0.0014)
Belgium	0.0047 (0.0012)
Italy	0.0036 (0.0010)
Spain	0.0033 (0.0009)
South Africa	0.0032 (0.0008)
Finland	0.0030 (0.0008)
United States	0.0023 (0.0006)
Canada	0.0020 (0.0005)

Source: NUS Consulting 2003/2004 International Water Report and Cost Survey (as quoted in *Global Water Intelligence,* September 2004).

TABLE 2	Price comparison of water versus other widely used consumer goods
Product	Average Price $/gal ($/L)
Tap water*	0.0026 (0.0007)
Gasoline†	2.20 (0.58)
Coca-Cola®‡	2.64 (0.70)
Organic milk†	4.25 (1.12)
Tide® liquid detergent‡	8.39 (2.22)
Imported beer†	12.00 (3.17)
Evian® bottled water‡	21.19 (5.60)
Peaberry Coffee® mocha drink‡	22.28 (5.89)
Pepto-Bismol®‡	58.52 (15.46)
Vicks Formula 44D® cough syrup‡	96.67 (25.54)
American whiskey†	150.00 (39.63)
Visine® eye drops‡	741.12 (195.80)
Revlon® nail enamel‡	983.04 (259.72)
Good French wine†	1,000.00 (264.20)
Chanel® No. 5 perfume‡	45,056.00 (11,903.83)

*AWWA and Raftelis Financial Consultants Inc.
†TechKNOWLEDGEy Strategic Group
‡AWWA

TABLE 3	Total US spending on water versus other products and activities	
Product/Activity	**Annual Expenditure $ billions**	**Comparison With Amount Spent on Water**
Viagra®	2	6%
Prozac®	3	9%
Cosmetic surgery	12	34%
Pornography	14	40%
Water	35	
Tobacco products	40	114%
Legalized gambling	68	194%
Alcoholic beverages	140	4 times
Military defense spending	558	16 times

Sources: American Society of Plastic Surgeons, CitizenLink, *Forbes Magazine*, the Pharmaceutical Research and Manufacturers of America, Reuters, the Statistical Abstract of the United States, and various company websites.

provide a much greater share of the world's population with clean, safe drinking water.

THE MONTHLY COST OF WATER VERSUS OTHER BASIC SERVICES

Another way to look at the relative cost of water is to review how much we typically pay for other basic services each month. The AWWA/Raftelis study cited previously suggests that the average US family pays about $20 a month for water. This compares with somewhere around $30 a month for Internet service, about $40 dollars a month for basic cable television service, $75 a month for telephone service, and $80 a month for electricity. Again, we pay much less for the service that—if push ever comes to shove—we clearly need the most.

A few additional examples serve to reemphasize and bring home the relative inexpensiveness of water. A quick check at your local nursery shows that chicken manure typically costs around $15 per ton, and potting soil (fancy dirt!) can cost as much as $2,500 per ton. By comparison, tap water goes for about $0.60 a ton.

US SPENDING ON WATER VERSUS OTHER PRODUCTS

Let's look at the cost of our water in another way—in terms of how much we as a society spend on water versus what we spend in other areas of life.

What do we really spend on water? The data are not exactly comprehensive, but according to AWWA, every day our public water utilities process approximately 38 bil gal (140 GL) of water, and the US Census Bureau estimates that the United States currently has a population of about 296 million people. That works out to

about 128 gal (484 L) of water per person per day. (Total water use, including untreated or less-treated water for agricultural irrigation and thermal power generation multiplies that per-capita figure by about a factor of 10.) If the average cost of water is $0.0026/gallon and if we each use about 128 gal (484 L) of treated water a day, it works out to a per capita cost of about $121 a year. If there are almost 300 million Americans, that works out to a total annual water cost of around $35 billion per year. According to a 2005 report from Environmental Business International and TechKNOWLEDGEy Strategic Group, the total annual revenues to water utilities in the United States are estimated at $33.8 billion. To help put this in context, Table 3 lists what the United States as a country spends on various other products, consumables, and activities.

WATER WASTAGE

Because water is so cheap, we tend to waste a lot of it. It is difficult to measure exactly how much water we waste—because this is obviously a somewhat subjective value judgment. Is watering a yard in Phoenix, Ariz., a waste of water? Is a 15-minute shower a waste of water? Is it a waste to wash your car once a week, and so on? However, one recent and comprehensive review pinpointed the United States as the most wasteful nation on earth in terms of water use. In a study by the Center for Ecology and Hydrology in the United Kingdom, the United States ranked last out of 147 countries in terms of efficient water use.

OUR IGNORANCE ABOUT WATER

Despite this veritable mountain of data (which basically show that water is still absurdly cheap relative to its true value), huge political controversies are often generated by municipal attempts to raise water rates by 10% or 20%. Town councils or mayors are regularly removed from office for raising, or threatening to raise, water rates—despite all of this evidence, staring us in the face, that water is obviously worth much more than what we pay for it. For most of us, even a large percentage increase in our water rates would be equivalent to no more than $10 or $15 a month. This is probably less than what many of us are already spending on bottled water (even though we have good, clean tap water available at one one-thousandth of the cost). According to the previously cited study by the Pacific Institute, if we took the dollars we currently spend on bottled water and used them instead to address those oft-cited infrastructure requirements, the infamous spending gap—the amount between the current levels of infrastructure investment and the levels that the US Environmental Protection Agency estimates are required to maintain our infrastructure—would almost disappear.

The United States is blessed with a wealth of water resources in most regions of the country, and we clearly have the innovative spirit and the technological wherewithal to figure out how to treat and transport water to those regions of the country with less water. Unfortunately, as a society we also tend to be characterized by an ignorant and careless attitude about water resources and water utilization in general.

We would all do well to regularly remind ourselves how valuable water really is. Think about those times your local utility has had to turn off your connection to do repairs and how difficult it was to get through the day without any water. Think about the last time you went hiking or camping and ran out of drinking water. Consider the fact that in many parts of Africa, women and children spend a good part of every day hauling water for the basic human needs of drinking, preparing food, and cleaning. Surely it's about time we realized that our water is worth a lot more to us than the price we currently pay for it.

Yes, water frequently falls from the sky. Yes, two thirds of the planet is covered by water. Yes, freshwater is abundant in many parts of the globe. But it's not always clean, it's not always located where we need it, and it costs the world hundreds of billions of dollars a year to collect, treat, store, and distribute that water. Sooner or later, we are all going to have to get used to reasonable and across-the-board water rate increases—especially those of us in the United States.

The Value of Water: What It Means, Why It's Important, and How Water Utility Managers Can Use It

BY BOB RAUCHER
(*JAWWA* April 2005)

There are good reasons for the increased focus on the value of water. Safe water—essential to life—is a highly valuable natural resource. Utility services also provide considerable value by reliably delivering ample quantities of a safe and aesthetically pleasing product to their customers' taps. Both aspects—water's role as a life-sustaining natural resource and the reliable delivery of service—form the basis of the value that water agencies deliver to their communities. The value of both aspects must be considered if we are to manage natural and utility assets in a wise and sustainable manner.

WHAT IS VALUE?

Despite the recent attention to value, a general vagueness about what the term means exists. Likewise, it is not clear what types of values apply to water, how great the values are, or how water utility managers can or should apply value-based insights. This article attempts to provide some answers to these questions. A recently completed AWWA Research Foundation report on this topic, *The Value of Water: Concepts, Estimates and Applications for Water Managers,* provides additional detail and guidance (see footnote on this page for availability information).

Most people understand that water is essential to life and therefore has a high value. Paradoxically, general ambiguity exists about how something so valuable and essential to life can be purchased at the tap at a price much less than a penny per gallon. As Adam Smith said more than 200 years ago, "How is it that water, which is so useful that life is impossible without it, has such a low price—while diamonds, which are quite unnecessary, have such a high one?"

It's also puzzling that 75% of water utility customers at times buy bottled water at a price up to 1,000 times the price charged for safe tap water. Further, concerns exist about how extremely difficult it can be for utilities to obtain modest price increases to cover necessary investments for regulatory compliance and infrastructure renewal. How can all of these realities coexist? What does this tell us about values? And what can and should the water profession do to help preserve and enhance the value of what it provides?

To answer these questions, we must begin by defining what we mean by value. *Webster's New World Dictionary* provides several useful definitions.

A quantitative measure. One definition is "the worth of a thing in money or goods," a quantitative measure, such as might be expressed in dollar terms. To economists, value is a manifestation of how much well-being a person derives from acquiring another unit of the good or service, because the person is willing to trade something of value in exchange for it (e.g., money, or an hour of labor). The value is thus how much the person is willing to pay for the good or service.

A quality-based approach. Another definition is "the quality of a thing that makes it more or less desirable or useful." This quality-based definition underscores the importance of attributes such as the location, purity, timing, and reliability of water. The value that utilities provide to their customers depends on the reliable delivery of a safe product.

Ethical or moral considerations. A third definition offered by *Webster's* is "beliefs or standards." In this context, value refers to core ethical conventions and morals that help govern or reflect a culture and society. For example, when a politician refers to "family values," he or she is not referring to a family's monetary worth or physical characteristics. The value reference pertains to higher-order principles that transcend the notion of markets or attributes. In this context, value refers to

If you would like to obtain a copy, please contact the AWWA Bookstore at 1-800-926-7337 or awwa.org/bookstore.

what may be collectively viewed as fundamental "truths" about how we as humans relate to one another and the world or universe around us. One example of such a value is the view expressed by some that "because water is essential for life, safe water should be a basic right for all people, regardless of their ability (or willingness) to pay." Another example is the use of water in various religious practices, such as baptisms and ritual bathing. In these practices, water is a vehicle for transcending or washing away sins or otherwise seeking spiritual purity and renewal.

Within the public policy-making domain, in which water resource issues typically are deliberated and decided, a broad view of value must be considered. In the context of public discourse and decision making, values must reflect what is important to appointed and elected officials, customers, stakeholders, and the public. These discussions need to accept a broad perspective of beliefs and perceptions about water values, embodying the types of value that might motivate or resonate with the general public, key stakeholder organizations, and those with water governance responsibilities. For example, local autonomy and a community's way of life may surface as key values that must be articulated and considered in some local water agency decision-making activities.

COST, PRICE, AND VALUE: NOT THE SAME

Part of the challenge in tapping the value of water is the confusion that arises because of three related but very distinct terms: *cost, price, and value*. Each term has a specific and distinct meaning, yet the terms are often used interchangeably, fostering unintended confusion.

• Cost refers to the expense of producing and delivering a unit of water.

• Price refers to the rate charged to a customer for the unit of water delivered. Usually prices (rates) are based on costs, because utilities strive to recover their capital and operating expenses (e.g., by developing rates based on cost of service).

• Value is a more ambiguous concept than cost and price. Utilities incur capital and operating costs, and set and apply prices; therefore, these costs and prices are easily identified and measured. Value is harder to measure. As Oscar Wilde said, "Some people know the price of everything and the value of nothing."

As noted earlier, economists define value as the maximum willingness to pay—in terms of money or other goods and services—that a person would exchange for an additional unit of a good or service. The WTP concept is what defines the "demand curve" for water (i.e., the demand curve reflects the WTP-based value of water to its users). In contrast, the utility's costs of providing that water is reflected in what economists refer to as the "supply curve."

If water were exchanged in what economists refer to as a perfectly competitive market, then the price of water would be determined by the point at which the supply curve (cost) and demand curve (value) intersect. In this economists' Nirvana, the market-clearing price would equal the value of the last (marginal) unit of water consumed, and the price simultaneously would also equal the marginal cost of supplying that last unit. Thus, for goods and services that are exchanged in free and open competitive markets, values can be deduced from the behavior of buyers and sellers, as reflected in market-driven prices and quantities of the goods exchanged.

But for goods and services associated with water, which for various reasons are not transacted in perfectly competitive markets, measuring value is a challenging proposition. Economists have developed several tools to estimate "nonmarket values," and past research sheds some light on the range of values that may apply to water. Some examples of value estimates are provided later in this article. However, by and large, little is readily known about how much value customers place on water and water utility services.

WATER SOMETIMES UNDERVALUED, BUT MORE OFTEN UNDERPRICED

What does this economist's discussion about cost, price, and value mean for water agencies and their managers? Because little information is available about what water and water services are worth, prices typically are set solely based on what can be readily observed—namely costs. Thus, value is not typically part of the price-setting paradigm.

The problem, then, is not necessarily that water is undervalued, though many customers may not fully appreciate the value of water and water service, but rather that water is often under priced. Water costs are the sole (or primary) basis in pricing water, and water costs typically are low in comparison to other goods and services consumers routinely purchase. Consumer water prices tend to be relatively low because of a number of factors.

Prices don't reflect opportunity costs. The cost of water as a resource is often not borne (or fully borne) by water providers and is thus not included in consumer prices. Utilities and customers typically bear out-of-pocket expenses for treatment, pumping, and other activities associated with delivering a safe product, but the natural asset used (i.e., raw source waters) is often obtained for free. This is akin to paying for gasoline based on the refining and shipping costs without bearing the cost of the crude oil. Even where water rights must be acquired, long-held rights were either established at no cost or purchased at past low prices; thus they have little impact on prices and do not reflect today's opportunity costs for those waters.

Prices are often based on average-versus-marginal costs. To meet growing demands, many utilities incur high marginal costs to add expensive new water to their supply mix (e.g., by buying expensive water rights, by expanding conservation programs, or by using high-cost technologies to render low-quality sources potable, as in desalination). These high marginal costs are diluted by the low-cost base, so customers whose prices are based on average costs do not see or pay for the true cost of additional water use.

Infrastructure renewal and other necessary investments have frequently been overly deferred. Such deferral means that the costs are not reflected in current average cost-based prices. Because marginal costs typically are incomplete or understated or their price signal is diluted, cost-based water prices are lower than they should be in economic terms. Further, there is no countervailing upward pressure on prices because values—which presumably are quite high—remain vague and outside of the price-setting calculus.

Affordable water is a social good. Because water service is essential to good health and basic sanitation, there is a concern that it be priced so as to be affordable to all. Affordability concerns thus can lead to underpriced water. However, other forms of low-income assistance are better mechanisms for protecting low-income community members than is underpricing water (AWWA, 2005).

The challenge thus contains two key elements:
- Full-cost pricing should be more seriously examined and implemented.
- Water professionals must find ways to better communicate the value of both water and water service provision. As a "silent utility," water agencies have become invisible, and the services they provide are often taken for granted. Because customers and governing officials do not often think systematically about the value of water or water-related services, they tend to focus on the more immediate, visible costs. It is up to the water utility profession to better understand water's value and to effectively communicate that value to the public.

HOW HIGHLY VALUED IS WATER?

To a residential customer, the value of water refers to how much the household would be willing to pay for an additional 1,000 gal (3,785 L)—or to avoid having the quantity of water they currently use reduced by 1,000 gal (3,785 L). Such residential household willingness to pay is not readily observed or measured.

A household's willingness to pay depends on whether one is considering the water used for essential purposes such as drinking, cooking, and bathing (i.e., highly valued uses), or if it's intended for discretionary purposes (e.g., lesser-valued applications such as hosing off the sidewalk). In much of North America, water typically is available at such a relatively low cost that cus-

tomers tend to use some water in low-value applications (e.g., inefficient turf irrigation practices). This means that much of what we can observe about household values pertains to water purchased at the margin for relatively low-value uses.

Despite the challenges in measuring values for such different uses, available empirical research literature results do provide some useful insights into the probable general magnitude of water values. More detailed reviews of this literature are provided in the forthcoming AwwaRF report, but a brief synopsis is offered here (all monetary values reported are in 2003 US dollars unless stated otherwise).

- Residential customers of municipal systems may value water at $1,400–$2,300/acre-ft ($1.14–$1.86/m^3). An acre-ft is about the amount consumed by two average American households per year. These monetary values are the equivalent of about $4.00–$6.00/1,000 gal ($1.06–$1.59/1,000 L). These estimates must be interpreted with considerable caution, as they are based on limited data from a small number of studies.

- Households also appear to place relatively high values on water supply reliability. Willingness to pay estimates to avoid drought-related or other restrictions on water use have been estimated to be $109–$421/household/year. This suggests a reliability value to residential customers that may translate to about $4,000/acre-ft ($3.24/m^3).

- Commercial, industrial, and institutional purpose values, for which data are limited, range from $28 to $804/acre-ft ($0.02 to $0.65/m^3).

- Agricultural irrigation values, based on empirical investigations, suggest a mean value of $21/acre-ft ($0.02/m^3) and $837/acre-ft ($0.68/m^3) for hops and potatoes, respectively. Mean values estimated for water applied to other crops tend to be within the lower portion of this range. Examples include $60/acre-ft ($0.05/m^3) for wheat, $134/acre-ft ($0.11/m^3) for cotton, and $242/acre-ft ($0.20/m^3) for vegetables.

- Instream water uses can provide considerable value. Some recreation-based estimates suggest $10–$770/acre-ft ($0.01–$0.62/m^3). Other recreational and nonuse values for instream uses have been developed, but do not always translate directly into meaningful dollar/acre-ft measures. For example, the whitewater rafting enabled by flows in Gore Creek, Vail, Colo., generates a value of more than $1 million/year to boaters and local economies. In peak-use summer periods, this translates to a value greater than $1,600/acre-ft ($1.30/m^3).

All dollar values provided here are estimates, often based on limited data, and are typically very sensitive to the location, uses, timing, and other site- and circumstance-specific factors. Therefore, the values provided are only illustrative, and any person using them should do so with caution; the values may not be suitable or accurate for the circumstances under consideration.

In addition to the values described earlier, other water sources and types of water-generated value exist. For example, the utility provision of reliable quantities of high-quality, reasonably priced waters can be instrumental in sustaining the economic base of a community or region. Water is also a valuable tool for managing growth in ways that are consistent with the community's broader objectives. It also has important cultural and spiritual values, as well as aesthetic ones (e.g., in instances in which water is used to provide green space, fountains, and other amenities).

ATTRIBUTES AFFECT WATER VALUES

The value of water depends largely on several factors. These include
- how the water is used,
- the quality of the water,
- the time and location at which water is available, and
  - water's relative scarcity.

Many people have an intuitive appreciation of water's considerable value. Most also recognize that the interaction of timing, location, quality, and end-use is integral to assessing water's value.

To a person stranded in a desert, a cup of freshwater has more value than a thousand gallons of runoff in a rain-swollen river. Water's value thus depends, in part, on the relative scarcity of water in a location (defined by supply relative to demand) and on the value of the activities that the water-use supports (e.g., the market value of additional crop yields enabled by irrigation or additional benefit to a community that has enough municipal water to support desired activities such as lawn watering and job creation).

Water quality, timing, and reliability are other attributes that play a large role in determining water's value. A gallon of high-quality water that keeps a human being alive may be "priceless" whereas the same gallon spread on the sidewalk from a misplaced lawn sprinkler may have no value. A gallon of untreated wastewater may have negative value because of the harm it causes when released into the environment. Likewise, an extra 100 acre-ft (123,000 m^3) of water left in a river at the right time and place may preserve an endangered aquatic species and enable downstream farmers to profitably grow their crops; if that same water were added to a flood-swollen river it might imperil lives and inflict considerable property damage.

To further complicate matters, different individuals hold different values and therefore might choose to use water resources in different ways. Thus the issue for water managers is not always limited to understanding the magnitudes of various values, but, in some cases, is also a matter of deciding whose values take prominence (i.e., who "wins" in a water decision).

For example, water can be

- extracted from rivers or aquifers for use by the residential and commercial customers of municipal or investor-owned community water systems,
- extracted for irrigation by farmers,
- left instream to support recreation (e.g., fishing and boating) and ecological purposes (e.g., providing critical habitat for fish and other species, some of which may be designated as threatened or endangered).

In any of these competing uses, the water generates values. The challenge for water resource professionals is to understand which use promotes the greatest overall value to society. This requires not only understanding the magnitude of the various possible values, but also insights on who gains and who loses (i.e., who benefits, and who bears the costs) from a water allocation or other management decision that affects water.

WATER VALUES BECOMING MORE IMPORTANT

The value of water and the value of providing water service will become increasingly important considerations moving forward. Most water utility service mangers and governing officials will need to raise prices to address infrastructure renewal, enhanced security precautions, and regulatory compliance demands. These costly investments are necessary to maintain reliable and safe water service, but the added costs will be hard for many governing officials and customers to accept. In order to avoid rate shock, managers and other water professionals must effectively focus communication efforts on the value of the service provided. By identifying and articulating such benefits, we can hope to steer the focus of public discourse from solely the cost to also the value provided by the expenditures.

A similar challenge is managing water as a finite natural resource. As water resources become increasingly scarce and pressures among competing water users grows, it becomes more important to understand the value of this essential resource. Questions to consider include:
- How much is water worth to residential and commercial customers of community water utilities?
- How much more might utility customers be willing to pay for more reliable or higher-quality water?
- How valuable is an asset such as a water right?
- Do the benefits of desalination or water reclamation outweigh the costs?
- How much compensation should be paid when surface or groundwaters become contaminated?
- Is the value of water greater when extracted for utility customers, when applied to cropland irrigation, or when left instream to preserve endangered fish?

With the challenges of measuring water's value, it is difficult to ensure that this precious resource is being allocated to competing (and complementary) uses in a manner that provides society the greatest possible value or well-being (what economists refer to as "maximizing social welfare"). As stewards of this essential resource, utility managers and other leaders in the water services profession face several critical challenges. Water utility managers need to deliver a product that protects public health, yet they must concurrently provide ample supplies at a cost that is affordable to low- or fixed-income households and that can sustain (or stimulate) their communities' economic prosperity. At the same time, utility managers must be mindful of how their decisions about acquiring and diverting water may affect broader social objectives, including values associated with instream ecological conditions, the economic viability of irrigated family farms in the region, or the long-term sustainability of groundwater extraction.

Therefore, we expect that water managers will increasingly need to integrate water values into their deliberations on supply options and management strategies. The value of water will become increasingly important to managers as they attempt to accommodate increased scarcity and cost into their communities' broader efforts to manage growth, promote sustainability, enhance equity, and balance competing demands.

CASE STUDIES REVEAL MORE ON WATER VALUES

Several case studies were developed as part of the AwwaRF project on the value of water. The case studies span the nation from Massachusetts (low- and no-flow conditions in the Ipswich River caused by excessive withdrawals) to California (the value of recycled water in San Diego). They also stretch from Florida (developing high-cost regional water supplies, including desalination, in the Tampa region) to Arizona (the value of in-stream and Phoenix-area extractive water uses for the Verde River). Additional case studies include the "triple bottom line" approach to measuring utility performance in San Francisco, the value of water for instream recreational purposes in the Colorado Rocky Mountains, and the cost of alternative water supply options for the water constrained city of El Paso, Texas.

CONCLUSIONS

The value of water and the value of water utility service provision are key concepts for today and the future. Though the exact meaning of value is not always clear, value is distinct from the cost of service or the prices that are typically based on those costs.

There are many types and levels of value provided by water supply agencies, and these need to be recognized in a more systematic and comprehensive manner. The dollar magnitude of these values typically is unknown or highly uncertain, and it is worth more research effort to more precisely identify and quantitatively measure these values. More important, utilities and water professionals need to think more systematically about what types and levels of value they provide to their communities. We need to work with customers, governing officials, and other stakeholders to begin to articulate what types of value are generated, why and how they are important, and to whom they are important. Communicating about values is an essential first step to addressing many of the emerging challenges of the water supply community faces.

ACKNOWLEDGMENT

The author would like to acknowledge the contributions made by his research partners, Project Advisory Committee members, and the participating utilities on AwwaRF project 2855. Their insights and efforts are reflected throughout this article. Thanks also to David LaFrance, Cheryl Davis, Marca Haenstad, Bill DeOreo, Annette Huber-Lee, Frank Blaha, and James Goldstein for material and case studies presented in this article.

REFERENCE

AWWA, 2005. *Thinking Outside the Bill: A Utility Manager's Guide to Assisting Low-Income Water Customers.* Denver.

Complementary Instream and Extractive Uses Generate Considerable Total Value for Local River Water

The Verde River is an important resource in north-central Arizona. It's Arizona's only designated Wild and Scenic River. In this designation, the river was found to contain outstanding scenic, fish and wildlife, and historical and cultural values. These and other nonconsumptive values such as recreation (e.g., fishing, boating, and hiking) and protection of threatened and endangered species are generally linked to the upper and middle reaches of the Verde River. The river also supports consumptive use values such as agricultural, municipal, and industrial use. The river is a major supply source for several cities in the Phoenix metropolitan area, at the downstream edge of the Verde watershed. Although some small municipal and agricultural uses are supplied in the upper Verde, the majority of the current consumptive use is located downstream of the nonconsumptive uses. As a result, instream and extractive uses generally do not compete for water.

A key issue for the Verde watershed is growth of extractive uses in the upper Verde watershed. The population of major cities and towns within the Verde watershed has more than doubled in the last 20 years and is projected to more than double again within 50 years. The region's population is projected to continue to increase in part because of its attractiveness as a retirement location. This includes growth of the cities of Prescott and Prescott Valley, which currently have no significant surface water source, relying on groundwater pumping for their supply and facing locally declining groundwater levels. Without another water source, increasing demand might force Prescott to look to the groundwater supplies that provide the base flow for the Verde River headwaters, as well as perhaps the Verde River itself.

As a result of this continuing and projected growth, management of the headwaters of the Verde watershed is of particular concern. Groundwater in Little Chino Valley is the source of springs that constitute the Verde's headwaters. Current groundwater extraction has resulted in groundwater declines in some areas and may reduce flow at the source of the Verde River headwaters. Future groundwater extraction may significantly affect the baseflow of the Verde River itself. Clouding the water management picture is that little is known about how much groundwater is actually in storage in many areas of the Verde watershed or about how much water use in the Upper Verde may affect continued water availability for instream and extractive uses below the headwaters in the Verde Valley.

This case study evaluated the value derived from flows in the Verde that originate at its headwaters (i.e., the springs fed by the groundwater in the Little Chino Basin). Results show that the combined nonconsumptive instream values calculated for the Verde River amount to a range of $48 million–$55 million/year. This value range includes greater than $35 million in instream or near-stream use values for recreational fishing, hunting, and wildlife watching, and $13 million–$20 million/year in nonuse values such as for preserving threatened and endangered species. A potential complete loss of baseflow in the upper Verde was estimated to reduce downstream municipal use values by $3.5 million/year. These values amount to a total ranging from $52 million to $58 million/year. This is an underestimate of total water values on the Verde River and does not include other recreation values (e.g., for swimming, boating, camping, picnicking, and sightseeing), historical and cultural values, and other nonuse values (e.g., existence and bequest motives) in addition to value from preserving threatened and endangered species and the Wild and Scenic River designation. This valuation also did not attempt to value the water used as it flows down the Verde River, before it reaches Phoenix, for irrigation, drinking water, power generation, and other industrial/commercial uses such as dairy, golf, sand and gravel, water bottling, and fish hatching.

This analysis shows that upper Verde instream uses are highly valuable, estimated to be worth more than $50 million/year. This value provides a benchmark for considering whether future possible extraction of water from the source groundwater basin and upper river may impose costs (foregone values) that outweigh the value that may be realized by new users; it is evident that if groundwater or upper river withdrawals do reduce instream flows, the impact on downstream values probably would be sizeable.

Developing High-Cost Regional Supplies to Replace Unsustainable Use of Local Groundwater

The greater Tampa Bay region, including Pinellas, Pasco, and Hillsborough counties, has relied almost exclusively on groundwater from the Floridian aquifer to meet its water supply needs. However, as populations have grown and groundwater extraction has increased to meet rising demands, a series of adverse environmental and economic consequences became apparent. These included appreciable declines in the lake and wetland levels, land subsidence (including damage to many houses and properties), and reduced flows (and increased salinity) in streams, rivers, and estuaries.

The economic value of the environmental and other consequences of overpumping has not been estimated but was significant. The adverse effects led regulators, local public officials, and local water agencies to explore ways to resolve the problem. Because of restrictions placed on groundwater pumping permits, a regionwide approach to developing alternative water sources was pursued. Tampa Bay Water was formed as a wholesale water supply agency to meet region-wide needs, using a combination of new desalination facilities and increased surface water storage.

The water developed by Tampa Bay Water will be relatively expensive, especially relative to the low cost of the previously relied on local groundwaters. However, the added cost of desalinated water—projected to be $2.60/1,000 gal ($0.69/1,000 L) but which may end up being considerably higher—is nonetheless considered a good investment for the region. The value of this water probably exceeds the cost when one factors in the damages (negative values) that would otherwise result from continued groundwater reliance (i.e., the avoided cost of overusing groundwater); the increased reliability of the regional water supply because of the drought-insensitive nature of desalinated water and the increased surface water storage capacity; and the region's ability to continue to follow its planned population and economic growth trajectory, which would have been severely constrained absent the development of alternative regional groundwater.

The Triple Bottom Line: Getting Beyond Cost and Revenues

Water utilities provide services that have value beyond the simple financial balance sheet of utility-incurred costs and revenues received from customers. A broader economic perspective of the total value includes social as well as environmental benefits. One approach to addressing this issue is the triple bottom line (TBL), which evolved from Australian applications.

TBL starts with the conventional financial bottom line used to record a utility's incurred costs and received revenues. This is the standard that accountants develop and auditors review. However, this bottom line does not include the other valuable contributions that a water agency provides to its community. Therefore, TBL adds a second bottom line to include social values, and a third bottom line to include environmental issues.

The case study (provided in full in a forthcoming AwwaRF report) examines the history and concept behind the TBL approach and provides examples of how TBL has been applied in four utilities spanning Australia, New Zealand, and the United Kingdom. Examples of the environmental and social bottom lines are provided for these overseas utilities. The case study then examines how TBL could be applied to the San Francisco (Calif.) Public Utilities Commission, which provides water for the city of San Francisco and 29 other Bay Area water agencies. Though the San Francisco Public Utilities Commission does not currently apply a TBL framework, the case study reveals how TBL could be developed and applied in the context of the key elements of the agency's capital improvement program.

Ultimately, the usefulness of the TBL approach is that it offers a template within which a utility can systematically identify and describe the broad range of values that the water agency provides to its customers and the broader community. It recognizes the out-of-pocket financial elements but then goes beyond those dollars and cents to help recognize important social and environmental values the utility furnishes. The key is not to dwell on trying to provide a monetized estimate of all the key values added; instead, the TBL becomes a forum for identifying and describing in qualitative terms what important values the utility action contributes. It is helpful for internal planning and review purposes but also useful as a mechanism for documenting and communicating to the public about the value of water and water utility services.

Under Valuing (and Over Extraction) of Zero-Priced Water

When scarce and valuable natural capital assets such as water are used without charge and with no institutional limits to govern who can extract them or how much can be used, then invariably the resource becomes overused. This is what economists refer to as "the tragedy of the commons," in which common property (shared) resources are not used or managed in a sustainable way to maximize the value generated.

Even though Massachusetts is considered a relatively water abundant and modestly growing state, compared with more rapidly growing and arid regions such as the American Southwest and California, it still faces sizeable challenges posed by increasing municipal water supply system demands for water. The Ipswich River, for example, has been affected by increasing groundwater withdrawals and direct river water extraction associated with watershed residential development. The problem has grown to the point that stretches of the river now face extremely low-flow and no-flow conditions during summer months. The absence of river water has several adverse consequences for the community, including losses in ecologic, recreational, aesthetic, and commercial values that would otherwise be provided by retaining instream waters.

The state and communities in the region are attempting to resolve the problems. The case study describes several options that are under consideration. These include encouraging conservation, importing water from the Massachusetts Water Resources Agency, and providing more in-basin return flows from treated wastewaters. The costs and benefits of these various options have not been fully estimated, though estimates exist for some components. The general perception, however, is that considerable value can be gained from programs and policies that ensure that suitable flows remain in the river year-round. The benefits include restoring lost or reduced services, such as recreational and aesthetic services, because of currently inadequate flows.

In addition to articulating some of the losses associated with overuse of water resources, this case study also reveals a fundamental problem that arises when neither economic signals nor institutional (i.e., political, regulatory) controls exist to manage use levels for a common property resource. Local groundwater and river water in the Ipswich basin has until recently been seen as a free good that is available to anyone with properties providing land-based access. If a price or fee were assessed on all users based on the volume of water extracted (i.e., users had to pay for the water resource itself), then the associated economic signals would help regulate use levels.

ABOUT THE AUTHOR

Robert S. Raucher is a founding partner at Stratus Consulting Inc. Raucher has been involved in the economics and risk management aspects of public water supply for 24 years. He is a member of AWWA, AwwaRF, the Water Environment Federation, the Water Environment Research Foundation, the WateReuse Association, and has served on the National Drinking Water Advisory Council. Raucher has been published in Journal AWWA, Land Economics, *and* Water Resources Research. *He is a three-time winner of* Journal AWWA's *Best Paper Award, Regulatory Affairs Division.*

No Doubt About Climate Change and Its Implications for Water Suppliers

BY JOHN E. CROMWELL III, JOEL B. SMITH, AND ROBERT S. RAUCHER
(*JAWWA* September 2007)

With the release of its Fourth Assessment Report: "Climate Change 2007," the Intergovernmental Panel on Climate Change (IPCC) has removed many doubts that previously shrouded both scientific and policy discussions of climate change. Mounting evidence about climate change and its effects made the situation much clearer to the scientists and government policy analysts from around the world who participated in a six-year process and arrived at the consensus presented in the report.

The World Meteorological Organization and the United Nations Environment Program (UNEP) established the IPCC in 1988. IPCC's role is to assess on a comprehensive, objective, open, and transparent basis the scientific, technical, and socioeconomic information relevant to understanding the scientific basis for the risk of human-induced climate change, its potential effects, and the options for adaptation and mitigation. The IPCC does not carry out research nor does it monitor climate-related data or other relevant parameters. Its assessment is based primarily on peer-reviewed and published scientific/technical literature.

A main activity of the IPCC is to provide at regular intervals an assessment of the state of knowledge on climate change. The First Assessment Report was completed in 1990, the second in 1995, and the third in 2001. Background reports are written by hundreds of scientists from around the world and are subject to extensive peer review. The "Summary for Policy Makers," typically a 20-page summary of the entire report that usually receives extensive attention in the press, is drafted by scientists but subject to approval by governments. In approving, adopting, and accepting reports, the IPCC makes every effort to reach consensus. If consensus cannot be reached, differing views are explained, and scientific and policy differences are distinguished. The IPCC's conclusions are not official until they have been accepted in a plenary meeting by the governments.

Thus governments cannot ignore the conclusions of the IPCC because they have endorsed them.

The IPCC's adherence to scientific rigor and openness, coupled with the objective of articulating the best possible consensus views, might result in a slower and more cautious process of deliberation than some would prefer, but it also gives tremendous weight to IPCC findings when they are finally put forward as consensus statements.

During the six years since the release of the Third Assessment Report, the scientific evidence regarding climate change has become much more compelling. Accordingly, the Fourth Assessment Report (IPCC, 2007a–c), released in 2007, contains consensus statements that are much more profound than those in previous reports. This article provides a brief review of some of the statements that pertain to water resources. These statements serve as a foundation from which to examine the implications of climate change for water suppliers.

IPCC REPORT REMOVES DOUBTS ABOUT CLIMATE CHANGE

The Fourth Assessment Report made headlines because the IPCC made very strong statements that left no room for doubt that global warming is producing long-term effects on natural systems and that anthropogenic sources are a likely cause. The summary statements that captured attention were:

• Warming of the climate system is unequivocal, as is now evident from observations of increases in global average air and ocean temperatures, widespread melting of snow and ice, and rising global average sea level.

• Most of the observed increase in global average temperatures since the mid-twentieth century is very likely due to the observed increase in anthropogenic greenhouse gas concentrations.

• Observational evidence from all continents and most oceans shows that many natural systems are being affected by regional climate changes, particularly temperature increases.

• A global assessment of data since 1970 has shown it is likely that anthropogenic warming has had a discernable influence on many physical and biological systems.

• Anthropogenic warming and sea level rise would continue for centuries due to the time scales associated with climate processes and feedbacks, even if greenhouse gas concentrations were stabilized.

The first three of the preceding conclusions had been made in previous IPCC assessments, although with less confidence. This time the authors felt there was much stronger evidence. In addition, the conclusions tying human emissions to these impacts are new.

This IPCC assessment made it clear that continued warming is inevitable. Six scenarios of greenhouse gas emissions were considered. All would result in an acceleration of the rate at which global temperatures and sea levels are rising.

CONSEQUENCES FOR WATER RESOURCES

The Fourth Assessment Report states with "high confidence" that among the effects of global warming on hydrologic systems the following are presently occurring:

• Increased run-off and earlier spring peak discharge in many glacier- and snow-fed rivers.

• Warming of lakes and rivers in many regions, with affects (sic) on thermal structure and water quality.

Turning to the future, the Fourth Assessment Report makes the following predictions regarding water resources:

• By midcentury, annual average river runoff and water availability are projected to increase by 10%–40% at high latitudes and in some wet tropical areas, and to decrease by 10%–30% over some dry regions at mid-latitudes and in the dry tropics, some of which are presently water stressed areas. In some places and in particular seasons, the changes differ from these annual figures.

• Drought-affected areas will likely increase in extent. Heavy precipitation events, which are very likely to increase in frequency, will augment flood risk.

• In the course of the century water supplies stored in glaciers and snow cover are projected to decline, reducing water availability in regions supplied by meltwater from major mountain ranges, where more than one-sixth of the world population currently lives.

• Coasts are projected to be exposed to increasing risks, including coastal erosion, due to climate change and sea level rise (0.2–0.6 m by 2100, and probably more).

The IPCC's more specific projections regarding future effects of global warming in North America include the following statements:

• Warming in western mountains is projected to

cause decreased snowpack, more winter flooding, and reduced summer flows, exacerbating competition for over-allocated water resources.

• Disturbances from pests, diseases, and fire are projected to have increasing impacts on forests, with an extended period of high fire risk and large increases in area burned.

• Moderate climate change in the early decades of the century is projected to increase aggregate yields of rain-fed agriculture by 5%–20% (in mid- and high latitudes), but with important variability among regions. Major challenges are projected for crops that are near the warm end of their suitable range or depend on highly utilized water resources.

• Cities that currently experience heat waves are expected to be further challenged by an increased number, intensity, and duration of heat waves during the course of the century, with potential for adverse health effects.

• Coastal communities and habitats will be increasingly stressed by climate change impacts interacting with development and pollution. Population growth and the rising value of infrastructure in coastal areas increase vulnerability to climate variability and future climate change, with losses projected to increase if the intensity of tropical storms increases.

BOTTOM LINE IMPLICATIONS FOR WATER SUPPLIERS: ADAPTATION

The IPCC concludes adaptation will be necessary to address the effects of warming that are already unavoidable because of past emissions. A further 0.6°C (1°F) increase in global mean temperature—relative to temperatures during the period from 1980 to 1999—is projected to occur by the end of the twenty-first century even if greenhouse gas concentrations remained at 2000 levels. The IPCC projects that the average rise in global temperatures could range from 1.1 to 6.4°C (about 2 to 12°F) by 2100 when examining a range of greenhouse gas emissions scenarios. Temperatures in the lower 48 states are expected to rise about a third more than the global average (Wigley, 1999).

One of the simplest ways to envision many of the implications of global warming for water resources is to follow the logic of what happens when water is heated; global warming will basically accelerate the pace of the hydrologic cycle. Consistent with this, forecasters predict effects for water resources in the arid western regions of North America that are different from those in the humid eastern regions. Recent trends indicate that many of these effects are already happening.

In the West, warming effects may be seen most prominently in reduced water supply capacity. Snowpack will be smaller and melt earlier, altering the recharge of surface water and groundwater sources. In addition to less rainfall in this region, droughts are expected to be more extensive, with more heat waves and dry days, accompanied by increases in evaporation and greater irrigation demands. When precipitation does come, it is likely to be more intense. The combination of earlier snowmelt and more intense precipitation will likely increase turbidity, sedimentation, and the risk of flooding in many areas.

In the East, evidence of warming will primarily come in the form of increased rainfall frequency and intensity. The increased rainfall intensity will likely produce increased runoff and erosion with resulting increases in turbidity and sedimentation. Related effects include direct flood damage to water and wastewater facilities, loss of reservoir storage capacity for flood control and water supply, and increased sewage overflows during wet weather events.

In both the East and the West, the changes in temperature and hydrology will produce changes in watershed vegetation and aquatic ecosystems, which in turn will have implications for water suppliers. One result will be changes in watershed conditions because of wildfires and pest infestations, both of which are likely to increase. With increased water temperatures and shallower reservoirs (because of lower base flows and sedimentation), eutrophic conditions will be more prevalent. Rising sea levels will pose the risk of flood damage to coastal water and wastewater facilities in both the East and the West, especially as a result of more intense coastal storms. Rising sea level, coupled with lower base flows in freshwater sources because of altered recharge, will also increase the salinity of both coastal aquifers and brackish surface water sources.

The individual effects of climate change are staggering enough, but they will also have compound effects on water resources because of the interactions of natural and human systems. The bottom line in water supply planning has always been a matter of coping with variability. With the coming changes in climate, there will be a heightened need to respond to increased variability. Global warming will change the variability of many key parameters affecting the quantity and quality of water that would normally be available at specific times and places. In addition, the capability to store water in various forms and the demand for water will be changed.

Portfolio approach to planning encouraged. Given the difficulty of predicting the magnitude of individual effects, let alone compound effects, IPCC and others have recommended that adaptation planning should employ a portfolio approach. This approach would maintain a maximum degree of flexibility within the portfolio by devising coping strategies to address an array of possible climate change scenarios that may affect the quantity and quality of supply sources, and the demands placed on them. Reevaluating the entire portfolio from source to tap with respect to adaptation strategies may seem like a daunting task; however, there are several guiding philosophies that can offer a pragmatic way to start.

A first practical step in evaluating the adaptation strategies is to conduct a "bottom-up" vulnerability assessment. Rather than starting with complex climate modeling, it is more tractable to begin with what a utility already knows about its own systems and water sources to determine the points at which water quality or water quantity changes would present a major challenge. This type of "threshold analysis"—based on a utility's current models and knowledge—can identify key climate-related tipping points (e.g., what reduction of instream flows would be most problematic given current supply strategies). Where possible, these thresholds could be verified through simulation modeling of altered operating regimes or through actual system tests. Once these thresholds are identified, climate expertise can be brought to bear in a focused manner to examine the potential process changes that could lead to these key tipping points being exceeded.

Once a water utility is equipped with a sense of its vulnerabilities, the IPCC and others recommend development of adaptation strategies by focusing first on ways to improve system flexibility and resiliency across the entire portfolio. In concert with this emphasis on flexibility, it may be prudent to adopt a step-wise approach that recognizes that irreversible choices or capital commitments might be easier to optimize within a portfolio as information improves over time.

TRIPLE BOTTOM LINE IMPLICATIONS: MITIGATION

In addressing water supply planning in the context of portfolio optimization, it is necessary to consider all of what should go into a sustainable solution. As has already happened in many water-short areas in both the East and West, it is possible to conceive of many ways to enhance the reliability of water resource management outcomes by essentially investing more energy to produce more water. But in evaluating these options, it must also be acknowledged that water utilities account for a significant share of total electric consumption and that power plant emissions account for a significant share of greenhouse gasses. There is no doubt, therefore, that water utilities need to apply a broader "triple bottom line" discipline to the design of adaptation strategies in order to balance the cost and reliability of water supply against social and environmental consequences. The key to implementing this approach in practice will be to apply it to overall portfolio outcomes and not just to the individual project elements. There are several areas in which the need for this broader thinking arises most visibly.

As urban and suburban areas continue to grow, water resource managers have already devised elaborate portfolio strategies to tap into multiple sources of supply and to make strategic investments in capabilities to move and store water. To wring every last drop from available supplies, flexibility in transmission and storage operations has been taken to great extremes in some water-short areas in order to stay within complex constraints imposed by environmental withdrawal limits and seasonal and annual swings in water availability. Although flexibility in transmission and storage will be valuable in re-optimizing current schemes to meet future challenges, some system features designed for the current understanding of climatic variability may not be reversible or easily adaptable under altered operating regimes that were never envisioned. Operating flexibility needs to be an even more important design consideration for future system improvements.

Balancing fiscal and environmental responsibility. The rising cost of electric power has caused many water utilities to reexamine their transmission and distribution operating strategies in search of ways to curtail electricity use during peak periods and conserve electric power. Some cost-saving strategies—such as using backup generators to serve peak-period loads (where not prohibited by air quality regulations)—may save money, but they do not reduce greenhouse gas emissions. On the other hand, renewable energy supply strategies such as solar- or wind-powered pumping, or in-line hydropower generation, may be more expensive initially but provide fuel-cost savings and avoid greenhouse gas emissions. The challenge is to integrate such strategies at the level of the overall portfolio to produce the best systemwide operating outcomes in terms of cost, reliability, and social/environmental consequences. For example, the reliability profile of a solar- or wind-powered option may be that it is only viable if a backup generator is available to fill critical gaps. Although this would produce some greenhouse gas emissions, overall it would be less than a conventional power supply strategy.

As many water suppliers are already aware, the cost and energy investment required to reliably meet water supply needs can rise exponentially when hard, unmovable constraints arise in one part of the portfolio and cause everything else in the strategy to be shifted to more extreme levels. Global warming has the potential to cause simultaneous adverse changes in both the quantity and quality of available water. These compounded effects could greatly weaken the foundation of some existing integrated supply strategies. It is hard to compensate for such things as a smaller snow pack, an earlier spring runoff, or increased salinity.

Many water suppliers in overconstrained settings have already begun to turn to energy-intensive membrane treatment processes to enable desalination of saline water sources and reuse of highly treated wastewater effluent. These processes make it possible to overcome a deterioration in the reliability of normal supplies by meeting part of the demand from sources that will be abundant under most climate change scenarios (i.e.,

yields from water reuse and desalination supplies are drought-resistant). Although the costs—especially the energy cost—of these technologies are significant, the triple bottom line must be evaluated in the context of the overall portfolio. If these technologies can plug a gap or shore up a vulnerability produced by climate change processes in a way that enables a broader scope for optimization across the entire portfolio, then they could play a critical role in making the broader strategy more sustainable. The economies of scale involved in such facilities also offer opportunities to address greenhouse gas emissions during the design phase of related power generation facilities and incorporate renewable energy sources into the strategy. Perhaps tomorrow's water can be decoupled from yesterday's energy, making the water sector a leader in the transition to carbon-neutral energy sources.

Demand management plays numerous roles. No portfolio of adaptation strategies to meet global warming challenges would be complete without including a comprehensive review of the water demand side of the equation. Warming processes will lead to altered demand patterns as a result of seasonal shifts in precipitation, more evaporation, more frequent heat waves, and more extensive droughts. From a triple bottom line perspective, conservation programs offer a bonus in reducing both water supply needs and energy use. Bolstering conservation incentives (and disincentives to outdoor water use) may become more essential as warming processes increase water demands, especially during peak demand periods when both water supply and electric power capacities are stretched to their limits. Any margin of relief from the more extreme water and energy supply strategies required to meet peak demands can significantly improve the triple bottom line of the overall supply strategy.

In many regions demand management is also related to population growth. Regional strategies for optimizing water supplies often meet with opposition in an indirect effort to control population growth. When broader regional portfolio optimization is blocked in this way, it can force individual jurisdictions to adopt more narrow and therefore less sustainable options because warming processes make it continually harder to meet demands with a limited range of adaptation options. At the other end of the spectrum, the energy intensity of moving water across distances and terrain probably limits the potential of broader regional approaches in terms of the triple bottom line criterion.

IPCC REPORT ALSO AFFIRMS HOPE

Despite the IPCC's sobering confirmations that global warming processes are well under way and will produce effects that are unavoidable over the next several decades, the Fourth Assessment Report also holds out hope in showing that actions taken to reduce greenhouse gas emissions over the same time period are well within our technical and economic capacity and should be capable of stabilizing the situation.

• Studies indicate that there is substantial economic potential for the mitigation of global greenhouse gas emissions over the coming decades that could offset the projected growth of global emissions or reduce emissions below current levels.

• In order to stabilize the concentration of greenhouse gases in the atmosphere, emissions would need to peak and decline thereafter. The lower the stabilization level, the more quickly this peak and decline would need to occur. Mitigation efforts over the next two to three decades will have a large impact on opportunities to achieve lower stabilization levels.

Adaptation by water utilities could ameliorate many adverse effects, although there will be costs to adaptations, and maintaining reliability may be more challenging. Many water utilities are already engaged in considering adaptation strategies and mitigation of greenhouse gas emissions. The state of California and municipalities from New York, N.Y., to Miami, Fla., to Los Angeles, Calif., to Seattle, Wash., are assessing their risks and developing strategies to lessen the effects of climate change. There is no doubt that it is time for all to rise to this call.

ABOUT THE AUTHOR

John Cromwell consults for utilities and performs national policy research for Stratus Consulting Inc. in Washington, D.C. He has bachelor's degrees in biology and economics as well as a master's degree in public policy analysis, all from the University of Maryland. Joel Smith is vice-president and Bob Raucher is executive vice-president, both at Stratus Consulting.

REFERENCES

IPCC (Intergovernmental Panel on Climate Change), 2007a (S. Solomon, D. Qin, M. Manning, Z. Chen, M. Marquis, K.B. Averyt, M.Tignor, and H.L. Miller, editors). Summary for Policy Makers. Climate Change 2007: The Physical Science Basis. Contribution of Working Group I to the Fourth Assessment Report of the Intergovernmental Panel on Climate Change. Cambridge University Press, Cambridge, UK and New York. http://ipcc-wg1.ucar.edu/wg1/Report/AR4WG1_Pub_SPM-v2.pdf/.

IPCC, 2007b (M.L. Parry, O.F. Canziani, J.P. Palutikof, P.J. van der Linden, and C.E. Hanson, editors). Summary for Policy Makers. Climate Change 2007: Impacts, Adaptation, and Vulnerability. Contribution of Working Group II to the Fourth Assessment Report of the Intergovernmental Panel on Climate Change. Cambridge University Press, Cambridge, UK, and New York. http://www.ipcc-wg2.org.

IPCC, 2007c (B. Metz, O.R. Davidson, P.R. Bosch, R. Dave, and L.A. Meyer, editors). Climate Change 2007: Mitigation. Contribution of Working Group III to the Fourth Assessment Report of the Intergovernmental Panel on Climate Change, Cambridge University Press, Cambridge, UK and New York. http://www.ipcc.ch/SPM040507.pdf.

Wigley, T.M.L., 1999. The Science of Climate Change: Global and U.S. Perspectives. Arlington, Virginia: The Pew Center on Global Climate Change.

This page intentionally blank.

Ten Primary Trends and Their Implications for Water Utilities

BY EDWARD G. MEANS III, LORENA OSPINA, AND ROGER PATRICK
(*JAWWA* July 2005)

Water utility future trends were identified in a reassessment of the 2000 American Water Works Association Research Foundation (AwwaRF) "Strategic Assessment of the Future of Water Utilities" project. This article describes the primary trends identified through an assessment of the literature, interviews with public water supply community leaders, and a futures workshop featuring futurists and scenario planning exercises. Many of the same trends at work in 2005 continue to be primary drivers of water utility strategy today, including aging infrastructure, financial constraints, succession planning, technology developments (including water treatment advances), and politicization of water. Trends that are comparatively new or more pronounced since the 2000 assessment include security (physical and information technology), climate change, growth in nongovernmental organization influence in water, total water management principles (e.g., watershed management), and energy risk.

The authors draw several inferences from the array of trends discussed here. On the basis of the findings of this assessment, the utility of the future will likely do three things really well.

• The utility of the future will be operationally efficient. The effective water utility will do this because active consumers and political interests will demand solid financial stewardship as water rate increases mount for repair, rehabilitation, and construction of water supply infrastructure. Operational efficiency will require a motivated and engaged workforce.

• The utility of the future will be engaged in the community. The effective water utility will never take its customers—or shareholders—and stakeholders for granted. Communications systems will be geared to creating a constant "dialogue" with the community and ensuring that there is philosophical alignment among the community, its elected leaders, and the utility's policies and actions.

• The effective utility will integrate water management, treatment, environmental sustainability, and public policy decisions in a fashion that leverages regional opportunities and relationships and economies of scale.

In order to identify and characterize future trends of importance to the public water supply community and begin a strategic discussion to address these trends, AwwaRF funded a study titled "A Strategic Assessment of the Future of Water Utilities" in 2000 (Brueck et al, 2002; Means et al, 2002a; Means et al, 2002b; Patrick et al, 2002; Dixon et al, 2001; Means, 2001; Miles et al, 2001).

The project approach included the development of three papers addressing societal, business, and utility trends. The papers served as background documents for a Futures Workshop in Denver, Colo., in June 2000, at which water utility leaders and stakeholders reviewed the papers, listened to futurists, debated the trends, and identified and ranked approximately 60 trends in terms of certainty and desirability. These trends were grouped into several potential future scenarios, which provided the basis for an American Assembly in Orlando, Fla., Sept. 20–21, 2000. The American Assembly identified and adopted specific strategies to respond to the trends as well as success attributes that future water utilities should have.

Based on an assessment of the trends and the benefits of a scenario-planning exercise, the participants identified nine competencies or attributes of excellence that were believed to best position a water utility to excel in the future. These attributes of excellence are

1. Maintain strong internal and external communication skills and programs.

2. Possess a consistent and strong view of the public stewardship responsibility for water supply (including quantity, quality, and fiduciary responsibilities).

3. Understand customers' knowledge levels and needs.

4. Seek innovation and improvement of utility products and services.

5. Run the enterprise like a business with a sharp eye on the "bottom line."

6. Use technology strategically.

7. Treat every employee with dignity and respect.

8. Maintain a strong working relationship with the governing board or council that clearly separates and institutionalizes management/governance responsibilities.

9. Create strategic alliances with other organizations in preparation for tactical response.

These nine attributes of excellence should be maintained or developed through utility strategic plans.

In recognition of the dynamic nature of the utility environment, AwwaRF commissioned a revisit of the 2000 effort. The 2005 report updates previously identified issues, introduces new trends, identifies potential implications of these trends, and defines potential response strategies for water utilities.

PROJECT APPROACH COMPRISED OF FIVE ELEMENTS

The 2005 study consisted of five elements: (1) developing and conducting interviews of managers to identify key issues and trends facing the industry today, (2) examining available literature/experience on future trends and scenarios, (3) developing the trend paper, (4) conducting an Expert Futures Workshop that included presentations by futurists, and (5) synthesizing the input from the workshop into "actionable" response strategies.

Project interviews. Telephone interviews were conducted with eight prominent water community leaders as a means of initially framing trends for inclusion in the paper. The key trends identified in the interviews are shown in Table 1.

Trend paper. The project trend paper was intended to be a critical briefing document to prepare participants for the Futures Workshop. The trend paper was derived from an evaluation of current published studies (in refereed and popular literature) identifying important trends and presenting the data behind each trend. Potential implications for water utilities were also included at the end of each chapter. Issues that were identified were vetted by the participants at the workshop described here.

Expert Futures Workshop. The futures workshop was held Nov. 30–Dec. 1, 2004, in Huntington Beach, Calif. The expert workshop was designed to gather the wisdom and expertise of the participants through breakout group discussions. The participants included 35 water professionals from across the United States, two AwwaRF project advisory committee members, and six project team members from McGuire/Malcolm Pirnie Inc. and Competitive Advantage Inc. The primary objective of the workshop was to identify 10 primary future trends and formulate the strategies to deal with each trend.

Day one of the workshop began with presentations from selected futurist speakers. The presentations were followed by breakout group discussions. The breakout group discussions were consolidated into consensus

TABLE 1   Utility trends identified in expert interviews

Societal	Business	Utility
Population/demographics	Employment trends	Regulatory trends
Environmental trends	Customer expectations	Political environment
The economy	Outsourcing/globalization	Rate sensitivity
Medicine/health trends	Technology (information technology [IT] and other)	Infrastructure aging
Energy policy/trends	Public confidence in financial market/ business ethics	Privatization
Terrorism/wars/ post-9/11 environment		Physical and IT security
		Workforce demographics
		Total water management
		Water resources/drought
		Treatment technology
		Regionalization
		Reuse

primary trends. Second-day discussions identified the implications and strategies to address the trends.

PRIMARY FUTURE TRENDS IDENTIFIED

The trend paper research identified 19 broad trend areas and potential implications (Table 2). Participants then identified 10 primary trends from the broad list. These primary trends and their implications are described in the following paragraphs. Because of space limitations, selected graphs are used to convey some of the trend data. The reader is encouraged to visit the AwwaRF website and review the detailed data in the trend paper posted there.

1—Population. The US population has increased at a steady rate over the past 40 years and is expected to maintain that rate into the next century. From 1960 to 2000, the population grew from 179 to 281 million, with the current population (September 2004) estimated at 294 million. US population is expected to increase by 50% between 2000 and 2050 (Figure 1). Population growth has occurred in the South and West, with the Northeast and Midwest remaining static. This increase in population strains supplies in dry areas and could place strains on rate bases in areas that need infrastructure upgrades.

In addition, the US population is aging—and living longer. Average life span has increased approximately 10 years for both males and females over the past 50 years contributing to the current concerns over the viability of the Social Security system.

Implications:

• Increasing use of marginal water supplies (affecting both cost and quality), including wastewater recycling and desalination (ocean and inland).

• Population growth in watersheds will represent additional contaminant loading, and shifts from agricultural to urban land uses will cause shifts in contaminant type.

• Water "transfers" will increase as agricultural lands are developed for urban uses and foreign competition encourages marginal farming operations to sell water. Protocols to facilitate transfers will be needed.

• Elderly consumers may be more interested in emerging contaminants from a health perspective because larger percentages of the elderly are immunocompromised. Additionally, the elderly tend to be more involved in the political process, attend public meetings, and vote.

• Rate increases will affect communities differently. Communities experiencing population reductions will have fewer rate payers across whom to spread fixed costs. Members of these communities will thus incur even higher water rates. Growing communities may be able to accommodate rate increase through rate-base growth.

• The cost of service and proper allocation of those costs will become bigger issues as rates climb.

• Diverse populations will necessitate the use of diverse communication methods.

Strategies. Utilities need to understand how the current population uses water and project how the future population will affect water use in their own service area as well as in the watersheds from which their water supplies are drawn. Engaging in broader watershed discussions will be critical for many utilities. Contiguous communities with shared values and/or issues should actively explore greater regional cooperation to leverage economies of scale.

A formal communication strategy to convey infrastructure needs to the decision-makers and the community should be developed. Decision-maker support and ultimately development of a community dialogue should be sought. Communication strategies will need to emphasize investment in infrastructure and future supplies and maintaining/increasing the quality of life in communities.

Utilities should use modern communication tools to maintain a dialogue with their diverse customer bases. Understanding their customer bases, expectations, and attitudes (especially those of the growing elderly population) will help to create regional forums for problem-solving and will enhance communication with new ethnic customers.

2—The political environment. The political environment is growing more complex. Large utility capital expenditures represent a source of political power. Term limits are being enacted in some communities. There is also an increase in nongovernmental organization (NGO) advocacy groups that may play a greater role in public policy decisions. These NGOs are generally well organized, technically capable, and willing to share information. General public participation will also play a larger role.

Implications:

• The significant infrastructure replacement and capital expenditures of the public water community will attract political interest.

• Rate increases will likely be politicized. Utilities may be forced to rapidly increase rates and fees (actions that are highly unpopular with customers) if long-term financial planning and rate-setting programs are not implemented.

• There will be a greater need to manage constituencies and develop new relationships.

• Utilities will need to be connected to their customers and treat them as both shareholders and customers. Utilities must understand where the public stands on issues. Improved customer communications/involvement may moderate some of the political pressures that affect water utilities.

• Although the privatization movement in the United States has apparently slowed down, some

TABLE 2	Future trends identified in research	
Number	Trend	Example Implications
1	Automation to reduce labor costs	Expansion of real-time monitoring Growth of staff operations and maintenance training requirements Increasing reliance on automation to manage complex data generation Repair that is less common; replacement saving labor
2	Population growth and demographic shifts	More growth continuing to be in the South and West Population increasing in urban areas Significant growth in Hispanic and Asian populations Population loss in some communities Increasing education levels resulting in increasing customer demands
3	Health	Increasing life expectancy Increasing immunocompromised population Public awareness of occurrence of trace contaminants in water
4	Medical	Chronic diseases as the leading cause of death in the United States Drug use increasing, with implications for occurrence in water and wastewater Increasing medical applications of nanotechnology
5	Regulatory	Effects of Disinfectants/Disinfection Byproducts Rule Shift to advanced water treatment (e.g., membranes, ultraviolet disinfection, chlorine dioxide, chloramines) Difficulty in coping with demands (especially in smaller utilities) Potential distribution system regulation based on security and aging infrastructure risk Energy and residuals management issues associated with advanced water treatment Increased constraints on operational flexibility Greater operating expertise required (operator training implications) Continued significant need for utility community involvement in regulations/legislation
6	Climate change	Changing conditions altering water supply Effects of sea elevation change on estuarine water intakes Dam design/water management strategies that must change where earlier runoff/snow pack melt occurs Need for water supply contingency planning
7	Total water management	Climate change that may alter water availability and contaminant loading Demand management/conservation that is necessary Increased use of marginal water supplies Watershedwide planning growth Solutions that require multiagency coordination and compromise Endangered species/amphibian decline that complicate water development Population growth and contaminant contribution to source waters/watersheds, which may spur additional discharge requirements
8	Workforce issues	Baby Boomer retirements that create opportunities for promotion and challenges for replacing skill sets; succession planning that grows in importance in the near term Engineering/scientific skills in short supply in some regions (especially where housing costs are high); competition for treatment operators/engineers Management of increasingly ethnic, gender diverse, and multigenerational workforce Growth of work/life balance issues Labor unions resisting work rule changes that threaten labor gains Management of increasingly ethnic and multigenerational workforce

Number	Trend	Example Implications
	TABLE 2	**Future trends identified in research, continued**
9	Customer expectations are rising	The necessity for utilities to understand customer knowledge level and needs
		Customer distrust/dissatisfaction that may complicate utility financing
		Customers paying for convenience/water quality in bottled water
		General utility need to improve customer communications
10	Information technology (IT)	An increasing focus on seeking information technology applications that will reduce cost
		Internet use maximized to communicate and learn
		24/7 communications and information access
11	IT security	Increased dependence on remote security
		Establishing secure databases to decrease vulnerability
		Need for mobile communications security
12	Drinking water treatment technology	Regulations that continue to affect treatment technologies and spur innovation
		Membrane technology and UV application grow
		Growth in treatment residuals management challenges
13	Energy	New treatment technologies that are energy-intensive
		Growth in energy optimization in utilities
		Increasing price of energy and instability of resources
14	Security issues	Financial systems that will require greater protection
		Physical security needs that require careful risk/cost balancing
		Information systems that are increasingly relied on and must be secured from internal and external threats
15	Economic	Rising interest rates
		Opportunities for savings through outsourcing
		The need for promoting benefits of careers within public agencies
16	Utility finance	A continuously aging infrastructure
		Required rate increases that are not affordable for some customers
		Rising operations and maintenance costs; public pension liabilities at issue in some communities
		Financial reserve policies increasingly important to help stave off reserve raids by cash-strapped local governments
		Areas with population decline facing a shrinking rate base
17	Political environment	Infrastructure replacement funding needs that will have political consequences as a result of need to increase rates and fees
		Control over operating costs that will be important to get rate increases
18	Regionalization	Needs of small utilities arguing for consolidation
		Improved coordination among utilities
		Increased watershed cooperation
19	Private sector participation	Water utilities that must keep public and private opportunities available
		Continues to drive innovation in water utilities
		Continues to provide a competitive tension in public water systems

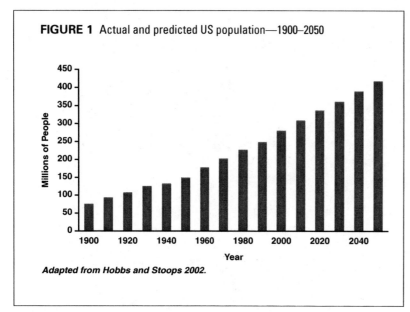

FIGURE 1 Actual and predicted US population—1900–2050

Adapted from Hobbs and Stoops 2002.

utility-governing entities are still evaluating the possibility of long-term public–private partnership agreements. Private operations will continue to provide tension, spurring public agencies to higher levels of excellence.

• Utilities will need to improve financial and capital improvement program transparency. Reserve policies will need to be clearly articulated and maintained to ward off attempts by financially strapped communities to tap water utility cash.

Strategies. Utilities should develop a clear strategic plan that establishes a course of action for the utility. The strategic plan should consider the nine attributes of excellence and reflect the external and internal trends shaping the utility. The strategic plan should integrate (or drive the development of) specific plans related to water supply, water quality, customer service, finance, and communications goals and objectives. Each of these areas is critically important to the future success of the utility and should receive specific planning attention. Collectively, these plans will lay out a purpose and vision for the utility that forms the basis of building and maintaining community trust and support. To this end, the plans should be developed in concert with the community.

Governance models must be structured and designed to overcome political problems. Clear separation of management and board responsibilities coupled with strong ethics programs will help minimize politicization of capital program expenditures.

Utilities must understand who their stakeholders are and develop and maintain relationships with them. Stakeholder input should help shape policy. These stakeholders include customers, the business community, NGOs, employees, vendors, and contractors. Development and maintenance of these relationships should be a managerial and board priority.

Active public outreach must be used to engage and inform the public and public officials on water issues. The utility should engage in a constant dialogue with the community that measures knowledge, values, and priorities. This information should be provided to the governing board to improve the alignment between the institution and the public it serves. At the national level, utilities should develop a strategy to engage elected officials in water challenges.

Utilities should consider regionalization and work to overcome natural political opposition to greater cooperation to spread the cost and benefits across a larger rate base. This opportunity is especially available to contiguous communities.

3—Regulations. Regulations will continue to challenge water utilities (e.g., disinfection by-products). Meeting the US Environmental Protection Agency's (USEPA's) stated goal of 95% of community water systems in compliance with health-based standards by 2008 will require steady progress. In order to reach this strategic target, small and very small water system regulatory compliance will be critical. Very small and small systems in the United States accounted for most of the Safe Drinking Water Act (SDWA) health-based violations (both maximum contaminant level [MCL] and treatment technique violations) in 2003 (Figure 2). The USEPA goal can either drive those systems to improve their facilities and operations in order to achieve compliance (through assistance from the Drinking Water State Revolving Funds [DWSRF]) or force more consolidation, privatization, or contract operations in order to achieve compliance. However, even though compliance has improved, point-of-use device and bottled water sales have increased.

Science and technology developments will continue to affect regulations. Improvements in analytical technology now allow the detection of contaminants in water supplies that were never anticipated. Low part-per-trillion detection of endocrine-disrupting compounds and pharmaceutical and personal care products are receiving increasing attention. Human health effects are not known, but effects on aquatic species have been documented.

Implications:

• Small-system compliance will continue to be a problem.

• Because large and very large system violations affect more people than small and very small systems violations, the USEPA and states may focus their compliance efforts on the large and very large systems in order to ensure that 95% of the population served by the community will receive water that meets drinking water standards by 2008.

• Increased water treatment will be necessary to meet higher regulatory standards.

• Installation of more sophisticated treatment may not coincide with public confidence, nor does it ensure relief from further regulations and increased costs.

• Increased regulatory requirements will continue to engage utilities in the regulatory/legislative process.

• The effect of emerging contaminants on human and aquatic life will continue to garner attention. Public understanding of and tolerance for risk will be tested. Older populations of immunocompromised individuals may become politically engaged in the debate about and regulation of emerging contaminants.

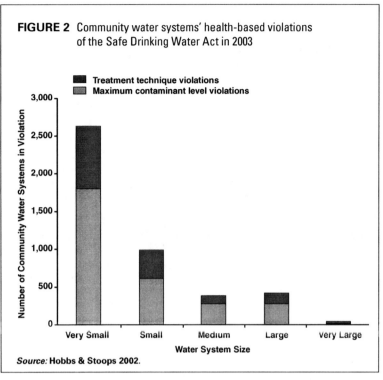

FIGURE 2 Community water systems' health-based violations of the Safe Drinking Water Act in 2003

Source: Hobbs & Stoops 2002.

Strategies. Strategies to address this trend include continuing to strategically engage regulators and actively participating in the legislative process (including lobbying). The public water supply community should take great care to protect the public health and be viewed as protectors. This is especially challenging given the limited financial resources available to many water utilities. The public water supply community should continue to emphasize peer-reviewed science in regulation-setting.

Influencing policy through coordination and cooperation with other utilities will continue to be important. The public water supply community should strive to speak with a cohesive voice on Capitol Hill, to educate decision-makers, and to advocate for public health. Utility support for research and development is vital. The public water supply community should strategically fund the science to support anticipated regulation.

The contaminant-by-contaminant regulatory approach continues to present incremental challenges to the public water supply community. Consideration should be given to moving to a more holistic compliance approach involving the application of treatment technology and monitoring.

4—Workforce issues. There are significant changes occurring in the workforce, including mass retirements of the Baby Boomers, increased use of technology, conflicting generational values, and growing ethnic and gender diversity. Ethnic diversity predictions are shown in Figure 3. In addition, growing job-skill requirements (e.g., multiskilled, technologically savvy workforce) and attrition and shortage of available technical talent in the

marketplace will drive competition for employees in the industry. Treatment plant operators and engineers will be in especially short supply.

In addition, good communications skills are going to be required as customer focus sharpens for many utilities.

Implications:

• Workforce expectations regarding work/life balance may change.

• Filling technical positions in engineering and treatment operations may become difficult and expensive. Medium and small systems may have a difficult time competing for talent in the marketplace. Wages for operators and engineers will likely rise as demand outstrips supply.

• The workplace will need to be sensitive to the needs of working parents and older workers.

• Individuals with good communication skills will be especially valued in the utility organization.

Strategies. Utilities will be driven to establish strong management and leadership strategies to adapt to a changing work environment. Utilities should consider developing a strategic human resources plan that defines the utility's functional needs and staff capability. The human resources plan could include

• creating a formal succession plan;

• modifying employment practices that discourage rehiring of retired employees;

• designing compensation packages and creative incentives to retain "the best and the brightest";

• developing a high school or university recruit-

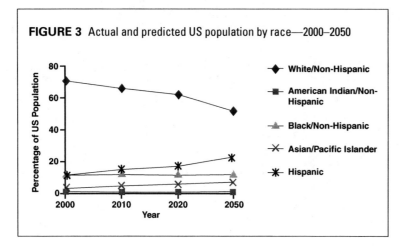

FIGURE 3 Actual and predicted US population by race—2000–2050

ment strategy to educate counselors on promoting the industry;

• creating incentives for retaining key staff past retirement age;

• rehiring retiring staff, when appropriate, to mentor junior staff;

• establishing workforce planning and development programs that strategically identify needs and then match the skills required to fulfill those needs;

• implementing career counseling, leadership, and skill training programs to respond to the new set of required utility skills (these could be cooperatively implemented across multiple utilities in a region); and

• developing and setting up apprenticeship programs.

Utility managers should also understand generational, ethnic, and gender differences of the workforce. Human resource policies should be tailored to reflect that understanding, providing workforce flexibility to recruit and retain employees. Recognizing that many new employees "work to live" as opposed to "live to work" may help to manage expectations, reduce workplace friction, and reduce attrition.

Utilities should continue to examine outsourcing/automation options. Identifying and retaining mission-critical jobs internally and shedding jobs and/or processes that are not mission-critical and that are more efficiently accomplished with outside resources should be common utility practice.

5—Technology. Technological advances will improve water quality, affect customer service, and reduce costs. Technology is becoming smaller, cheaper, and disposable. Lower costs, higher processing power, and higher labor costs are conspiring to drive automation of treatment technology and utility processes. Online monitoring technology is poised to expand greatly, as are mobile applications. Instant messaging and 24/7 communications are here and expanding rapidly.

Technology will continue to transform the utility workplace from traditionally labor-intensive (and high-cost) work activities to automated systems. Increasingly powerful supervisory control and data acquisition (SCADA), radio frequency identification tag technology, automated meter-reading systems, geographical information systems, and distribution system–modeling tools will find broader deployment in water utilities. The use of technology as a competitive tool will grow in order to manage staffing and labor costs. Electronic technician skill sets will grow in importance as will the need to protect the information technology assets of the utility from internal and external threats.

Implications:

• Electronics will not be repaired (because labor is expensive) but rather thrown away when possible. Technicians will be needed to do diagnostic work and to procure and install replacement equipment.

• Automation will grow in importance. Many utilities may take another look at the possibility of having minimally or unattended treatment plants.

• Real-time water quality monitoring systems will grow in importance as new analytical technology is deployed.

• Knowledge-management systems will manage complex multisource information and provide analytical and decision tools to water utility managers.

• Communications will be cheaper, faster, and 24/7.

• Instant messaging will be common among the workforce.

Strategies. Utilities should track and understand technology development and apply it strategically. All areas of utility business and operations should consider the applications, benefits, and challenges of information technology and automation. These developments should be incorporated into a strategic technology plan, and the plan should be modified periodically to reflect new developments. The plan should convey

• how information technology can lower utility costs;

• customer preferences for information and interaction with utilities via information technology;

• automation initiatives (including treatment plant automation) to reduce costs associated with labor, chemicals, and energy;

• how to use SCADA systems and new data communication technology to monitor and control remote facilities; and

• how to use the Internet to improve efficiency in purchasing, research, training, and customer and employee self-service, thereby allowing utilities to operate as a business and to be perceived by its customers and employees as efficient.

It will be important for the public water supply community to commit to research and development in order to foster innovation and new technologies. This could involve establishing strategic technology initiatives to develop specific products of interest to the public water supply community. Establishing links between universities and the private sector to drive research and development and encourage adequate health-effects research by others should be encouraged.

6—Total water management. Significant population growth, climate change, increases in the number of impaired US waterways, species loss, waste disposal, and other issues are leading toward a need to plan and manage resources collectively in order to avoid unanticipated trade-offs. Balancing water and wastewater generation, cost, environmental effects, population growth, demand management, and watershed implications will challenge traditional water utilities that have grown accustomed to operating in relative isolation. As communities grow together, integrated planning and decision-making will become more important. Managing the environment (including water) will require more cooperation and development of a common regional planning vision.

Implications:

• Population growth will require additional water resources. Many of these resources will be developed from currently impaired water supplies (wastewater, brackish groundwater, seawater).

• Population growth could increase pollutant loading in current source watersheds unless that growth is managed carefully.

• Development in suburban and agricultural regions will have environmental and water resource trade-offs. Regional approaches to managing the trade-offs may become more politically acceptable as communities reach build-out. Political resistance to regionalization and consolidation may lessen.

• Climate change (e.g., snowmelt and runoff) will affect different regions differently. Profound effects are predicted for some areas.

Strategies. Watershed-based resource management approaches should be adopted. An effort must be made to further understand European and Australian strategy models. Watershedwide planning and/or control approaches, including source water protection and integrated resource plans, can help manage environmental trade-offs and reduce costs related to various water resource development strategies. These cooperative initiatives will require technically competent water utility diplomats. Cooperative efforts must include all significant water users and wastewater and nonpoint source dischargers, including water/wastewater utilities and agricultural, environmental, and significant commercial/industrial entities. Planning should specifically address demand-management measures such as conservation and understand and incorporate the environmental attitudes of younger generations. Total water planning should also consider the effects of climate change on water supply and include contingency planning.

Regionalization should be encouraged to improve decision quality (e.g., in consideration of trade-offs). This includes promoting coordination between water and wastewater divisions within the utility.

7—Customer expectations. Consumer satisfaction with tap water is relatively high in the United States. Increasingly, however, segments of the population—particularly consumers on the East and West coasts and those under age 30—seek bottled and/or home-filtered water as their primary drinking water source.

Consumption of bottled and filtered water varies widely by region, with 80% or more of the population on the West Coast, in the Southwest, and in part of the Southeast drinking bottled and/or filtered water. In contrast, in the Midwest, such consumption in some areas is limited to only 20% of the population.

Other highlights from the AwwaRF research include

• Eighty-six percent of Americans are concerned about their tap water.

• By 2003, 48% of the American public was using either a point-of-use or point-of-entry device or bottled water at home (up from 41% in 2001).

• Safety was the primary motivator for those who drank home-filtered water.

• Those who drank bottled water cited taste, safety, or healthiness as their motivation for drinking this product.

Bottled water use continues to grow in terms of per-capita consumption and total sales (Figure 4). Statistics reflect a 10% annual growth in bottled water sales. There is some concern that public trust and confidence in water utilities may diminish as water rates rise and as customers' perceptions of the safety of tap water remain cautious.

AwwaRF research (Tatham et al, 2003) revealed three key expectations of water utility customers: (1) they expect their water to be safe to drink, (2) they expect their water to come out of the tap when they turn it on, and (3) they expect that their water bill will be fair and accurate. The research supported the assertion that people who are informed about quality and safety-related issues are more likely to report that they are satisfied with their water utility.

Research also showed that utility managers tend to overestimate consumer satisfaction and underestimate the level of importance consumers place on tap water safety. In addition, most customers cannot remember receiving a Consumer Confidence Report; these reports are not deemed to be effective in conveying information.

Customer service is an area that can be improved.

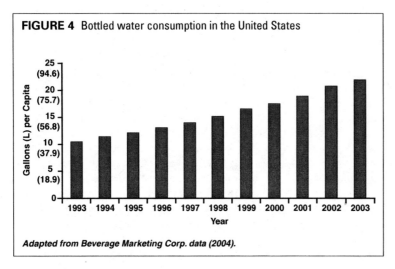

FIGURE 4 Bottled water consumption in the United States

Adapted from Beverage Marketing Corp. data (2004).

Implications:

• Consumers buy bottled water for its convenience, a preference for aesthetics, and concern over the safety of tap water. Should concern over safety translate into loss of trust in public water supplies in general, there could be implications for rate-payer support for rate increases if (1) consumers are not drinking tap water and therefore not interested in improving the system and (2) there is distrust that the utility can manage money effectively (because they are not managing water quality effectively).

• The SDWA now requires the assessment of costs and health benefits when drinking water regulations are established. MCLs are set to protect the health of individuals and are based on an assumed consumption of 2 L/d (0.5 gpd) of tap water over a lifetime. Bottled water and/or point-of-use device deployment is widespread in many communities; the actual exposure of consumers to tap water may be substantially less than the consumption the MCLs were based on, and therefore the "benefits" of the MCLs may be overestimated.

• There is a need for state-of-the-art outreach methods to communicate to the public the importance of water quality. New methods of conveying water quality and risk information are needed, particularly given the generally low science and technology aptitude of the American public.

• Utilities should embrace understanding customer needs, desires, and best methods of communication. As customers become better informed, they will likely become engaged in water policy issues.

• The role of public interest/advocacy and special interest groups in providing information to utility customers may grow in communities where utilities fail to provide needed information.

Strategies. Utilities should provide consistent and accurate information to internal and external stakeholders to foster communication and trust. The information should be readily available to all stakeholders. Utilities should take advantage of the Internet and other information technologies to make information available about water quality, compliance with water quality standards, budgets, status of capital budgets, and other aspects of utility operation. Utilities should also be cognizant of cultural nuances in how information is disseminated.

Utilities should actively engage their stakeholders. Utilities should work to increase public understanding of water quality health risks, infrastructure replacement needs, water supply reliability, and other poorly understood areas. Partnerships with independent parties should also be fostered. Aligning the utility with the needs and values of the community will ultimately result in less conflict and generate support for the utility's initiatives. Improving communications with stakeholders may require upgraded customer service systems and employee models for many utilities.

8—Utility financial constraints. The challenges of replacing and repairing infrastructure will strain many systems. Resistance to rising rates will require key cost-containment strategies. Federal subsidies will be static at best. Interest rates are rising.

Implications):

• Local sources will continue to provide the bulk of water utility funding. Federal and state governments are not likely to put markedly higher-than-usual amounts of funding into the public water supply community.

• Capital will become more expensive, putting increasing pressure on water rates to finance infrastructure.

• Difficult priority-setting and political conflict may occur in cash-strapped cities where enterprise fund contributions to the general fund have historically provided city revenue for nonwater services.

• Development of prudent water system reserve funds may make an attractive target for cash-strapped governments.

• Cities may increasingly work together to build regional infrastructure.

Strategies. Water utilities should formalize their financial policies. This includes documenting financial targets in their mission statements and having policies on rates and financial returns that ensure ongoing financial health. Water utilities with high debt ratios and/or those not earning above their cost of capital, need to look at creative strategies to improve their financial decisions. Also, water utilities should establish reserve policies for unforeseen events (such as a percent of replacement asset value). Rate risk associated with climate change (e.g., water sales fluctuations) should be considered.

Utilities should formally document infrastructure and rate needs and these needs/plans must be consistently communicated to stakeholders. Modern asset-management strategies can help in this regard. Rate-increase plans should be accompanied with measurable cost-containment strategies to remove the potential for politically damaging charges of inefficiency. Alternative funding mechanisms (e.g., public–private partnerships) should not be ruled out.

Public water agencies should consider better matching their high-risk costs as a fixed component of rates to reduce sales volumes risk.

9—Energy. Energy cost and supply reliability will become major issues for utilities. Despite ongoing efforts to encourage conservation and concerns over energy efficiency and potentially adverse environmental effects from harmful emissions, total energy consumption in the United States is projected to grow 21% by 2025 (DOE, 2004). Worldwide energy demand is projected to grow by 43% during the same period, with energy demands in India and China being the highest and increasing 100%. Electricity "deregulation" is fragmented, and its future is uncertain.

Implications:

• Energy use is—and will remain—a large component of the cost of water. Higher costs will increase the need to focus on energy efficiency.

• The price of energy will likely trend higher and be accompanied by increasing price instability. It will be difficult for water utilities to predict energy prices accurately and important to pass the additional costs through to consumers effectively.

• Energy shortages may occur with increased frequency, and water systems will need increased capability to switch to alternate or backup energy supplies.

• Energy efficiency in plant operations will be increasingly important to lower costs and risk of interruption. Equipment replacement efforts will increasingly consider energy efficiency and perhaps de-emphasize capital costs.

• New energy-intensive treatment technologies may not achieve their expected potential, despite their advantages.

Strategies. Utilities should develop an energy plan that includes aggressive energy-conservation strategies. These energy plans should also address a long-term goal of self-sufficiency.

• Utilities should develop alternative and backup energy capabilities (e.g., solar power, windmills).

• Energy costs should be a pass-through item in rate structures.

• Relationships should be fostered with energy providers in order to secure contracts for power delivery.

• A primary goal for water utilities should be to get off the power grid during peak use and possibly consider selling energy from generators during this time.

• Utilities should develop and ensure redundancy where appropriate in backup systems and understand where critical vulnerability points are.

• Utilities will have to examine the wisdom of water industry involvement in deregulation and encourage AWWA to work together with the National Association of Regulatory Utility Commissioners.

10—Increased risk profile. The terrorist attacks in September 2001 brought an obligation for water utilities to look beyond the established conventional threats to drinking water supplies and quality. Water utilities have always faced the potential for some level of malevolent attack from such threats as vandals and criminals and have responded with appropriate measures. However, after the terrorist attacks the importance of such measures was underscored and reinforced by state and federal legislation. Overall utility risk related to information technology, physical security, climate change, and litigation is increasing.

Implications:

• The need for improved physical security has been accepted by drinking water utilities following the September 2001 terrorist attacks.

• Utilities now need to examine the nature and capabilities of the real threats faced by the water industry.

• Utilities are becoming increasingly vulnerable to security breaches on information technology systems by both internal and external threats.

• Risks and cost trade-offs are not well understood by a traditional mindset.

• Utilities will require a sophisticated workforce to understand the risks and use the required tools for responding to these risks.

Strategies. Utilities need to revisit their risk-management approach in light of new internal and external risks. Successful utilities will develop a strategic plan that aligns the utility with the community. The strategic plan should consider growth, water supply, water quality, human resources and organizational efficiency, capital planning, financial planning, communications, and physical and information technology security. Transparency of planning will be increasingly important. Contingency strategies should be considered in each of these areas.

Specific risk issues related to staffing, training, health and safety, litigation risk, climate change and water supply risk, security, and aging infrastructure should be identified. These will vary by utility. The insurance and legal representation needs of the utility should be revisited and tailored to the current risk. A mix of reliance on internal and external legal resources should be carefully considered given the increasing legal complexity of water issues in general.

Utilities will need to invest in infrastructure. This investment should be guided by formal asset management plans that articulate capital improvement spending and financial plans. These plans must be based in a solid understanding of population growth trends and infrastructure repair and replacement needs. Specific methods for managing risk should be considered in each of these areas. Utilities should understand the implications of the current global capital market and translate that information into how it would operate in a high-capital-cost environment.

Utilities should develop and define risk-management success metrics and performance indicators. These performance indicators will help validate security response to various scenarios. This will allow the utility to optimize risk-management expenditures.

Utilities should stay engaged in federal security discussions and institutionalize security as a full-time function within the utility.

DISCUSSION

In the 2000 study, much utility attention appeared to be focused on providing customer satisfaction. Other trends of concern included water utility restructuring, private sector competition, water resource constraints, infrastructure needs, finance constraints, environmental regulations, and stakeholder relations. These same trends are still in play in the 2005 study.

Several trends seem to be gaining in importance. The rising risk profile of water utilities is more apparent now. This appears to be partly driven by increased sensitivity to physical and information technology security in the wake of the September 11 attacks. Although terrorist threats were identified in the 2000 study, the participants then felt that the likelihood was so remote as to make it relatively unimportant from a strategic viewpoint. What a difference a year makes. Similarly, rashes of computer viruses, spyware, identity theft, and computer hacking have sensitized the public water supply community to the potential threat.

Total water management ranked high in importance among the participants in this trend assessment. The strategic importance of water and environmental planning on larger scales than traditionally practiced appears to be growing. Optimizing resource decisions to reduce costs increasingly requires crossing broader geographic and policy boundaries (e.g., water, wastewater, growth, discharge limits, watersheds, land-use planning).

Environmental activism may also increase as the population in the United States increases and quality-of-life issues become more important. A "sustainability ethic" for water-poor and water-rich communities may develop in which the goal is to leave the smallest "environmental footprint" through environmentally sensitive water policies. Some of this can be seen in the expansion of NGOs as advocates for local watershed protection.

Continued tightening of drinking water regulations

will drive debate about the feasibility of maintaining high-quality water from centralized sophisticated treatment plants through distribution systems. Maintaining the quality of water during distribution will be a particular challenge as US infrastructure ages. Water utilities have characterized this issue as one that will have a significant effect on the water industry.

It is also likely that technology developments and regulatory interest will drive online monitoring of water quality in distribution systems.

Infrastructure management will continue to be a pivotal issue for water utilities. Water infrastructure needs in the United States will require significant investment of capital, causing large increases in consumer costs in affected communities. Improvements in pipe-replacement technology and other asset rehabilitation and replacement techniques may reduce the degree to which rates rise to support this investment. Nonetheless, many utilities will be hard-pressed to raise sufficient capital to fund their infrastructure needs and will turn to state or federal assistance. Managing infrastructure is ultimately tied closely with the ability to raise capital. Financial constraints will require innovative financing strategies (including the use of private capital). It may well drive greater regionalization and consolidation.

CONCLUSION

The strategic future of water utilities rests in the confluence of many significant trends, only several of which were discussed here. Trends can act in a cumulative, synergistic, or antagonistic fashion. Although all crystal balls are flawed, it is reasonable to draw several inferences from the array of trends. To that end, the utility of the future will likely do three things really well:

1. It will be operationally efficient. The effective water utility will do this because active consumers and political interests will demand solid financial stewardship as water rate increases mount for repair, rehabilitation, and construction of water supply infrastructure. Operational efficiency will require a motivated and engaged workforce.

2. It will be engaged in the community. The effective water utility will never take its customers or shareholders and stakeholders for granted. Communications systems will be geared to creating a constant "dialogue" with the community and to ensuring that there is philosophical alignment among the community, its elected leaders, and the utility's policies and actions.

3. It will use total water management principles to conduct water planning in a holistic fashion. The effective utility will integrate water management, treatment, environmental quality, and public policy decisions in a fashion that leverages regional opportunities and relationships and economies of scale.

This study is the equivalent of a general "environmental scan" that can be used by any utility in helping to identify future trends that will affect their utility and community. Water utility managers can use these trends as guidelines, adding local and internal trends to the strategic planning process. By anticipating these trends, water systems will be better able to position themselves to best serve their customers and communities in the future.

Future articles in this series will explore the study's findings regarding water resources, technology, regulatory, and consumer trends.

ACKNOWLEDGMENT

The authors thank the Awwa Research Foundation for funding and Project Manager Linda Reekie for making this work possible. The authors also acknowledge the guidance of Andrew DeGraca, John Huber, and Rosemary Menard of the Project Advisory Committee. The authors thank the utility managers who gave generously of their time and expertise in the Futures Workshop as well as the experts who were interviewed at the start of the project. This work reflects their collective wisdom and active engagement. Important logistical support by Gloria Rivera, Ryan Reeves, and Nicole West is acknowledged and greatly appreciated.

ABOUT THE AUTHORS

Edward G. Means III is vice-president of McGuire/ Malcolm Pirnie, Irvine, CA. Means has 26 years of experience in water utility management, water resources, and water quality. Roger Patrick is the president of Competitive Advantage Consulting Ltd. and is a specialist in improving the management of water and wastewater utilities. Lorena Ospina is a scientist with McGuire/Malcolm Pirnie.

REFERENCES

Beverage Marketing Corp., 2004. Beverage Marketing's 2004 Market Report Findings, New York.

Brueck, T. et al, 2002. Facing the Workforce Reality: Making More Out of Less. *Jour. AWWA*, 94:6:34.

Dixon, L. et al, 2001. The Challenges of a Changing Customer Base. *Jour. AWWA*, 93:9:46.

DOE (US Department of Energy), 2004. Energy Information Administration, Office of Integrated Analysis and Forecasting. International Energy Outlook.

Hobbs, F. & Stoops, N., 2002. Demographic Trends in the 20th Century. Census 2000. Special Reports. US Dept. of Commerce. Economics and Statistics Administration, US Census Bureau.

Means, E.G., 2001. Meeting the Future Head On. *Jour. AWWA*, 93:7:36.

Means, E.G. et al, 2002a. The Coming Crisis: Water Institutions and Infrastructure. *Jour. AWWA*, 94:1:34.

Means, E.G. et al, 2002b. Drinking Water Quality in the New Millennium: The Risk of Underestimating Public Perception. *Jour. AWWA*, 94:8:28.

Miles, J. et al, 2001. Planning for the Effects of Climate Change. *Jour. AWWA*, 93:10:34.

Patrick R. et al, 2002. Alternative Service Delivery Models. *Jour. AWWA*, 94:4:44.

Tatham, E.; Tatham, C.; & Mobley, J., 2003.Customer Attitudes, Behavior, and the Impact of Communications Efforts. AwwaRF, Denver.

This page intentionally blank.

How Big Is The "Water Business"?

A DIFFUSE AND DISPARATE INDUSTRY LIKE THE WATER AND WASTEWATER TREATMENT BUSINESS IS BOTH A FRUSTRATION AND A CHALLENGE TO THE MARKET RESEARCHER.

BY STEVE MAXWELL
(*JAWWA* January 2005)

The business that most of us rather casually refer to as the "water industry" is, in reality, a confusing array of fundamentally quite different niche businesses—businesses that can't really be classified under any one single heading. Obviously when it's difficult to define or draw boundaries around an industry, estimating market size and growth characteristics is likewise going to be difficult, if not impossible.

The water industry consists of numerous and very different sectors and types of players—new technology developers, established product manufacturers and integrators, specialty chemical producers, measurement and monitoring firms, manufacturer's representatives who sell equipment to the end user, engineers and consultants of all stripes, laboratories, contract operating service providers, and many others. These companies' only real similarity may be their ultimate end goal—helping to provide clean water.

Trying to accurately describe the water business brings to mind the old fable in which three blind men try to describe an elephant and come to startlingly different conclusions. Observers of the water and wastewater industry may not be blind, but it is still very difficult to accurately define and assess the size of this overall business. One thing, however, is increasingly clear to all observers, regardless of their vantage point—the water industry "elephant" is a large and growing business whose true economic significance is only beginning to be realized.

Individual perspectives on what comprises the "water industry," and, thus, how big it is, are typically influenced by that part of the business with which each individual is most familiar. Although the overall business seems to be gradually consolidating and coalescing into more of a unified or single industry, it is still composed of different sectors that each have their own growth, profitability, and strategic profiles. Lumping too many different segments together into one large whole tends to mask important trends and developments within a specific part of the business. For example, a 5% overall industry growth rate may disguise the fact the

one large and mature sector in decline is masking rapid innovation, economic vitality, and explosive growth in another, perhaps smaller, sector.

Over the past few years, there have been countless studies conducted regarding the future of the water business and the size of the market. As concerns have grown about the future availability of water resources, a number of academic institutions, industry groups, government agencies, and associations (including this one) have begun to make detailed economic studies and projections about the future of the water business. All of these have indicated a consistent and growing marketplace for the overall business, but specific predictions have been all over the board. Just a few of the more notable studies have included the following:

• In 2000, in a report entitled *Clean and Safe Water for the 21st Century* the Water Infrastructure Network estimated that the United States needed to spend an additional $24 billion a year—$480 billion over the next 20 years—in order to maintain and expand the water and wastewater distribution infrastructure.

• In 2001, AWWA estimated that some $250 billion would need to be spent over the next 30 years, or about $8 billion a year, in order to maintain just the drinking water side of US infrastructure.

• At about the same time, the US Environmental Protection Agency estimated drinking water infrastructure spending needs at about the same level—$160 billion over the next 20 years.

• In 2002, the Water Environmental Federation put forth its own estimate of approximately $300 billion over 20 years, or about $15 billion per year, for both water and wastewater infrastructure.

• Later in 2002, the Congressional Budget Office weighed in with the highest figures yet—$800 billion over the next 20 years, or $40 billion per year, for water and wastewater infrastructure expenditures.

These are only a few of the studies conducted by various trade and governmental organizations during the past few years. Although their estimates vary, these studies uniformly project future spending requirements far in excess of today's level of infrastructure spending. It also seems that the estimated gap between current investment rates and required future spending gets larger with each successive study. In addition, most of these studies have only attempted to estimate the infrastructure requirements of the municipal water and wastewater treatment and distribution systems in the country—not even attempting to take into account the (separate) industrial water and wastewater needs of the US economy.

Because there are so many ways to analyze, slice, and dice the water industry, it is critical that market researchers be thorough and clear about their assumptions when the current status and potential future growth of the industry are analyzed. As is the case in any industry in which there is a general lack of detailed or accurate economic data, certain market estimates seem to gradually evolve over time into conventional wisdom. One of the most widely cited estimates of the overall water business is developed by Environmental Business International (EBI). To develop its market estimates, EBI conducts surveys of companies in several different service or product sectors of the market and then totals and extrapolates the survey figures to result in a "supply-side" segmentation of the overall market, as shown in Table 1.

This $100 billion rough figure is increasingly becoming the conventional wisdom in terms of the overall size of the US water business. Estimates vary, but this is probably as good as any, and a number of interesting factors can be gleaned from Table1. Note that the fees we all pay to utilities for our drinking water and sewage services make up about two thirds of the overall annual spending on water in the United States. The products and services that industry and municipalities buy to

provide water and wastewater services comprise the other third of the US water business. Emphasizing the point made earlier, note the range of projected growth for these different sectors of the industry—varying from a 0% to 10% per annum growth rate.

Other estimates of the US market tend to be fairly close to these figures. For example, the market research firm Frost and Sullivan (F&S) recently estimated the US market at $105 billion, with an overall growth rate of 7%. That study segmented the water market in a slightly different way—residential point-of-entry/point-of-use products were broken out separately in a segment estimated at $1.4 billion per year. Interestingly, more detailed studies of individual segments of the market often yield different figures than do those of more general overviews. For example, the market research firm Freedonia has projected the US market for water treatment equipment and supplies at about $4.5 billion—apparently considerably smaller than the figures estimated by either EBI or F&S. On the other hand, the market research firm McIlvaine and Company recently estimated the global water treatment products and services market at $122 billion per year. Although definitional differences and assumptions clearly explain some of the discrepancies between studies, estimates of individual markets clearly range widely.

Although the domestic market is clearly the world's largest and most developed, opportunities also abound for water companies around the rest of the world. Even though it is difficult to adequately estimate the size of the US market, it is much tougher to get a handle on the world market size. Estimates of the overall world market seem to center around a figure roughly three to four times the size of the US market. Again, you have to wonder whether such figures are based on real research and data or whether they simply represent a convergence toward some conventional wisdom formed in response to the general paucity of reliable data. In any event, different parties have pegged the business at around this level. General Electric has indicated that it sees a world water market of $360 billion, and when Siemens recently acquired US Filter, it estimated a world market of $400 billion. The publication *Global Water Intelligence* recently laid out a set of assumptions that estimated a global market size of $540 billion per year; another private consultant set the figure at $655 billion per year. Perhaps these estimates are so much larger because many of these geographic markets are at much earlier and more rapid stages of growth than those in developed countries.

So, in the final analysis, what can you make of all of these widely varying estimates of the future market? How big *is* "the" water market? One obvious conclusion is that, even when we define things rather precisely, we really don't have a very good idea of exactly how big this market is. However, perhaps the

TABLE 1   The US water industry

Business Segment	2003 Revenue $ millions	2004–2006 Growth %
Water treatment equipment	8,860	4–6
Delivery equipment	8,880	2–3
Chemicals	3,660	0–1
Contract operations	2,290	6–10
Consulting/engineering	6,090	5–6
Maintenance services	1,640	3–5
Instruments/monitoring	800	5–7
Analytical testing	480	2–4
Wastewater utilities	30,780	3–4
Drinking water utilities	32,650	3–4
Total US water industry	96,130	

Source: Environmental Business Journal 2003.

more important question for most vendors is: "Does it really matter?" With a market somewhere in the range of the figures mentioned here, and given the types of underlying factors and challenges that are clearly driving this market, many firms are not too worried about the details.

Many observers and analysts have predicted an exploding rate of the growth for this industry in the near-term future; however, at least in most sectors of the business, the situation has been more one of lower but consistent growth. Historically—at least over the past decade or so—the total water industry has experienced mid-single-digit rate growth, say 4%–6% per year. There is little doubt that fundamental supply and demand considerations, whether expressed in a quantitative or qualitative manner, clearly suggest continuing growth into the long-term future. One way to summarize the likely future of this business might be to refer to, and slightly amend, another old fable—the race between the tortoise and the hare. From time to time the water industry may produce a few "hares"—businesses that will grow at 10%–15% for a few years (such as the membrane filtration sector of the business). But by and large, it seems more likely that the water business will consist of a number of dependable and relatively speedy "tortoises"—sectors that will perhaps grow a bit slower but that will be steady and dependable businesses with consistent growth well into the long-term future.

This page intentionally blank.

The Less Glamorous Side of the Water Industry. . . Where Most of the Dollars Will Be Spent

STEEL AND CONCRETE PIPE, PUMPS AND VALVES, FITTINGS, AND STORAGE TANKS—RATHER THAN HIGHER-PROFILE AREAS SUCH AS MEMBRANE AND ULTRAVIOLET TECHNOLOGIES—ARE EXPECTED TO GARNER THE LION'S SHARE OF INFRASTRUCTURE SPENDING BY WATER UTILITIES DURING THE NEXT 20 YEARS.

BY STEVE MAXWELL
(*JAWWA* November 2006)

Trying to keep up with the forecasts of future water needs and capital expenditure requirements can quickly set your head to spinning: $300 billion here, $400 billion there, estimates as high as $1 trillion dollars—it seems as if there is no end to the amount of money that we are going to have to spend to maintain and expand our drinking water infrastructure—and that's just in the United States.

Driven by this presumably continuous and rapid future growth, investors of all stripes are falling over each other trying to get into the water business. Most of these investors, both strategic and financial, want to buy into the exciting and rapidly growing areas of the exploding water business—membrane filtration or ultraviolet treatment technologies, the contract operations business, or other subsectors that are thought to represent high profitability and double-digit growth potential over the long-term future (see January 2006 Market Outlook.)

But guess what? A high percentage of these projected dollars are going to be spent on considerably less glamorous sectors of the business—on steel and concrete pipe, pumps and valves, fittings, and storage tanks. This may not be the "sexy" side of the business, and the companies in this sector don't seem to be attracting as much attention from the Wall Street analysts, the strategic industrial buyers, or the private equity players looking to invest in the water industry. However, this is where the lion's share of the dollars are going to be spent in the future.

USEPA ESTIMATES 20-YEAR NEED

Let's take a look at some statistics. As mentioned, scholarly studies suggest water infrastructure capital needs will range anywhere from $100 billion to $1 trillion over the next two decades. However, for a good point of reference let's use the widely cited and authoritative *Drinking Water Infrastructure Needs Survey and Assessment,* recently issued by the US Environmental Protection Agency (USEPA), which pegged the capital need at $276.8 billion over the next 20 years. Approximately $165 billion of this total is designated as current needs—improvements or upgrades to older and deteriorating systems that are needed right now in order to ensure a continuity of water supply and the maintenance of minimum water quality standards. The remaining $112 billion of the estimate is made up of needs that will gradually develop in the future

• as currently satisfactory systems continue to age and deteriorate,

• as new systems must be built to satisfy a growing and shifting population, and

• as improvements to existing systems are required to comply with new water quality and security regulations.

The size of this future capital requirement continues to grow, even by USEPA's own estimates. The $276.8 billion estimated in the most recent study compares with 1990s studies that suggested that the total need was about $165 billion. Many observers think that USEPA's estimate is, if anything, conservative.

Several interesting and illuminating ways to break down and segment this total capital expenditure need exist. One method is to estimate requirements in terms of the size of an individual drinking water system—how many people it serves. It is well-known that there are approximately 52,000 community water systems (public water systems that deliver water to the same population year-round) and that a small number of these serve the great majority of the US population. USEPA estimates that "large" and "very large" systems (those serving more than 10,000 people) supply slightly more than 80% of the total US population; within that group, just 372 systems serve 120 million people in the major urban centers of the United States. Table 1 summarizes how this total capital expenditure of $276.8 billion breaks down as a function of utility size. (There are some 107,000 additional systems, not shown in the table, that supply water but not on a year-round basis or to the same population—systems in such places as gas stations, campgrounds, and schools.)

It is notable that about 45% of the total capital expenditures will be made by only about a thousand systems, each of which serves more than 50,000 people. From the perspective of water technology and service vendors, this allows for a strong marketing focus.

More interesting, however—for vendors in the water industry and for investors looking to enter the business—is a breakdown of this total investment by type of product or service sector, as shown in Table 2.

More than 66% of this total amount, some $184 billion over the next 20 years, is going into transmission and distribution (pipes, pumps, hydrants, meters, and valves). Furthermore, an additional 9%, or about $25 billion, will be spent on storage. Total treatment expenditures—of which the popular purification technologies represent only a small part—constitute only about $53 billion, or less than 20% of the total. Simply put, over three quarters of the total spending will go into the nonglamorous and often overlooked infrastructure equipment side of the business—the one

sector that no one has seemed very excited about.

It is interesting (and perhaps revealing about the general level of sophistication and knowledge of most investors) that this basic fact does not seem to be very widely recognized or understood. Nor has it seemingly been translated into the general valuations of companies in the infrastructure equipment sector. For example, three leading public companies in this sector—Northwest Pipe Co., Mueller Water Products Inc., and Ameron International Corp.—are trading at average multiples of earnings before interest, taxes, depreciation, and amortization that range from 7 to 8, whereas the water utility and treatment companies are typically trading at multiples in the 10–12 range.

Finally, when these total expenditures are segmented in terms of geography, the focus of US spending will be

---

TABLE 1   Summary of US capital expenditure needs by system size

System Size	Number of Systems	Total Capital Need $ billions
Large community water systems (serving more than 50,000 people)	1,041	122.9
Medium community water systems (serving from 3,300 to 50,000 people)	7,638	103
Small community water systems (serving fewer than 3,300 people)	43,039	34.2
Other water systems and needs*	NA	16.6
Total US need	NA	276.8†

Source: USEPA 2003.
NA—not applicable
*Includes certain not-for-profit noncommunity systems, Native-American village water systems, and costs associated with certain newly proposed and promulgated regulations
†Numbers may not total because of rounding.

Table 2   Breakdown of US capital expenditure needs by project type

Project Type	Total Capital Need $ billions	Portion of Total Capital Required %
Distribution and transmission	183.6	66.3
Treatment and purification	53.2	19.2
Storage	24.8	9.0
Source water development	12.8	4.6
Other	2.3	0.8
Total US need	276.8*	100*

Source: USEPA 2003

*Numbers may not total because of rounding.

concentrated, as might be expected, in the areas with higher populations, in the older and more industrialized cities of the Northeast, and in the southern states (where most of the current population shift is occurring). Highest expenditures are expected to take place in California, Texas, Florida, Illinois, Michigan, New York, Pennsylvania, and North Carolina.

THE NEEDS ARE LARGE AND ON THE WAY

No matter how you slice or dice it, the vast nature of our future needs is pretty clear. As it was put succinctly in the 2001 AWWA report, *Dawn of the Replacement Era: Reinvesting in Drinking Water Infrastructure*— "replacement needs are large and on the way."

Unfortunately, too many utilities and cities are taking a dangerously reactive approach to actually making these expenditures. Many financially strapped agencies upgrade their treatment and distribution infrastructure only as it truly begins to fail, or they are applying temporary fixes to problems that should be solved by more thorough, ongoing reinvestment. It is widely assumed that this sort of delaying or last-minute spending is due to a true lack of funding. However, as is increasingly being pointed out by industry observers, many cities have the money but simply not the political willpower to spend on infrastructure reinvestment when they should.

In the final analysis, however, this head-in-the-sand tendency on the part of many utilities will only delay eventual spending. Future spending on water infrastructure is a classic case of "when, not if." As Andy Seidel, the president and chief executive officer of Underground Solutions Inc., pointed out at a recent water investment conference, delays will only have the effect of creating a larger pent-up demand for infrastructure products and services. There is a huge and overflowing "reservoir" of need behind a "dam" that will eventually break, with huge infrastructure demands spilling out into the market at some point in the future. Infrastructure companies— pipe and tank manufacturers, pipeline rehabilitation companies, pump and valve distributors, and a broad array of infrastructural services providers—face an inexorably growing demand over the longer-term future. Although these types of companies may operate at somewhat lower margins than do other sectors of the water business, it seems inevitable that they will experience future growth opportunities unparalleled in the broader water industry.

REFERENCE

USEPA (US Environmental Protection Agency), 2003. Drinking Water Infrastructure Needs Survey and Assessment, EPA 816-R-05-001. Ofce. Ground Water & Drinking Water, Washington.

This page intentionally blank.

Strategic Planning in the Water Industry: Making it a Truly Useful Process

STRATEGIC PLANNING IS ONE OF THE MOST CRITICAL BUSINESS FUNCTIONS IN ANY ORGANIZATION. IT IS AN ONGOING PROCESS—AND A NEVER-FINISHED TASK—THAT MUST BE REGULARLY REVISITED, REFINED, AND UPDATED IN ORDER TO HELP YOU BE SUCCESSFUL AND ACHIEVE YOUR KEY OBJECTIVES OVER THE LONG TERM.

BY STEVE MAXWELL
(*JAWWA* January 2004)

In a rapidly changing and highly competitive industry like the water business, it is critical to regularly assess the marketplace and the competition and to continually figure out how to best mesh your strengths and weaknesses as an organization with the challenges and opportunities presented by the marketplace. A key attribute of the successful executive leader is having the ability to lead his or her management team through this kind of thinking exercise, to be explicit in the statement of the organization's objectives, to detail the means by which those objectives will be achieved, and to be able to articulate them on paper. Unfortunately, a quick poll of both private and public organizations suggests that formalized and truly useful strategic planning programs are still not all that common in the water business.

Many organizations in the water business—both private companies and public agencies—seem to believe they cannot afford to invest the time, effort, or money necessary to build and institutionalize a formal strategic business planning function. Some organizations even seem to believe there is no particular need for strategic planning, especially when things are going well right now; they say they know what they're doing and that whatever happens in the future is just going to happen. Then there are those managers and chief executive officers who claim that they personally conduct all the requisite strategic planning in their own heads—if and when they need to.

Part of the reason for this gap may be the fact that many of the senior managers in this industry are engineers or scientists who never really received formal training in business management and who may be generally unaware or skeptical of the value of long-term strategic planning. Unfortunately, many firms that have tried to initiate a formal planning process have ended up dumping the process a few years later because they didn't think it was very useful when, in reality, they may have just not developed the right kind of program.

However, as discussed here, strategic planning is a critical communication and objective-setting exercise that must involve numerous stakeholders and different levels of people throughout the organization in order to be successful. In short, although the evidence shows that few tasks are more critical for long-term success and growth, too many organizations end up fighting today's fires rather than spending a little bit of time planning for the future.

Many of us are only too well aware of the common complaints about strategic planning—that it is somehow an "academic" or "paperwork" exercise, divorced or detached from the day-to-day urgencies and realities of running an organization. Some argue that longer-term planning efforts are a waste of time because it is impossible to predict the future. Unfortunately, too many of us *have* been involved in strategic planning processes that were poorly managed, never implemented, or simply irrelevant. Too often, after a hopeful golf weekend or retreat in the mountains, the strategic planning process has resulted only in a thick bound volume that gathers dust forevermore on the bookshelf.

Strategic planning certainly can be an irrelevant and academic process—*but it doesn't have to be*. In fact, many well-intentioned corporate efforts at formalizing and instituting planning processes could be more successful and useful over the long term if they were much more straightforward and simply stuck to a few key rules and guidelines. There is nothing magical or ephemeral about the strategic planning process. In fact, the best strategic planning exercises are often quite easy and simple.

In a nutshell, strategic planning can be thought of as

consisting of four simple steps, as highlighted in Figure 1.

1. assess the "external world"—market opportunities and competitive threats in the marketplace—and determine where the greatest opportunities are;

2. assess the "internal world" —frankly and critically evaluate your own strengths and weaknesses as an organization;

3. match your key strengths as an organization to those most relevant opportunities identified in the marketplace; and

4. articulate that "match" or plan to your key stakeholders and put in place a tactical approach, schedule, and set of responsibilities for achieving that plan.

This approach, sometimes referred to as a "SWOT analysis" (the analysis of strengths, weaknesses, opportunities, and threats), is the foundation of most strategic planning efforts. This process can be a relatively simple "seat-of-the-pants" process, or it can be incredibly detailed and complex and supported by reams of market and competitive research data—either way, it is an important exercise. It can yield insights, strategic opportunities, and directions for the organization, as simplified in Table 1.

Once established, a strategic plan should be re-evaluated and updated on at least an annual and preferably a quarterly basis—a well-written plan will allow the user to check progress on previously established goals and determine if and where changes or corrections need to be made. As a more regular and formal strategic planning process is developed and implemented and as it begins to be more broadly accepted and utilized in an organization, several benefits will begin to emerge. First, the planning process will get easier as it becomes more institutionalized—once a framework plan is in place, it becomes easier to adjust and modify that plan, as opposed to starting from scratch. Second, as the process becomes understood by more people, a better common understanding of the firm's overall direction and vision should start to become clearer across the organization. Informational needs and/or organizational changes will begin to be more obvious.

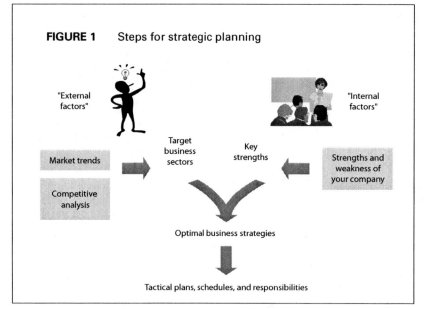

FIGURE 1   Steps for strategic planning

Internal Issues → ↓ External Issues	STRENGTH	WEAKNESS
OPPORTUNITY	This area represents achievable opportunities that should be emphasized.	Issues in this area should be de-emphasized, unless new strengths can realistically be built.
THREAT	Work to secure your strengths in this area; establish contingency plans in case of greater competition.	Steer away from this area, as success is highly unlikely; focus efforts on areas in which success is more likely.

TABLE 1   Key issues to evaluate in the strategic planning process (SWOT* analysis)

*SWOT—strengths, weaknesses, opportunities, and threats

The real benefit of the strategic planning process is not just the process of strategic thinking and debate, the clearer objectives, or the more formal plan document that results. Often, the real benefit is in the interaction and communication that occur when various leaders of the organization get together to talk about strategy and direction. The planning meetings may be the only time that many of these people have to get together and talk through the broader future directions and strategic alternatives faced by the organization, and significant and productive new understandings and personal relationships may be formed. Often, this communication is just as valuable as the plan process itself.

We would all do well to remember that when things are going well, it may not be simply because our management is that much smarter or our execution that much better than the next guy's—it may just be that we're being lifted up by the same tide that is lifting all the other boats in the business. The trick is figuring out how to continue to perform well even when the tide starts to go out—how to succeed and continue strong performance even as others in your business may begin to suffer. Good strategic planning can help to confer those competitive strengths and advantages.

The winners of the future are likely to be those organizations whose leaders have had the commitment and vision to consistently monitor the external market environment and the courage to adjust their paths as they move into the future, even when current cash-flow or pressing business problems may make it difficult to take the long view.

In Article 9, I will highlight 13 specific steps and guidelines you can use to make sure your strategic planning process is as effective and useful as possible.

This page intentionally blank.

13 Ways to Make the Strategic Planning Process More Useful

ARTICLE 8 SUMMARIZED THE OVERALL PREMISE AND APPROACH TO STRATEGIC PLANNING FOR WATER-RELATED ORGANIZATIONS. IN THIS ARTICLE, WE EXAMINE SOME GUIDELINES FOR BUILDING THESE PLANS.

BY STEVE MAXWELL
(*JAWWA* March 2004)

s you begin to develop a truly useful and sustainable strategic planning process, there are a number of key guidelines of which you should be aware.

1. Plan the planning process. Decide at the outset what it is you want to accomplish, who is going to be involved over what time period, and what the end product should look like. This may sound obvious, but it often isn't. If you are just starting a new process, it may be wise to keep your expectations fairly modest to begin with. How detailed will the plan be? What period into the future will it span? Will it cover general objectives and directions for the firm, or will it be a more specific, department-by-department breakdown of detailed tactics for achieving those broader objectives?

2. Strategic timing is pivotal. The timing of the effort, vis-a-vis other corporate processes, is critical. The strategic planning process is most effective if it is conducted in advance of and tied directly to the organizational budgeting process. It is important that the financial staff of the organization be directly involved in the process. The strategic planning process should essentially be seen as setting the broad guidelines within which the more detailed financial budgeting process can be conducted. Indeed, it is difficult to understand how any organization can even try to effectively budget without at least some strategic assumptions behind the process.

3. Make sure the right people are involved. The strategic planning process should include most of the key executive management staff. If the strategic plan is drawn up in a vacuum by the CEO or the planning department, it isn't a very valuable process, and it's not likely to be followed or given much attention by others. It is especially important to involve the sales and business development leadership of the organization because they are often the people in closest touch with trends and developments in the outside marketplace.

4. Top management must lead and champion the process. Many executives in the water business and other indus-tries pay lip service to being strategically oriented and "market-driven," but in reality too many fail to practice what they preach. If top management doesn't take the strategic planning function seriously, it's unlikely that anyone else will either.

5. Make sure the data you're using are as accurate and timely as possible. One of the biggest risks inherent in the planning process is that you are evaluating your situation and planning your future based on faulty data. As a sage once said, "It is difficult to make predictions, especially about the future." Obviously, none of us can know the future with much certainty, but try to make sure that your assumptions about competitors, clients, services, the economy, and so on (see Figure 1 in Article 8) are as complete as possible, and recognize that it may be better to draw no conclusions than to draw conclusions based on clearly faulty data. This is an area in which many firms could do much better—the significance of accurate market research cannot be overemphasized.

This data accuracy aspect of the strategic planning function can be very challenging in the water industry because of the accelerating pace of change—in terms of economic forces, driving regulations, technology, and the competitive landscape. Many sectors of this industry are changing so fast that they are plagued by a lack of good market and competitive information, making it even more challenging to position and prepare for the future.

6. Be as specific as possible in the delineation of your goals. This is perhaps the most critical prerequisite of a meaningful and useful plan. Strategic plans that consist solely of "motherhood and apple pie" objectives may look and sound nice, but they don't really accomplish much. Many strategic plans tend to be far too general— "We want to be the leading firm in our industry," "We want to exceed the expectations of our clients," and so on—without any mention of the metrics or means of even measuring such lofty goals. A good strategic plan should function as a detailed (if evolving) road map of where the business is headed. What are the main objectives of the firm, what are the specific subtasks required to accomplish those objectives, what routine and more day-to-day efforts will have to happen, who is responsible for which task, when should it be started, when should it be finished, and so on.

7. Objectives and goals should be ambitious, but realistically achievable. The strategic objectives and financial goals of the company should be ambitious, but they should also be realistically achievable. Many strategic plans tend toward superlatives—oftentimes involving growth assumptions or profitability rates that are clearly unachievable. If the plan proposes objectives that are obviously unattainable and if there is no follow-up after such lofty goals are set, then the rank and file of the organization is probably not going to take the whole process very seriously. In short, the strategic plan should be objective, and the specific goals should be achievable given a strong effort.

8. Know exactly who's doing what by when. As you develop specific plans and objectives, make sure you specify who is responsible for doing what and by when. Many firms put together good planning documents with reasonable goals and objectives that everyone agrees on, but they don't really allocate the responsibilities very clearly. This is one of the most important points to understand in getting a strategic planning process up and running successfully—specifying not just where you want to be, but nailing down the details of how you are going to get there.

9. Recognize you're probably not going to accomplish everything. This realization is especially true if you are just starting up a formal planning program. Focus your efforts on a few key areas, and work to accomplish those first. Walk before you run, and show progress on a broad level before trying to address more specific or detailed areas.

10. Consider outside help. It's not a bad idea to get some outside assistance to help you implement and facilitate the process as you go forward. There are many strategic planning and management consultants around the water industry who can assist you in evaluating your needs and getting a more formal strategic planning program started. Sometimes outside facilitators can help to force a more objective and critical self-analysis and can also help to keep the process on schedule when key participants have many other responsibilities and fires to fight on a daily basis.

11. Quantity does not equal quality. This truism is especially apt in the strategic plan document. Although some firms have produced multivolume tomes, the most useful strategic plan documents can usually be adequately expressed in 10–20 pages of text or less. The document must be readable and understandable, and although we have stressed detail as a critical component, it should not be one huge spreadsheet. A key value of a strategic plan is to promote internal communication and discussion of the company's current position and future objectives.

Thus it should be the type of document that can be readily absorbed by many different kinds of people within the company.

12. Get the word out to all employees. After the planning cycle is completed, make sure the key results of the plan are effectively communicated properly to the "troops," i.e., the rest of the relevant people in the company and most particularly those who did not have a specific role in the process. Some firms have had good success and "buy-in" by giving the broad employee base a chance to participate in the process via an intranet or mail survey, for example.

13. A strategic plan is never "finished." Keep in mind that a strategic plan must necessarily be an evolving and changing document—internal capabilities or directions may change over relatively short time periods, and circumstances in the outside world can certainly change dramatically and unexpectedly. For example, many firms had a very distinct change in their strategic plans and directions two and a half years ago following the unpredictable events of September 11. We are planning in an environment where things are constantly changing. We also must constantly change and adapt to keep up with the outside environment.

In order to be successful, water utilities and other organizations must build a clearly defined strategy, based on an in-depth knowledge of the outside world (particularly your customers) and the strengths of the organization. That strategy must be clearly communicated to employees, customers, and other stakeholders. The strategy must be refined and continue to evolve over time as circumstances in the outside world change. Remember that ultimate success will be achieved not by simply having a strong strategic plan and vision but by successfully implementing that plan. As Larry Bossidy, the former CEO of Allied Signal, said in his recent book, execution is the real job of running an organization—not simply formulating a vision and leaving the work of carrying it out to others.

Preserving Sustainable Water Supplies for Future Generations

BY PAMELA P. KENEL AND JAMES C. SCHLAMAN
(*JAWWA* July 2005)

The most widely accepted concept of sustainability is development that "meets the needs of the present without compromising the ability of future generations to meet their own needs" (World Commission on Environment and Development, 1987). Sustainability, as applied to water resources, is a multidimensional way of considering the interdependence among natural, social, and economic systems in the use of water. Accordingly, enhancing and preserving ecological systems and ensuring social well-being and security while accommodating economic vitality are objectives that must coexist to achieve sustainability (Kranz et al, 2004).

Some contend that the problems of sustainability, including overconsumption, overpopulation, fossil fuel use, and species destruction, are not mainly technical. Additionally, these problems involve complex and interactive systems in ways that are naturally unpredictable (Prugh et al, 2000). There is growing recognition that our institutions were established based on historical water needs that focus on physical, chemical, engineering, and other traditional water concerns as nearly discrete components. If water availability, water quality, land use, physical, chemical, and ecological characteristics must be considered simultaneously in varied geographical settings, we need to devise ways of relating these factors (USGS, 2005a).

The Sustainable Water Resources Roundtable was created as a US government–sponsored group under the Advisory Committee on Water Information to promote information exchange among governmental, industrial, environmental, professional, public interest, and academic group representatives and to ". . . serve as a forum to share information and perspectives that will promote better decision-making in the [United States] regarding the sustainable development of our nation's water resources."

The roundtable proposes that sustainable development of water resources

- involve policies, plans, and activities that improve equality of access to water;
- include limits and boundaries of water use beyond which ecosystem behavior might change in unanticipated ways;
- entail consideration of interactions occurring across different geographical ranges—global, national, regional, and local; and
- challenge us to fully assess and understand the implications of the decisions made today on the lives and livelihoods of future generations, as well as the natural ecosystems upon which they will rely.

From a water supply perspective, perhaps sustainability can be defined simply as the utilization of water sources while ensuring that we do not affect the ability of future generations to use the same sources. The AWWA Water Resources Division's ad hoc committee on sustainability offers the following working definition: "Sustainability is the necessary planning and management of water resources to provide an adequate supply of high-quality water while providing for the economic, environmental, and social needs of current and future generations."

Regardless of the definition, the concept of sustainability clearly requires a long-term view of water systems and must therefore factor into long-term water supply planning. A drinking water provider's or supplier's primary objective is to ensure that a plentiful supply of high-quality drinking water is available to the system's customers. To accomplish these objectives, today's water provider also needs to stay familiar with changing and tightening regulations, keep abreast of improving and innovative technologies, and continue to learn about emerging issues of concern. Thus the water provider's job continues to grow in complexity. Ensuring adequate water supplies for current and future customers requires concentrated attention to water system planning, management, and operations as well as to more efficient water use.

LONG-RANGE PLANNING

The factors likely to affect long-term water supply planning include population growth, economic development, regulations, environmental policy changes, financial conditions, and weather extremes. Additionally, many pollution threats may affect the quality and long-term usability of a system's sources.

Population growth and economic development. As economic prosperity continues to fuel development of lands that at one time may have buffered our sources of supply, it is becoming increasingly apparent that nontraditional management techniques need to be considered in order to maintain high water quality and minimize future degradation of these sources. The renovation and revitalization of older urban areas may provide additional opportunities for use of new and innovative techniques and strategies to control and treat runoff while providing aesthetic benefits.

Balancing economic growth and development with water resources considerations is essential. Attention should be placed on the available quantity and quality of current sources to ensure their sustainability as drinking water sources for the future. This means that we must protect existing sources and prevent their future degradation as well as plan for protection of future sources. Many source water protection activities undertaken by public water suppliers and their watershed partners are practices that ensure the sustainable quality of those drinking water sources.

The experience of those who have invested in wastewater treatment plant improvements to remove nutrients from effluent shows that further significant investments will not provide a proportional amount of water quality benefit. Nonpoint sources of pollutants—runoff from urban, agricultural, and construction site areas—are principle pollutant sources and become primary targets for additional control.

What is not sustainable? Nonpoint source discharges, especially from urban and agricultural land activities, are significant contributors to surface water impairment. Continued degradation of water quality may affect the long-term suitability of drinking water sources. Because point source discharges have been brought under control over the past three decades, it has become broadly acknowledged that nonpoint sources of pollution play a dominant role in degrading surface water quality. In May 2001, the US Environmental Protection Agency (USEPA) published a report detailing the progress made in safeguarding water resources in the United States (USEPA, 2001). Nonpoint

source pollution was identified as the leading threat to US waters, requiring increased USEPA focus on programs to address pollution from farmland, animal feeding operations, septic tanks, urban and suburban developments, and forestry operations. The control of nonpoint sources is a significant issue of concern for many drinking water suppliers.

What can be expected as land uses shift from agricultural or undeveloped to more urban? In the Chesapeake Bay watershed, located in Pennsylvania, Maryland, Virginia, New York, Delaware, West Virginia, and the District of Columbia, the US Geological Survey has documented land use change as the primary factor causing water quality and habitat degradation (USGS, 2005b). Current USGS data-gathering efforts are focused on documenting changes in land cover and land use that relate to sediments, nutrients, and toxins associated with urban, suburban, and agricultural lands in the large watershed.

In Philadelphia, drinking water is withdrawn from the Delaware and Schuylkill rivers. Upstream of the city's intake in the Schuylkill River lies 2,000 sq mi (518,000 ha) of watershed and more than 180 tributaries. The river responds dramatically to frequent rainfall events. More than 30 years of water quality monitoring has provided data for an evaluation of observed water quality trends and for correlation to changes in the watershed land cover. Recently, a build-out analysis was performed to predict the water quality that may be expected after 50 years of additional development in townships outside the city. Over the past decade, the water quality has degraded at a rate that can be tracked to development in the watershed. There is concern that the quality of the source will continue to degrade as the watershed population increases to more than 3 million inhabitants and the amount of impervious area in the watershed doubles.

"We have seen these water quality [effects] in the past 10 years, and because of these [effects], our water treatment options are limited. Imagine the problems 20 or 50 years from now, with the anticipated level of development and regulations affecting us and our treatment goals," said Philadelphia Water Department's Watershed Protection Manager Chris Crockett. "It will be extremely difficult to meet our needs in a cost-effective manner."

Water quality degradation can be cumulative, especially in reservoir settings where sediments, nutrients, and other pollutants may build up and contribute to more rapid eutrophication of the drinking water supply. Therefore, protection of water quality and management of land-based activities that have the potential to contaminate supplies are particularly important. Increasingly stringent regulations imposed on drinking water treatment require a more critical consideration of water source protection.

Changing regulations and environmental policy. Regulations can prevent use of the most efficient ways to reduce emissions—thus increasing costs and discouraging cost-saving innovations (OECD, 2004). Tightening regulations and environmental policies, such as those relating to endangered species and wetlands protection, creates challenges for drinking water suppliers. In some cases, pollutant source regulations, such as total maximum daily loads (TMDLs), may provide opportunities for water suppliers to protect drinking water sources from point and nonpoint source discharges. On the other hand, if a regulation is a mandated percentage reduction or low end-of-pipe concentration, it is possible that costs for control will increase significantly without a comparable water quality benefit.

Even for nonpoint source discharges, the situation may be complicated and not easily mitigated. The interrelationships between water, air, and land are complex. Continued water quality monitoring and modeling of the Chesapeake Bay region have revealed that airborne nutrients are responsible for 25% of the nitrogen contribution to the estuary. Further, as the percentage of nutrients directly discharged is reduced, the airborne percentage becomes even larger.

Water demands and revenue trends. Although per-capita water consumption has leveled since the 1970s, the total consumption of water is increasing as the population continues to expand. Per-capita consumption has been reduced in recent years by local ordinances established to conserve water during periods of drought and by the development of long-term water conservation programs. Regulations requiring technological changes such as low-flush toilets and reduced-delivery showerheads have further helped to reduce per-capita consumption.

Although water conservation is being implemented in many areas, the overall water consumption trend continues to increase. As a result, new source waters are being investigated to offset the demand. However, new sources are more expensive to develop because they are either less accessible than already tapped sources or they are not of high quality. If the demand for water increases while the development of new sources slows, costs will increase. Urban areas and industrial users will have the ability to pay more for water than rural and agricultural users. Therefore, a new balance will need to be struck between rural and urban users to offset the economic advantage to urban users.

Extreme natural events and climate change. Long-term water supply planning issues must consider extreme natural events, at least in a probabilistic sense. Historical weather extremes as well as possible global climate changes should be a consideration in long-range plans. Most scientists agree that the burning of fossil fuels during the past 150 years and clearing of forests are major causes of the identified global warming trend. The trapping of greenhouse gases, such as carbon dioxide

and methane, in Earth's atmosphere has caused surface temperatures to rise. Warmer global temperatures are likely to produce extreme weather such as droughts, intense rainfall, and intense hurricanes. In California's request for public comment on its most recent Water Plan Update (State of California, 2005), climate change is identified as a consideration. The California Department of Water Resources writes that the magnitude of potential effects on California's water resources infrastructure depends on the degree of warming experienced (Roos, 2005). Although the degree and significance of climate variability that might result from warming is highly uncertain, the results of increasing temperatures that may affect water supplies can be predicted. Shifts in precipitation patterns, intensity, duration, and type are expected. The potential effects include changes in snowpack accumulation and melting, runoff patterns, water demands and requirements, sea-level rise, floods and droughts, and surface water temperature. For instance, levels of snow in the mountains of western states will rise, and the land area covered by snow will decrease, resulting in decreased snowpack.

Concerns about changes in snowpack are already news. In February 2005, the snowpack in Washington state was affected by warm-weather and near-record low snowfall. The Natural Resources Conservation Service reported that Washington's snowpack was only 26% of the historical average for that time of year (NRCS, 2005b). The prospect of global climate change and the unpredictability of the potential effects are highly unsettling considerations. Although the magnitude, timing, and direction of the climate-induced changes in a region's water supplies are unpredictable, the weather change is certain to affect the hydrologic system as well as the demand for water.

It is generally accepted that the increasing levels of carbon dioxide in the Earth's atmosphere are causing a global warming trend. Shifts in precipitation patterns because of warming currents in the Atlantic and Pacific oceans have already been observed with resulting changes in seasonal weather patterns. The El Niño and La Niña phenomena affect the regions of North America in different ways, some dry and some wet. So, global climate change will influence stream flow patterns, and the frequency of these climate events may be increasing.

As precipitation patterns across the globe change as a result of climatic shifts and global warming, so will watersheds. Extended droughts may reduce the "safe yield" expected from water sources, and what might have been sustainable for the past 10 years may not be in the next decade. Droughts can dramatically affect runoff water quality because sediment and contaminants can be transported much more easily when vegetation is decreased. Extreme weather events, although difficult to predict, have predictable outcomes. Long-term droughts, reductions in mountain snowpack, death of vegetation, and saltwater intrusion from rising seas are all possibilities from current global warming trends and will increase the demand for new sources of water in the affected areas. If future weather patterns can be deduced, long-term sustainability plans demand that efforts be made to quantify and predict the range of possible outcomes to the changing environment.

WATER QUALITY PROTECTION

Sustainable source water quality requires that both current and future supplies be protected. In many cases, it is a challenge to maintain the high quality of existing sources, and others are in need of improvement because of quality degradation over time. It is imperative to collect water quality information now to be able to track changes in water quality over time, as well as to provide a basis for implementation of quality improvement strategies.

Source water assessments. Since passage of the 1996 amendments to the Safe Drinking Water Act, protection of drinking water sources has become a focus of USEPA, state drinking water programs, public water suppliers, and water consumers. Providing safe water is a comprehensive and integrated endeavor involving water protection and treatment from the water source to the consumer's tap. The required Source Water Assessment and Protection (SWAP) programs have, in some cases, inspired and empowered water suppliers to implement new and creative source water protection programs. This innovation is in part thanks to a few available funding mechanisms from which some water suppliers have secured financial assistance and incentives for their watershed protection activities.

The prevalence of geographic information systems has enabled the straightforward delineation, identification, and evaluation of watershed protection areas. It has become much easier to communicate with drinking water customers and watershed residents (who may not be users of the drinking water system) about the need to protect the quality of the water source. The foundation of effective source water protection is high-quality source water assessment. The assessment delineates the source of the public drinking water, identifies potential sources of contamination, determines the drinking water source's susceptibility or vulnerability to contamination, and makes the information available to the public and local decision-makers.

Although source water assessments are helpful in identifying threats from existing potentially contaminating activities, it is very difficult to protect drinking water sources from activities or entities that do not yet exist. Therefore, vigilance is needed, and when a potentially contaminating activity is identified in the planning stages, a methodology for evaluation and decision-making, as well as a public process of consideration, is necessary.

Long-term water quality monitoring and analysis of reservoir supplies. Predicting changes in source water quality is an important factor when determining long-term sustainability goals, a process essential for long-term planning. Evaluation of a reservoir's current trophic status coupled with predictions of the future status (extrapolated from current reservoir trends) will help identify the reservoir's long-term water quality prognosis. If analysis determines that the reservoir's quality is at risk, measures to mitigate the risk and improve water quality can be implemented before the problem becomes critical. In-reservoir remediation techniques to improve water quality may include aquatic vegetation harvesting, dredging, circulation, aeration, and nutrient inactivation. Reservoir and inflow quality information are the basic data needed to determine an appropriate course of action with the highest likelihood of success. In many cases, treating the lake water is only one strategy in a multistrategy approach. It is also important to appreciate the influence of the direct watershed and how it may affect the reservoir over time. When the effects of the watershed are considered, a different set of measures for protection of the drinking water source is required.

WATERSHED APPROACHES

Integrated water resource management at the level of natural hydrologic units is an essential step for improving water use efficiency, protecting aquatic ecosystems, and reducing conflicts among water users (Frederick, 2002). As competition for water rises, all users within a hydrologic unit or watershed become increasingly interdependent.

- Upstream users affect downstream users.
- Groundwater use affects surface water supplies.
- Surface water supplies affect groundwater supplies.

Most effective protection strategies are based on a watershed approach to managing the water supply. Effective source water protection also requires the support of the community because protection measures may involve voluntary actions, best management practices, or local zoning issues. To educate the community on the importance of such measures, the results of the source water assessments must be publicized.

Further, the case must be made to link drinking water protection actions with watershed protection actions. Historically, water programs were developed to protect separate parts of the ecosystem or separate uses of its resources; however, this fragmented approach can be a barrier to public health protection. Rivers, streams, and groundwater that serve as sources of drinking water also have ecological value, and their functions cannot be separated. Therefore, ensuring that our institutional programs work together is critical.

To adequately define management strategies, a watershed plan is usually developed. This plan should define the following: participants, issues, constraints, alternative management programs, recommendations, and an implementation plan. It is important to start with a plan so that goals and objectives are defined as the community evaluates the most effective organization for implementing the plan.

The next step is to identify and develop the mechanisms that will be used to manage the area to be protected. Several mechanisms for managing the protected area include watershed districts, legislation, ordinances, land acquisition, and collaboration. A description of these mechanisms is given in the sections that follow.

Watershed districts. The watershed district concept breaks regions and municipalities into small groups that include all stakeholders in a given watershed boundary. Conceptually, the districts help define which stakeholders are interrelated. The districts lead to a more comprehensive and effective governance of the system by compartmentalizing all of the stakeholders for a given watershed into one representative body, which can then make logical and appropriate management decisions for the system as a whole.

Legislation. Some states and communities have enacted separate legislation requiring the protection of water sources. The legislation can designate the responsibility for protecting a water source to an existing entity or require creation of a new entity with the legal ability to enforce the legislation. Generally, the responsible entity has the power to review and approve proposed development within the protected area.

Ordinances. Many communities manage their source water protection areas through city or county ordinances. These are appropriate for managing protection areas within jurisdictional boundaries. Neighboring jurisdictions may adopt the same ordinance to accomplish regional protection.

An ordinance should have language specifying allowable and prohibited land uses within the source water protection zone. For example, many source water protection ordinances limit or forbid the storage of hazardous materials and place restrictions on the location of businesses that use these materials within the overlay district. An ordinance should also include procedures for reviewing proposed projects within a source water protection district to verify that the project is consistent with the ultimate goal of the ordinance. This might include requiring applicants to submit geotechnical and hydrological analyses to determine the potential effects on water quality, as well as spill-control plans for businesses performing potentially contaminating activities. Finally, language explaining the mechanisms for enforcing the ordinance requirements, including the civil and criminal penalties that may apply for failure to obey, should be included.

Another method of managing source protection areas is to acquire land upstream from the water source.

Limiting use of lands can also be accomplished through permanent easements that prevent potentially harmful development while allowing those land uses that will not adversely affect water supply sources.

Collaboration. Virtually all source water protection plans involve outreach to the residents of the protection area. Many communities are not able to implement watershed districts or rely solely on ordinances as described. The primary method for such communities to manage their protection areas is outreach and education efforts with other communities, residents, and businesses in their watershed. These efforts can be effective at achieving many positive controls on development and activities within the protection area. The key is to build public support and understanding of the importance of drinking water supply protection.

Typically these collaborations involve the formation of a steering committee or other formal group that jointly defines goals, objectives, and standards for development and land use to protect water supplies. These groups also typically develop ordinances for each participating entity that will protect their supply sources.

Outreach and education. Even those communities that are able to implement watershed districts or ordinances use public outreach and education to build awareness. Outreach and education are important throughout the development of the watershed district or ordinance. Involving the community can help define goals and will build support for regulations. Once the district or ordinance is in place, public outreach, education, and involvement help to ensure that management practices are implemented and to identify where regulations are not being followed. Virtually all management strategies require collaboration.

LINKING QUALITY AND QUANTITY OBJECTIVES

During the planning process, consideration of factors that affect quality and quantity of the source will help utilities move toward more sustainable solutions. Evaluating the use of advanced technologies, incorporating other stakeholders in the planning process, examining opportunities to optimize and make more efficient use of existing facilities, and incorporating reliability and security features are some additional factors to consider.

Technological advances. Advances in technology have allowed the use of waters that would have been unavailable using conventional treatment technologies. Membrane and other desalting technologies allow the use of brackish and saline waters as possible sources of potable water. Water reuse and recycling can reduce the overall consumption of water by removing some demands from the potable system. The use of highly treated wastewater as a source for irrigation, industrial applications, or environmental enhancement could be viewed as creating a new source of supply. Strategies such as aquifer storage and recovery reduce the influence that diurnal and seasonal demand patterns have on the sizing of primary source, treatment, and transmission infrastructure. Strategic location of natural underground storage and recovery facilities can provide additional system reliability in time of emergency. Source water protection requires the evolution of an expanding array of management best practices to minimize the effects of nonpoint and point source pollutant loadings on source water quality. In addition to the technological advances that directly affect available quantities and quality of drinking water sources, information management tools allow for better real-time system operations and decision-making. Geographic information system tools and databases coupled with sophisticated modeling software enable more comprehensive analyses of a system of interrelated components.

Information management. A decision support system (DSS) enables planning, operations, and management of water supply systems by providing information and knowledge in a more responsive and effective manner. A DSS is a computer-based system that integrates information stored in databases with process information and plan-evaluation tools and presents it in a single platform.

An example of an organization that has implemented a DSS is Tampa Bay Water. As a regional wholesale water supplier in the Tampa Bay region, Tampa Bay Water provides treated water to member utilities who serve more than 2 million people in Hillsborough, Pasco, and Pinellas counties. Before 2002, Tampa Bay's system depended solely on groundwater from area wellfields. Since then, surface water, desalinated seawater, and additional groundwater supplies have been added to the agency's sources of supply. Planning and investigation of additional supplies are ongoing to support future increases in water demand and continued reduction in groundwater supplies, as well as to further diversify supplies. Through partnerships with other agencies throughout the region, Tampa Bay Water is integrating the use of reclaimed water into the water cycle. As diversity and number of supply sources increase, so too does the need for coordination, management, and operation of the sources. To aid in decision-making, Tampa Bay Water has invested time and resources in developing a custom DSS. This DSS involves the coordinated use of data collection systems, a database system,[1] and applications such as hydrologic and forecasting models and operational reporting tools.

Alison Adams, source selection and environmental programs manager for Tampa Bay Water, has facilitated the development of the DSS. Her work is ongoing with colleagues throughout the organization to develop applications and reports that allow better management of the existing water sources while adding new sources to the system. Tampa Bay Water views supply diversity as

a tool to ensure the long-term sustainability of its water resources.

"The benefits of supply diversity include improved availability, reliability, and reduced vulnerability of existing sources," said Adams. "We are also able to provide balance between our environmental and human needs while allowing proactive management of our water resources." The DSS is a tool that enables the authority's supply diversity goals to be implemented.

Security, reliability, and redundancy. Having multiple sources or source locations from which to withdraw water improves the reliability of the system by increasing the probability that supplies will be available. When multiple sources are available, the decision about which source to use can be based on water quality and other system benefits, in addition to whether the source is available.

John Witherspoon, retired manager of water treatment and supply for City Utilities of Springfield, Mo., chairs the AWWA Source Water Protection Committee and the Missouri Section's committee. During the past several years, Witherspoon has performed 30 risk assessments of systems of all sizes. Generally, he has found that surface waters are less susceptible to long-term effects than is groundwater because of flushing and dilution that occurs naturally. Multiple wells and well fields are important to ensure reliable and sustainable groundwater sources. Even with surface waters, a second source, dual intakes, or an interconnection with another system is needed for emergencies. However, having multiple sources among which to selectively switch is preferable.

"Source redundancy is paramount to ensure sustainable water supplies. Although our water systems focus on providing redundant treatment facilities, the need for redundancy of sources cannot be overstated," Witherspoon said.

Multi-objective planning. Multi-objective planning is a logical approach for developing strategies to pursue complex goals. In the context of conservation planning, three elements are crucial. A set of goals must be established, actions and resources that can help achieve the goals must be defined, and predictions of the various consequences of each action should be quantified (Rozdilsky, 2002). The various stakeholders should establish the objectives and goals necessary to attain water supply sustainability. According to the World Commission for Water in the 21st Century (2000), stakeholder decisions should be participatory, scientifically informed, and made at the lowest possible level. Municipal and watershed boundaries can act as logical divides for stakeholder inclusion and provide a way to understand the local demands and interrelationships. Based on the system, its demands, and the multi-objective goals, supply and demand management can be implemented in a logical and holistic way. Management techniques could include optimizing existing sources, changing operational practices, implementing local/regional conservation practices, and local/regional drought contingency planning.

An example of multi-objective planning is demonstrated by the efforts of the East Bay Municipal Utility District (EBMUD) in California. The utility supplies 1.3 million people with potable water and serves the wastewater needs of 640,000 people. The utility's Long-Term Water Supply Management Program, adopted in 1993, set goals to reduce average water demand by 48 mgd (182 ML/d) through 2020, to reduce water use by 25% during periods of drought, and to incur zero net demand on the system from new residential development. To achieve those goals, the utility developed the Water Conservation Master Plan in 1994, which identified viable forms of water conservation and water recycling. The utility is also using development fees and financial incentives to fund conservation measures and manage demand. EBMUD plans to add new groundwater sources and source water diversions to better manage its resources. The steps taken by EBMUD demonstrate a comprehensive approach to addressing the complex issues facing various water supply and water resource entities.

Strategies with multiple benefits. As we search for strategies that provide multiple benefits for our communities and water resources, a holistic and integrated approach is needed. Competition for scarce resources is a fundamental economic aspect of life. Quality of life in large measure depends on an adequate supply of freshwater, because water is critical for the health of humans and ecosystems. Likewise, water is an important element influencing economic and recreational activities. When the water resource is scarce, trade-offs among these uses occur. Identifying strategies that lessen the negative effects of those trade-offs by providing multiple benefits is a positive planning objective.

One example of a holistic strategy is the use of stormwater management to protect drinking water sources. This approach makes sense from an economic, ecological, and water quality perspective. Other integrated approaches include the use of reclaimed or recycled water to provide environmental and ecological enhancements.

Water supply plus environmental benefits. The degradation of aquatic and related terrestrial ecosystems and depletion of groundwater supplies pose long-term problems for meeting both instream and withdrawal water demands and limiting flood damage. An example of this problem can be seen in Las Vegas, Nev., where urban development has increased the stormwater runoff and peak flow to the nearby Las Vegas Wash. This, coupled with year-round treated wastewater flows, has resulted in significant channel-cutting in the wash and downstream water quality degradation in Las Vegas Bay, located in the

northwest corner of Lake Mead. Lake Mead is the main source of potable water for Las Vegas. As a result of the channel downcutting, vegetation along the wash has diminished, encouraging further erosion and downstream water quality degradation. However, the water and wastewater purveyors in the valley who are members of the Southern Nevada Water Authority (SNWA) have taken steps to reverse this trend. Gary Grinnell, chair of AWWA's Water Reuse Committee and senior civil engineer for the Las Vegas Valley Water District, stated that "It is important for the valley purveyors to act as good stewards of our water resources and the environment." Through the cooperative efforts of the SNWA purveyors, reducing flows in the wash by using reclaimed water instead of potable water for irrigation, constructing erosion barriers, and planting vegetation to reestablish wetlands, the objective to restore the wash is being accomplished. Through time, with the comprehensive efforts of the SNWA, better optimization and utilization of the water resource will help sustain the urban water supply and water quality demands while positively effecting the environment.

CONCLUSION

Water supply sustainability is ensuring that high-quality water supplies are readily available for generations to come. Agencies responsible for providing safe and reliable drinking water supplies face increasing challenges as they plan for future community needs and are given the task of ensuring the sustainability of their water resources. In 1994, AWWA endorsed the long-term goal of total water management—to ensure that water resources are managed for the greatest good of people and the environment. The total water management philosophy is well suited to a discussion of sustainability for drinking water supplies. As Bill Maddaus, former chair of AWWA's Water Resources Planning and Management Committee, has said, "To accomplish sustainability, we need to do more with less." This simple statement deserves further contemplation.

The current challenge is to consider the various components of the water cycle in a holistic manner, viewing wastewater, stormwater, and other forms of water as resources that can be applied in appropriate situations and settings. Those settings may seem to fall outside the bounds of drinking water and drinking water sources, but the complex interrelationships among water resources must be recognized and integrated into long-term planning and decision-making processes. As agencies serving the public good, water utilities have a unique opportunity to take a leadership role as good stewards of the available water resources. Providing safe water for human use is a primary mission that cannot be segregated from other water use considerations. Rather, an integrated consideration of water resources may identify opportunities that ultimately enhance the reliable quantity and quality of drinking water sources. The application of the drinking water community's abilities to innovate, to use technology for beneficial use, to conserve water assets, and to see opportunities in trying situations is needed to address the issue of sustainability. If sustainability is the necessary planning and management of water resources to provide an adequate supply of high-quality water while providing for the economic, environmental, and social needs of current and future generations, it is clear that the necessary water utility role is broad and meaningful.

FOOTNOTES

[1]Microsoft SQL Enterprise, Microsoft Corp., Redmond, Wash.

EBMUD Makes Strides in Ensuring a Sustainable Water Supply

BY LORI STEERE

Mother Nature grabbed the East Bay Municipal Utility District's (EBMUD's) attention with a severe drought from 1976 to 1977. Pardee Reservoir, with a maximum capacity of 198,000 acre ft (244.13×10^6 m^3), in March 1977 reached its lowest level since it was first filled in 1930. Storage fell to only 47,000 acre ft (57.9×10^6 m^3), less than 24% of capacity, with the water level 112 ft (3,414 cm) below the spillway.

This short, severe drought, followed by a longer-term drought from 1986 to 1992, kicked the Oakland, Calif.-headquartered EBMUD into high gear to find ways to reduce potable water demand and diversify its water supply portfolio. The goal was to ensure adequate and reliable water supplies to meet the district's customer water needs well into the 21st century.

The district looked for ways to develop a balanced supply portfolio that included new supplemental surface and groundwater supplies during droughts. In 1993, the board of directors adopted a Water Supply Management Program (WSMP) that also set two key goals to reduce demand on EBMUD's limited drinking water supply: to conserve a total of 34 mgd (129 ML/d; recently increased to 35 mgd [133 ML/d]) and to recycle a total of 14 mgd (53 ML/d) by 2020. EBMUD, which serves a population of 1.3 million people in a 325-sq mi (84,175-ha) service area, is well on its way to achieving both goals.

EBMUD, a leader in California water conservation, offers a large number of residential and nonresidential conservation services and programs that collectively have contributed to conserving some 17 mgd (64 ML/d) of potable water as of this year, almost halfway to the WSMP's 35-mgd (133-ML/d) goal. EBMUD has had conservation programs in place since the mid-1970s. The 35-mgd (133-ML/d) goal is above conservation savings achieved before adoption of the WSMP in 1993. Additionally, the 17 mgd (64 ML/d) savings does not include total water savings from the mid-1970s to the present, but rather savings as a result of new or enhanced conservation services and programs intended to help achieve the WSMP goal.

Excluding the 5.8 mgd (22 ML/d) of recycled water used at the district's main wastewater treatment plant for irrigation and various industrial purposes, recycled water projects currently serving five customers offset potable demand by 2.8 mgd (10.6 ML/d).

Additionally, the first phases of two multiphased recycled water projects are now under construction and when built out will serve a few hundred customer sites and offset potable demand by a total of 4.9 mgd (18.6 ML/d).

One project is in partnership with the Dublin San Ramon Services District (DSRSD). The DSRSD/EBMUD Recycled Water Authority was created in 1995 to help develop this joint project, known as the San Ramon Valley Recycled Water Program (SRVRWP). This cooperative, regional-benefit project will serve portions of southern Contra Costa and eastern Alameda counties in the eastern part of the San Francisco Bay area. This project is planned to serve DSRSD with 3.3 mgd (12.5 ML/d) and EBMUD with 2.4 mgd (9.1 ML/d) of recycled water for irrigation use. Construction of the initial phase of the SRVRWP is almost complete, and first recycled water deliveries are expected in the summer of 2005.

The second project under construction is EBMUD's East Bayshore Recycled Water Project (EBRWP), with first deliveries expected in the summer of 2006. This project ultimately will serve portions of five cities along the eastern shoreline of San Francisco Bay and will save 2.5 mgd (9.5 ML/d) of potable water. The EBRWP is a multiple-use project that will provide recycled water for irrigation, wetlands restoration, and various industrial and commercial uses, including toilet and urinal flushing in a 22-story office building in Oakland, the first dual-plumbed high-rise in the San Francisco Bay area.

Planning for future projects, including a satellite recycled water treatment plant and a proposed recycled water project to produce highly purified boiler feedwater, is focused on helping to meet the 14-mgd (53-ML/d) goal.

Although a limited potable supply drives EBMUD's long-term sustainable-supply effort, some key financial measures that help drive the district's ambitious recycled water program include:

· new development helps to generate funding for the program;

· EBMUD payment of the cost to retrofit existing customers to use recycled water, provided those retrofits are cost-effective; and

· the Non-Potable Water Use Incentive Rate. This new rate, adopted by the EBMUD Board of Directors in 2004, prices recycled water at 20% less than the potable water rate, and customers identified for recycled water use who do not permit the timely retrofit of their site will pay 20% more for all potable water used at their site until the site is retrofitted and has started using recycled water.

Desalination, while not included in EBMUD's 1993 WSMP, offers another potential asset in a diversified and balanced water supply portfolio. Not too long ago, getting freshwater from salty sources was a too-costly dream. But great strides have been made in both technology and regional cooperation that could make that dream a reality. EBMUD has joined with the San Francisco Bay area's other largest water suppliers—San Francisco, Santa Clara, and Contra Costa water districts—to seriously explore a regional desalination project that would benefit 5.4 million residents served by these Bay Area water districts.

The four partners predict the Bay Area Regional Desalination Project could turn out 120 mgd (454 ML/d) of drinking water by treating salty water. This added water supply would help meet local needs in emergencies such as earthquakes or a severe drought, make it easier for agencies to perform maintenance on current facilities, and increase local water supply reliability.

Preliminary studies assessed the financial feasibility of a regional project and looked at possible locations. Three sites appear best suited to house a desalination plant. One site is in Pittsburg near the Mirant power plant, another in Oakland at the foot of the

Bay Bridge, and a third in San Francisco's Oceanside area. These sites are typical of the ones where desalination plants are found around the world: near power plants and with access to a place where salty wastes could be safely released.

Just recently, EBMUD received a $250,000 state grant on the partnership's behalf, which will help fund a more detailed assessment of planning and building desalination facilities. A detailed feasibility and environmental study, including a thorough public review of the proposal, is planned for 2006. If the project makes sense, it will be designed and constructed—at a cost ranging from $450 million to $700 million—by the end of 2009.

Desalination illustrates the great strides the world is making toward finding environmentally sensitive approaches to dealing with a shortage of drinking water. Closer to home, it also demonstrates the increasing regional cooperation that is necessary to meet California's drinking water needs.

Stewardship in the Spotlight

BY IRA S. RACKLEY AND DAVID R. YANKOVICH

As noted by the Greek philosopher Heraclitus, "You cannot step into the same river twice, for other waters are continually flowing in." Much to the dismay of those responsible for sustaining water resources and water quality, variable water quality is one of the few aspects of life that never changes.

In southern Nevada, two cities and a county have adopted a consensus approach to protecting water quality. The Las Vegas Alternate Discharge Program is a stellar example of communities pulling together to sustain water resources and protect the environment.

Observing too much potential for change in the composition of Lake Mead water, Clark County and the cities of Las Vegas and Henderson, Nev., joined together to form the Clean Water Coalition (CWC). A natural symbiosis among Lake Mead, the Las Vegas Wash, and the Colorado River system led the CWC to take action, which included retaining Black & Veatch Corp. to help the coalition control the challenges of variable water quality.

Las Vegas, Henderson, and Clark County operate wastewater treatment plants with a current combined capacity of approximately 180 mgd (681 ML/d). The effluent is discharged to Lake Mead through the Las Vegas Wash, an ancient natural channel that serves as the only drainage for the entire 1,600-sq mi (414,400-ha) Las Vegas region. A major source of drinking water in the desert region, Lake Mead is a near-perfect demonstration of an effective water cycle; the CWC wanted to ensure continuation of that cycle.

As the region continues to grow and the discharge volume increases, concerns have surfaced that the plume from the highly treated effluent will someday affect the water quality of Lake Mead. Although the effluent is highly treated at the wastewater facilities, water treatment is most cost-effective when it processes fully diluted water. "It was time to start considering alternatives," said Doug Karafa, CWC program administrator.

Lake Mead, created in 1936 with the construction of Hoover Dam, is fed by the Colorado River. With a capacity of almost 30 mil acre-ft (36 × 10⁹m³), it is the largest artificial lake in the United States, a regional recreational treasure, and an integral element of the extensive Colorado River system.

Although the region had implemented aggressive effluent reuse and water conservation measures over the years, continued industrial growth and a booming population in Las Vegas—coinciding with the renewal of important discharge-limit permits—spurred a wide-ranging set of studies and planning activities. Numerous federal, state, and local planners and regulators, citizen groups, water and wastewater utilities, downstream water agencies in the states of Arizona and California, and other public and governmental agencies have actively participated in the development of a plan for the conveyance and discharge of the highly treated wastewater. Each entity is responsible for preserving the rights of its stakeholders, and many have common jurisdictions. This called for an integrated solution that encompasses an array of environmental and technical issues, including water and wastewater treatment, water quality and flow management, environmental impacts, public participation, and substantial agency interface. The solution needed to be flexible, responsive, and, above all, committed to sustainable and adaptive management of this valuable resource.

"Our need is clearly to return the portion of the effluent not devoted to reuse to the Colorado River system in a manner that is implemental, cost-effective, environmentally sound, and publicly and politically acceptable," Karafa said.

The geographic and technical scope of the Alternate Discharge Program, the large number of stakeholders, and the CWC's commitment to an enduring legacy of water quality has resulted in an unusually large number of complex project components encompassing many aspects of watershed management. Chief among these is the Systems Conveyance and Operations Program (SCOP), which consists of integrated elements designed to govern the effluent discharged to the wash and eventually to Lake Mead. The effluent interceptor and the lake conveyance system pipelines allow for flow diversion to both the wash and Lake Mead (depending on both the regulated seasonal discharge permit limits and lake conditions), giving the CWC plants more control over the discharged effluent.

An environmental impact statement (EIS) now under development will provide the basis and understanding to identify alternatives with the common goals of better managing the effluent discharged to the Colorado River system. A Citizens' Action Committee (CAC), made up of public officials, technical experts, and members of the public with an interest in the outcome of the project, was formed to help obtain public input on EIS alternatives. The CAC has submitted recommendations to the CWC board based on meetings conducted during project development and results from several technical memoranda prepared for the SCOP.

REFERENCES

AWWA, 2001. *Water Resources Planning, (M50)*. AWWA, Denver.

AWWA, 1994. Total Water Management. www.awwa.org/Advocacy/ govtaff/totwapap.cfm, accessed 2005.

Beck, M.B., 2002. Sustainability in the Water Sector, Reflections Within on Views From Without. Proc. 2002 IWA Conference on Sustainability in the Water Sector, Venice, Italy.

Frederick, K., 2002. Handling the Serious and Growing Threats to Our Most Renewable Resource—Water. Issue Brief 11. Resources for the Future, Washington.

Kranz, R. et al, 2004. Conceptual Foundations for the Sustainable Water Resources Roundtable. Universities Council on Water Resources, *Water Resources Update*, 127:11.

Maddaus, W.O. et al, 2004. Innovative Water Conservation Eliminates Water Supply Impacts, Enabling Sustainable Housing Development. Proc. 2004 AWWA Ann. Conf. Orlando, Fla.

National Council for Science and the Environment, 2004 (C.M. Schiffries & A. Brewster, editors). Water for a Sustainable and Secure Future: A Report of the Fourth National Conference on Science, Policy, and the Environment. Washington.

NRCS (Natural Resrouces Conservation Service), 2005a. *Record-low Snowpack Measurements Confirm Water Worries*. Press Release.

NRCS , 2005b. Washington State Basin Outlook Report for February, 2005. www.wcc.nrcs.usda.gov/cgibin/ bor2.pl?state=wa&year=2005&month=2&format=text.

OECD (Organisation for Economic Co-operation and Development), 2004. Implementing Sustainable Development, Key Results 2001–2004. Paris. www.oecd.org/dataoecd/25/35/ 31683750.pdf.

Prugh, T. et al, 2000. *The Local Politics of Global Sustainability*. Island Press, Washington.

Roos, M., 2005. Accounting for Climate Change. California Water Plan: Update 2005. www.waterplan.wter.ca.gov, accessed June 3, 2005.

Rozdilsky, I., 2002. Multi-objective Planning; an Introduction for Laymen. www.princeton.edu/~irozdils/multiobj.html, accessed June 3, 2005.

State of California, 2005. California Water Plan, Department of Water Resources Planning and Local Assistance. www.waterplan.water. ca.gov/, accessed June 3, 2005.

USEPA (US Environmental Protection Agency), 2001. A Watershed Decade. EPA 840-R-00-002, Washington.

USGS (US Geological Survey), 2005a. Sustainable Water Resources Roundtable. http://water.usgs.gov/wiicp/acwi/swrr, accessed June 3, 2005.

USGS, 2005b. Land Use and Watershed Characteristics. http:// chesapeake.usgs.gov/landcover.html, accessed June 3, 2005.

World Commission on Environment and Development, 1987. Our Common Future; The Brundtland Report. United Nations, New York.

World Commission for Water in the 21st Century, 2000. World Water Vision Commission Report, A Water-secure World: Vision for Water, Life, and the Environment. World Water Council, London.

ABOUT THE AUTHORS

Pamela P. Kenel is a senior water resources engineer with Black & Veatch Corp., Gaithersburg, MD. Kenel has a BS degree in civil engineering from Virginia Polytechnic Institute and State University (VPI), an MS degree from the University of Maryland, and she is undertaking advanced studies in environmental engineering at VPI. As chair of AWWA's Water Resources Division, Kenel is an advocate for further association work on the topic of sustainable water supplies. *Jim Schlaman is a water resource engineer at Black & Veatch. He has been involved in stormwater, combined sewer overflow, and water quality projects since he joined the company in 2001. He holds a BS degree in biological systems from the University of Nebraska–Lincoln with an emphasis in water and environment and is currently pursuing his MS in civil engineering at the University of Kansas.*

Meeting Customer Expectations in a Fluid Utility Environment

BY ROGER PATRICK AND EDWARD G. MEANS III
(*JAWWA* September 2005)

Water utilities are in the midst of significant change. Although the degree of pressure for change varies by utility, the industry as a whole faces a complex set of market, cultural, technological, and political pressures such as a shortage of infrastructure funds, political gridlock on water allocation, and growth in water-short regions. Preparing for change requires understanding these trends and their implications, and managers and governing boards that foresee and act on such trends will be best positioned to successfully lead their utilities.

In 2000, The AWWA Research Foundation (AwwaRF) commissioned a study to characterize trends, determine their meaning for water utilities, and plan the best response. Titled "A Strategic Assessment of the Future of Water Utilities," the study tapped the experience of water industry leaders and futurists and identified and documented key trends that have been used by water utilities and associations in their planning processes.

Four years after the initial report, in 2004, AwwaRF commissioned an update to the original study. "An Update of the Strategic Assessment of the Future of Water Utilities" was intended to address the acceleration of many key trends and the emergence of factors such as better information on climate change, crises in energy markets, and terrorism. This project used new trend research, futurists and future scenario development, and engaged national water utility leaders to debate these trends and develop strategies for future success.

A detailed paper documenting trend data in key areas of concern for water utilities was then developed. Key areas include population and demographics, health and medical advances, regulations, climate change, total water management, employment and workforce trends, customer expectations, information technology, drinking water treatment technology, energy, automation, information technology security, physical security, current economic issues, utility finance/infrastructure, politics, regionalization, and private-sector involvement in water.

This article was the third in a series presenting the results of the project (The first article, *Ten Primary Trends and Their Implications for Water Utilities*, is Article 5 in this collection. The second article, *Population Growth and Climate Change Will Post Tough Challenges for Water Utilities*, is article 12 in this collection.) and summarizes the findings on customer expectation and service trends.

BACKGROUND

Where do customers get their information on water-related matters? In a study conducted for the National Environmental Education and Training Foundation (NEETF; 1999) it was reported that 61% of respondents obtained their information on water quality from the media, whereas 34% obtained information from their water supplier. A subsequent NEETF-commissioned study (Coyle, 2004) found this trend had intensified; more children (83%) get environmental information from the media than from any other source, and for most adults the media is the only source of environmental information.

How accurate is this information? NEETF found that although people in the United States count on safe drinking water, few know where their water comes from, what may be threatening its quality, or what actions are appropriate to protect its source. Findings also indicated that environmental literacy in general and knowledge of water in particular are very poor.

In addition, the research showed that people think they know more about the environment than they actually do. Data show a steady pattern of environmental ignorance even among the most educated and influential members of society; the research showed little difference in knowledge levels among the average American and those who sit on governing bodies, town councils, and in corporate board rooms.

What information are people hearing in the media? Unfortunately, some articles provide an unfavorable portrayal of the public water supply and what is in it. Some of the

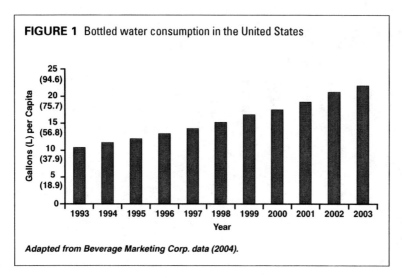

FIGURE 1 Bottled water consumption in the United States

Gallons (L) per Capita

Year

Adapted from Beverage Marketing Corp. data (2004).

more prominent contaminants discussed include methyl tertiary butyl ether, arsenic, lead, chromium 6, and perchlorate.

The cumulative effect of these news items may include the massive shift in drinking water consumption patterns: half of the US population now drinks bottled and/or filtered water regularly. The cost of this shift is high, with the amount spent on bottled water and point-of-use/point-of-entry (POU/POE) devices now approximately the same as the total revenue of United States water utilities. That is, Americans as a whole are now "paying twice" for drinking water.

RESEARCH ON CONSUMER EXPECTATIONS AND SATISFACTION WITH PUBLIC WATER SUPPLY

According to Rambo et al (2003) water utility customers expect

• a focus on their needs (e.g., being listened to);

• courtesy and respect, (e.g., being called by their surname);

• knowledgeable representatives but who do not pretend to know something;

• safe water (although taste is not equated to safety; i.e., even if people do not like the taste, they do not automatically conclude it is unsafe);

• complete restoration of property when any on-site work is done; and

• communication with a live person rather than use of an automated system.

This research also shows that consumer satisfaction with tap water is relatively high in the United States. Yet increasingly, segments of the population—particularly consumers on the East and West Coasts, and those younger than 30—seek bottled and/or filtered water as their primary drinking water source.

Consumption of bottled and filtered water varies widely by region, with 80% or more of people on the West Coast, in the Southwest, and in part of the South-

east drinking bottled and/or filtered water. In contrast, in the Midwest, such consumption in some areas is limited to only 20% of the population.

Highlights from the 2004 AwwaRF research project also indicated:

• In the United States, 86% of the population was concerned about tap water.

• By 2003, 48% of the US public was using either a POU/POE device or bottled water while at home (up from 41% in 2001).

• Safety was the primary motivation for drinking filtered water.

• Taste, safety, and healthiness were the motivating factors for drinking bottled water.

In fact, only about 5% of the 55,000 water treatment systems in the United States ever exceed any US Environmental Protection Agency maximum contaminant level (MCL) standard. Furthermore, 75% of all MCL violations are a result of monitoring/reporting practices and are not necessarily related to actual drinking water quality. It is a paradox that despite the safety of the nation's drinking water, the sources from which the public receives its information are generating increasing distrust of the public water supply.

The AwwaRF research also revealed that utility managers tend to overestimate consumer satisfaction and underestimate the level of importance consumers place on tap water safety. Additionally, most customers cannot remember receiving a Consumer Confidence Report (CCR), indicating it is an ineffective tool for conveying water-quality information.

Trends in bottled water consumption and home treatment. From 1993 to 2003, US per capita consumption of bottled water grew from 10 to 23 gal (38 to 87 L) as shown in Figure 1. By 2003, the total volume of bottled water sold in the US exceeded 6 bil gal (23 GL), continuing the steady 10% annual growth trend seen during the previous decade. At estimated retail prices, the amount paid by US consumers for bottled water is now similar to the total annual revenue of US drinking water utilities. The recent AwwaRF research also found that people do not generally object to the price of bottled water (Mackey, 2003).

Although this volume is insignificant compared with water sales by utilities, it indicates a willingness to pay, at least for some consumers, for better tasting water and/or an increased sense of health and safety. Further, the growth trend may not be over for some time: although the United States is the world's largest market for bottled water, its per capita consumption ranks only thirteenth, with Italians consuming 48 gal (182 L), Mexicans 42 gal (159 L), and the French 39 gal (148 L) per capita annually, respectively (Beverage Marketing Corp., 2004). It may be useful to point out these facts to municipal officials who worry about rate increases.

How does satisfaction with water utilities compare with other organizations? Customer satisfaction with goods and services purchased in the United States today is the highest it has been since the second quarter of 1995, according to a 2004 report from the American Customer Satisfaction Index (ACSI). The report indicated that overall satisfaction had increased 0.3% to 74 (out of 100) continuing a two-year upward trend. Similarly, Rambo et al (2003) found that consumers gave water utilities an overall score of 73 for service, putting them roughly on a par with the average private sector organization and slightly higher than the average federal agency (which scored 71, according to the ACSI). However, the taste of tap water was criticized, reinforcing other research on why people drink bottled water and/or use POU/POE treatment.

What are customers looking for from the utilities that serve them? It is known that utility customers place a high importance on

- accurate, understandable bills and convenient payment options;
- accessibility by phone and resolution of issues on the first call;
- quick responses and attentiveness to customer problems;
- knowledgeable, friendly employees; and
- value for money.

In addition, organizations with effective customer service typically place an equal emphasis on customers and the employees who serve them. This is because in competitive markets, customers most often leave because of perceptions of indifference displayed by the organization's employees. Other issues are not usually critical as long as employees are trying to help. In turn, employees leave organizations primarily because of a poor relationship with their supervisor and also commonly because of problems with balancing the demands of work and home, the lack of meaningful work, the level of coworker cooperation, and the level of workplace trust. Therefore, leaders in customer service typically place a high emphasis on such employee factors, not just factors that directly affect customers.

RESEARCH IMPLICATIONS

Consumer education and perceptions of drinking water quality. As already mentioned, the general population, including those who sit on governing bodies and town councils, has limited knowledge of water-related matters, and much of what individuals think they know is inaccurate.

Furthermore, there is a growing level of distrust within the US population regarding tap water safety.

This sentiment is fueled by media reports that are not counterbalanced by CCRs or other utility-generated communication. Without reversing this trend, the reputation of being the public provider of clean, safe water will likely continue eroding, and the water sector's ability to influence water policy on things such as infrastructure spending and rates will diminish.

The challenge for the water industry now is to come to grips with this combination of public ignorance and the powerful influence of the media. The implications for funding requests are obvious: the water industry must do a better job of consumer education and outreach. In the case of governing boards and others in decision-making positions, better education is imperative.

The growth of POU treatment and bottled water consumption also implies that consumers are willing and able to pay for perceived better-quality water. The industry needs to inform customers that paying reasonable rates is a less-expensive alternative to paying for bottled or home-treated water.

Additionally, Tatham et al (2004) has found that customers who say they are "very informed" on water quality issues are more likely to be satisfied with their utility's performance. Informed customers, in turn, are less likely to report errors on their bill, problems with taste and odor, or water safety concerns. These findings suggest that communicating with and educating customers may increase customer satisfaction as well as counter media inaccuracies.

Customer service and satisfaction. A 2001 ACSI report showed that water utilities' customer satisfaction may increase by emulating practices of leading private-sector organizations. It found that the IRS had improved its satisfaction rating by adopting some private-sector techniques such as actively listening to customers and then responding by changing processes, reallocating resources, and focusing technology to enhance the overall tax-filing experience.

J.D. Power and Associates' 2003 Gas Utility Residential Customer Satisfaction Study found that customer satisfaction is highest among those who believe their gas utility performs well in providing pricing and payment options. Although 75% of consumers pay their gas utility bill through the mail or by visiting their utility, 90% of customers would prefer to pay through an alternative method, such as the Internet, a payment agency, or automatic deduction from their bank account. If similar conclusions can be inferred from this research, improving payment options will boost customer satisfaction. Energy utilities also frequently mention achieving high levels of customer satisfaction as a key strategy to obtaining rate increases. This suggests that investments in customer service improvements are investments in a utility's sustainability.

STRATEGIES FOR IMPROVING CUSTOMER SERVICE

Improve communication with customers and other stakeholders. The water industry needs to become more active in providing accurate water quality–related information to the public to counter misperceptions and misinformation. This is too great a task for individual utilities and needs to occur at the industry level. In addition, individual utilities should enhance their CCRs as a mechanism for providing relevant water information in lay terms and building confidence in the public water supply.

Water utilities must also keep their community leaders abreast of water-related issues so that reasonable funding requests are approved, issues are kept in perspective, and misinformation is countered. Water utilities should recruit informed people to sit on their governing bodies and educate them on relevant issues to ensure decision-making occurs in an informed atmosphere.

Improve customer service. Water utilities must maintain a high level of customer service in order to build confidence in the organization. They should also strive to keep up with customer service best practices from other industries. In particular, water utilities should organize processes around customer needs rather than internal convenience. Examples include resolving problems in a timely manner by providing customer service representatives (CSRs) with the tools to resolve most problems on the first call, and by providing convenient appointment windows.

In order to keep pace with rising standards for customer service, water utilities should use technology to improve service while reducing costs. It is encouraging to see that technological advances are occurring at some water utilities despite a "cash-strapped" environment.

With the many high-tech tools that are now available, water utilities should focus on using the Internet and information technology to support high levels of service while reducing costs. Leading energy utilities have made technologies such as web self-service and interactive voice response (IVR) a focus of customer-service improvement efforts. Energy utility customers now expect to be able to independently access information regarding accounts, view and pay bills, prepare a home energy audit, and obtain information on energy-saving appliances or home modifications.

In turn, energy utilities have used the expansion of payment options and self-service information as an opportunity to close business offices and reduce operating costs. Water utilities can now offer Internet billing and flexible payment options such as direct debit, online credit card payments, check-by-phone, and payment through third-party agents such as banks. In addition, water utilities can use IVR systems to provide billing and payment information, directions and office hours, and options such as the ability to extend payment due date, enter a meter reading, and make changes to property ownership information.

Maintain a fair rate structure. Water utilities should charge a fair price for providing water that meets customer expectations and only subsidize low-income customers. Ideally, water utilities should also condition their customers to expect small, regular rate increases to prevent the need for rate "hikes." Water utilities must also provide creative ways for customers to manage their water usage and bills, especially as rates rise. Giving customers greater control of their bills improves satisfaction.

Improve customer contact handling. Research shows that customers are happiest when their questions and issues are resolved on the first call, even if the average time per call is increased (Patrick & Kozlosky, 2005). Leaders in customer service provide comprehensive information to CSRs, including instantaneous meter reading where automated systems are being used, as well as access to other systems such as work-management systems. The average wait time before a customer is helped (i.e., time on hold) is also an important piece of the customer service puzzle, and the acceptable wait time seems to be dropping. Commercial call centers aim for very rapid answer speeds—in seconds—and leading utilities aim to answer calls in 20 seconds or less.

Other actions water utilities can take to improve the customer's phone experience include better selection and training of CSRs, close monitoring of call-handling quality, feedback and rewards to CSRs, allowing CSRs to work at home, and consistently making CSRs feel valuable.

Pool CSRs with over-the-counter and back-office personnel. This strategy of pooling the human potential of various departments improves service while reducing cost, especially for small and medium-sized utilities. These utilities have the same personnel responsible for handling many or all customer calls, billing, collection, outage reporting and dispatch, and back-office work. In other cases, part-time agents—including agents working from home—help provide coverage during breaks, training sessions, and peak work times.

Track performance using commonly accepted measures. By using standard industry performance measures to both track internal trends and compare performance with other organizations, utilities can better identify areas to improve and justify the associated costs. A set of more than 100 such measures is available at www.wisebenchmarks.org.

Implement automatic meter reading. Utilities are seeing both tangible and intangible benefits from automatic meter reading (AMR) technology. Depending on the specific technology chosen, benefits can include the elimination of manual meter reading, the elimination of estimated reads (which reduces the call center workload), the ability for high bill complaints to be resolved by the CSR while the customer is on the phone, the ability to read meters as frequently as desired, the improved accuracy of large-meter reads, the elimination of manual checks, the automatic detection of service theft, and the early detection of leaks on customers' premises.

Some of the newer AMR technologies offer more features and are more financially suitable than the earlier drive-by systems. Future uses of AMR technology could see time-of-day pricing and options to help consumers manage their (larger) water bills.

In the future, water utilities could also share an AMR system with local energy utilities where possible, which would spread the cost of implementing the upgraded technology over a number of service providers and would offer even more customer service features.

Provide special services for industrial and commercial customers. Business customers have less patience than residential customers with waiting time, and they also have specialized needs. Customer service leaders strive to answer calls from business customers with no waiting time. Leading utilities are providing business customers with dedicated CSRs and special toll-free numbers connecting them directly to business specialists that function like phone-based key account representatives. Business customers are also more amenable to the use of voice mail than they are to waiting in a phone queue, so providing that option is valuable to them as long as they know they are dealing with specialists.

Other special services for business customers include web-based methods for builders and contractors to submit requests for service to new homes, easy ways to check the status of service requests, updated common job status information, and response to issues or concerns related to the installation of services.

Put simply, larger water utilities should set up dedicated services and access for business customers, whereas smaller water utilities should consider providing such services in conjunction with other local or nearby utilities.

REACTION FROM INDUSTRY LEADERS

At a workshop of industry leaders conducted in December 2004, the reaction to these trends could best be summarized as "regulatory compliance does not equal customer satisfaction."

There was also an acknowledgement that the water industry needs to do a much better job at public relations, including communicating the value of water and the value of services provided by water utilities. Also discussed was the fact that events and customers' reactions to issues (real or perceived) with their drinking water were outstripping water utilities' ability to respond. Accordingly, leaders also discussed the need to involve the community in relevant issues and decisions. In addition, it was recognized that communication should be tailored to the community and its subgroups, including ethnic groups and customers with special needs such as immunocompromised individuals.

Participants expressed the value of comparing customer satisfaction among water utilities (and recognized that customers already do this). In particular, it was recognized that customers make decisions on whether the water is safe partly by assessing other service aspects such as the condition of facilities and the responsiveness of utility personnel.

REFERENCES

ACSI (The American Customer Satisfaction Index), 2004. US Customer Satisfaction Highest Since 1995, Fueling Consumer Demand. www.theacsi.org/press_releases/0204q1.PDF.

Beverage Marketing Corp., 2004. Beverage Marketing's 2004 Market Report Findings, New York.

Coyle, K.J., 2004. Understanding Environmental Literacy in America. National Environmental Education and Training Foundation (NEETF), Washington.

J. D. Power & Associates, 2003. 2003 Gas Utility Residential Customer Satisfaction Study. www.jdpower.com/awards/industry/pressrelease.asp?StudyID=795.

Mackey, E.D. et al, 2003. Consumer Perceptions of Tap Water, Bottled Water, and Filtration Devices. AWWA Res. Fdn. (AwwaRF) Denver.

Patrick, R. & Kozlosky, C., 2005. Benchmarking Water Utility Customer Relations Best Practices. Submitted to AwwaRF, June 22, 2005. Denver.

Rambo, E. et al, 2003. Developing Customer Service Targets Through Assessing Customer Perspectives. AwwaRF, Denver.

Tatham et al, 2004. Customer Attitudes, Behaviors, and the Impact of Communication Efforts. AwwaRF, Denver.

ABOUT THE AUTHORS

Roger Patrick is the president of Competitive Advantage Consulting Ltd., Santa Fe, NM. He has 15 years of experience as a management consultant helping public and private organizations improve their performance. Patrick is a member of AWWA and the Australian Water Association. He received his BE degree in chemistry from the University of New South Wales in Sydney, Australia, and his MBA from the university of Western Australia in Perth. Ed Means III is senior vice-president of McGuire/ Malcolm Pirnie and has 26 years of experience in water utility management and water quality.

Population Growth and Climate Change Will Post Tough Challenges for Water Utilities

BY EDWARD G. MEANS III, NICOLE WEST, AND ROGER PATRICK
(*JAWWA* August 2005)

The AWWA Research Foundation (AwwaRF) recently revisited its 2000 study of water utility trends in order to track the developments and challenges that have emerged over the past five years. This article is the second in a series presenting the findings of the project, "An Update of the Strategic Assessment of the Future of Water Utilities." (The first article, *Ten Primary Trends and Their Implications for Water Utilities,* is Article 5 of this collection. The third article, *Meeting Customer Expectations in a Fluid Utility Environment,* is Article 11.)

To launch the update, project team members prepared a detailed issue paper on trends. The paper was designed to serve as a briefing document for participants of the Futures Workshop held Nov. 30–Dec. 1, 2004, in Huntington Beach, Calif. The purpose of the workshop was to

• assess the significance of the trends,

• develop strategies to help prepare for the future, and

• test those strategies against several future scenarios.

Workshop participants included water professionals from across the United States, AwwaRF project advisory committee members, and project team members from McGuire/Malcolm Pirnie and Competitive Advantage Consulting. Together they identified the key trends and developments driving the increasing application of total water management principles in water resource development and management. Two of those developments—population growth and climate change—are discussed in this article.

POPULATION DEMOGRAPHICS WILL CHANGE HOW THE WATER INDUSTRY OPERATES

US population and demographic shifts have been identified as major trends that will affect how the water industry operates in the next 20 years. The number of people and their place of residence, age, education, and income level are important factors that water industry leaders must take into account as they plan for the future. In addition, water utilities must consider the increase in water demand created by this population growth.

Population growth will continue. The population of the United States is expected to increase by 50% by the year 2050, reaching 309 million by 2010, 336 million by 2020, and 419 million by 2050 (Figure 1). Most of this growth will continue to occur in the western and southern United States and in urban areas. As a result, absent demand management, water demand will increase most rapidly in these two regions (Figure 2). It is predicted that by 2025 the South and West will be home to nearly two thirds of the nation's population. Furthermore, population growth appears to be centered in areas that currently have water supply constraints. The US Bureau of Reclamation has identified 25 western "hot spots" where water conflicts are expected over the next 25 years (Bureau of Reclamation, 2003). The report cited four primary tools to address this issue: (1) conservation, efficiency, and water transfer markets; (2) collaboration; (3) improved technology; and (4) removal of institutional barriers and an increase in interagency coordination.

As the US population increases, so will water utilties' customer bases. Expansion of customer bases allows growing communities to spread rates across a broader

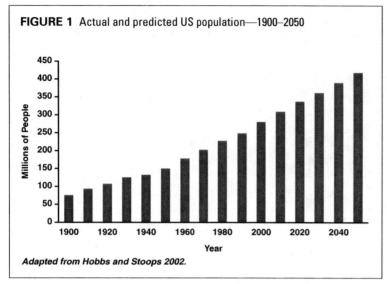

FIGURE 1 Actual and predicted US population—1900–2050

Adapted from Hobbs and Stoops 2002.

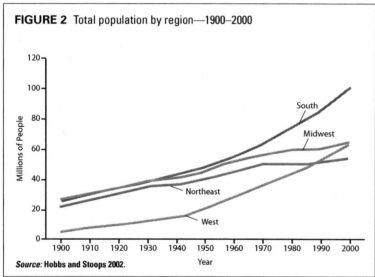

FIGURE 2 Total population by region—1900–2000

Source: Hobbs and Stoops 2002.

base. However, the capital programs necessary to provide facilities to accommodate that growth will raise water rates significantly in many communities. Whereas communities with burgeoning populations will have to find the resources to fund new facilities, areas with static or shrinking populations face the challenge of a diminishing customer base, resulting in a limited rate capacity to replace aging water infrastructure.

Accommodating burgeoning populations will require adequate housing and transportation systems, and this growth will occur in the watersheds of many water systems. Utilities will need to consider the contaminant loading from additional human activity and manage this loading to minimize deleterious effects on water quality and treatment. In addition, development of agricultural land to accommodate housing may free up water for urban use and facilitate water transfers from willing sellers to buyers, but it will also result in a shift

of contaminant loading from agricultural contaminants to urban contaminants.

Customer base is aging. During the last 50 years, life expectancy in the United States has steadily increased (Figure 3), a trend that is expected to continue into the next century. As a result, a larger portion of the US population will be older. The number of people age 65 and older will begin to increase rapidly in 2011 when the first members of the Baby Boomer generation reach age 65 (Figure 4). By 2030, the number of Americans age 65 or older will have more than doubled to 70 million (one in every five Americans). In contrast, the number of people younger than 20 is expected to increase only slightly, although the overall number of people age 18 to 64 will continue to increase. The aging populace could be an important consideration for the water industry because older people may be more sensitive to contaminants. Also, because many older people live on a fixed income, they may be more resistant to rate increases. Plus, seniors tend to vote and be more politically active than younger generations.

Nation's ethnic makeup will continue to shift. Between 1980 and 2000, the Hispanic population in the United States more than doubled and this trend is projected to continue. Over the next 25 years, most of the nation's population growth will occur in the Hispanic population, which is predicted to increase by 37 million people, compared with an increase of just over 13 million people in the White, non-Hispanic population, 15 million in the Black population, and 12 million in the Asian population. The ethnicity of the water utility customer base will affect what and how drinking water information (e.g., Consumer Confidence Reports) is disseminated.

Income gap will expand. The level of water customers' personal income will affect the types of houses and appliances purchased and possibly the type of landscaping installed. Along with the cost of water, income will affect how much water people buy. Recent trends indicate that the gap between the rich and poor in the United States is growing. During the last two decades, the earnings of Americans with the highest incomes have grown faster than those with the lowest. From 1980 to 2001, the average income of the top-earning 5% of US households grew from $66,617 to $260,464 (adjusted for inflation). During the same period, the household income of the lowest-earning 20% of US households increased by only $5,653. As a result, utilities may be confronted by an increasing gap between people who are willing to pay more for water and those who will resist rate increases. Rate structures will need to consider this disparity and ensure that those least able to pay can obtain basic water services at an affordable rate.

Population is becoming more educated. Recent trends indicate that US education levels are rising. The percentage of the population with college degrees or with some college education has increased in the past two decades. College graduates and people with some college education grew from 32% to 52% between 1980 and 2002. During the same period, the portion of the population with no high school degree shrank from 31% to 16%. As the population becomes more educated, customers are likely to demand more and better information from their water utilities.

Knowledge can help utilities manage population and demographic changes. To prepare for future customer bases, water utilities must be familiar with current customers and their water use. A utility must understand the nature of its community's population growth (e.g., land-use types, density, landscape trends, income trends) and the way population growth will affect current water use. Recycling and conservation programs should be a featured aspect of integrated resource plans for water utilities contemplating development of new surface water supplies. These programs diversify water supply and help inoculate utilities against criticism about the environmental effects of new water development. Reducing unaccounted-for water losses will take on more importance, especially in communities with deteriorating infrastructure.

Communication with consumers will be key. State-of-the-art communication tools can be used to convey information to consumers and receive feedback on community values, water needs, and willingness to pay. Focus groups and surveys can help utilities understand their consumers, and education programs can help consumers understand the intricacies of water supply and quality challenges. Education programs should focus on educating

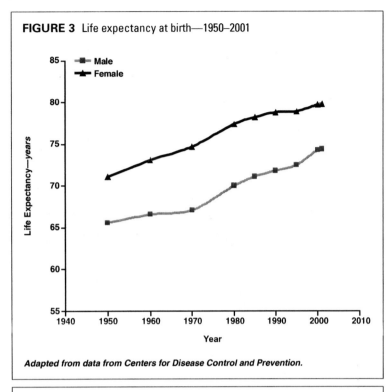

FIGURE 3 Life expectancy at birth—1950–2001

Adapted from data from Centers for Disease Control and Prevention.

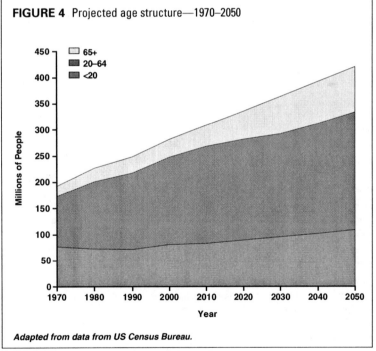

FIGURE 4 Projected age structure—1970–2050

Adapted from data from US Census Bureau.

not only homeowners but also planners, environmental groups, government officials, and children.

SHIFTS IN CLIMATE REQUIRE PREPARING FOR THE UNPREDICTABLE

Climate change has the possibility to exacerbate the pressures that population growth will cause on water

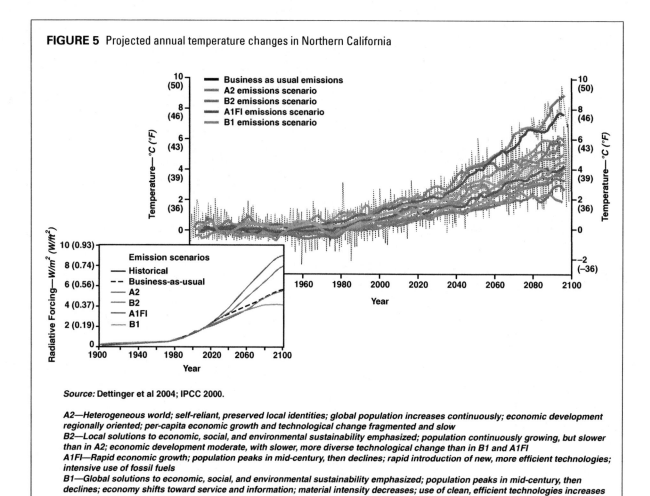

FIGURE 5 Projected annual temperature changes in Northern California

Business as usual emissions
A2 emissions scenario
B2 emissions scenario
A1FI emissions scenario
B1 emissions scenario

Emission scenarios

Historical
Business-as-usual
A2
B2
A1FI
B1

Source: Dettinger et al 2004; IPCC 2000.

A2—Heterogeneous world; self-reliant, preserved local identities; global population increases continuously; economic development regionally oriented; per-capita economic growth and technological change fragmented and slow
B2—Local solutions to economic, social, and environmental sustainability emphasized; population continuously growing, but slower than in A2; economic development moderate, with slower, more diverse technological change than in B1 and A1FI
A1FI—Rapid economic growth; population peaks in mid-century, then declines; rapid introduction of new, more efficient technologies; intensive use of fossil fuels
B1—Global solutions to economic, social, and environmental sustainability emphasized; population peaks in mid-century, then declines; economy shifts toward service and information; material intensity decreases; use of clean, efficient technologies increases

supplies. Many different climatological models are available, all of which project rising temperatures (Figure 5).

The potential effects of global warming on the United States are uncertain and vary by region. For example, there is no consensus on how global warming will affect average annual precipitation. The variability in annual precipitation is projected to increase, however, meaning wet years will be wetter and drought years will become more severe. Storm events may be more infrequent but more intense and may come during different times of the year than has been typical. As a result, water infrastructure, such as dam spillways, may prove to be underdesigned to deal with future climate variability.

The warming trend may also result in snowfall melting earlier, which will cause stream flows to rise earlier and high flows to dissipate faster. The water providers most affected will be those who rely on spring snowmelt to fill their reservoirs with a supply for the summer months. If current reservoir management practices continue, the water that melts earlier in the year will be released for flood control. With less snow melting in the spring, less water will be available to capture

for municipal use. Lower summertime stream flows could affect endangered species (whose health, in turn, affects water diversion permitting) as well as increase contaminant loading on water treatment plants. Utilities may need to revise their water resource management practices or consider contingency plans to deal with the loss of current water resources.

Many climate-change models predict higher sea levels as a result of melting polar ice caps and potentially increased runoff volumes. The rising sea levels may increase saltwater intrusion into coastal groundwater in some areas. In addition, increased sea levels could render estuary water intakes inoperable during certain tides. Regulatory compliance could become more of a problem because of increases in bromide (a disinfection by-product precursor) attributable to seawater intrusion and the increase in source water contaminant concentrations attributable to low summertime stream flow. Changing flow patterns could also cause turbidity spikes at intakes of drinking water treatment plants.

Strategies can help utilities weather climate change. The strategies for handling climate change are similar to those for dealing with population growth.

Utilities must understand the ways in which their water supplies are vulnerable to changes in precipitation volume, duration, or timing. This kind of deep understanding requires careful assessment and/or modeling of watershed precipitation that takes into account climate models. Such modeling is especially important given the variations in precipitation that local geologic features can cause. Conducting sensitivity analyses and contingency planning for water resource loss are additional ways that utilities can tackle the uncertainty of climate change. The implications of climate change should be expressly considered in utility integrated resource plans.

Shifting climate holds promises as well as challenges, however. One opportunity that few water utilities have considered is the selling of carbon dioxide reduction credits to participating nations under the Kyoto Accord. Conservation programs that save hot water generate carbon dioxide credits that have cash value.

TOTAL WATER MANAGEMENT HELPS WITH FUTURE CHALLENGES

The trends in population increase and climate change can be summarized as follows:

• Public water suppliers are being forced to meet increasing water demands as the US population increases and more people connect to public water supply systems. Public water supply withdrawals more than tripled between 1950 and 2000 whereas the US population did not even double (Figure 6). Public water supply withdrawals are expected to continue to increase over the next two decades as the population continues to grow.

• Climatological influences have significantly affected water utilities in recent years and will continue to do so. Nearly half of the continental United States has experienced drought conditions during the past few years (Mehan, 2003), which has resulted in regionalized depletion of short- and long-term drinking water supplies. Some researchers believe that much of the recent Western regional growth occurred during a wetter-than-average time period. Continuing research into drought cycles over the past 800 years suggests that the 20th century provided the West with more water than is "normal" (Johnson & Murphy, 2004). Thus the amount of actual drinking water resources available to utilities in the future may be significantly less than expected.

New challenges call for new approaches. Historically, water utilities have focused water planning efforts

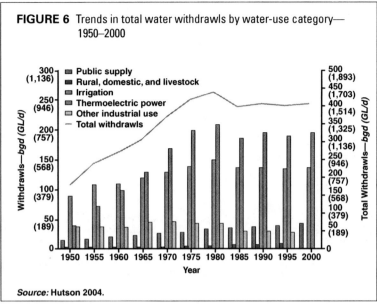

FIGURE 6 Trends in total water withdrawls by water-use category—1950–2000

Source: Hutson 2004.

on obtaining additional surface water or groundwater sources to meet the increasing demands associated with population growth and economic development (AWWA, 1995). These traditional approaches have been augmented by conservation programs, wastewater reclamation, and development of alternative supplies (Maddaus & Maddaus, 2001).

Water conservation typically is one of the first efforts that water utilities implement to address current or future water shortages (AWWA, 1995) and can lead to a 10%–30% reduction of per-capita water consumption within 10 to 20 years (AWWA, 2004; Maddaus & Maddaus, 2001). A 30% decrease in water consumption across the entire United States would result in a savings of more than 5 bgd (19 ML/d), equating to a daily savings of roughly $11 million (AWWA, 2004). Furthermore, these alternative water supplies have become increasingly viable as the costs of developing traditional sources of supply rise and alternative technologies become cheaper. For example, the cost of desalination has decreased in recent years because of advances in treatment technology (Figure 7).

Managing supply risk by diversifying alternatives has driven the development of conservation, recycling, conjunctive-use projects, and brackish and ocean water desalting. Integrated water resource planning considers these and other supply alternatives. Expanding integrated water resource planning to consider watershed management, managing stormwater systems to maximize water yields, and developing source water protection programs are hallmarks of total water management.

Total water management looks at "the big picture." AWWA has defined total water management as the effort of the "water supply industry to assure that water resources are managed for the greatest good of the

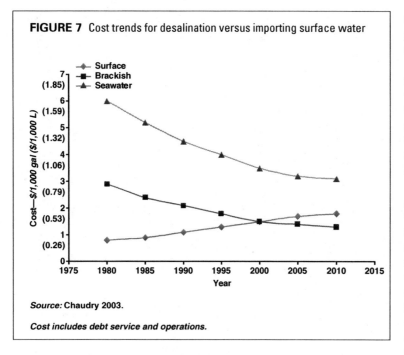

FIGURE 7 Cost trends for desalination versus importing surface water

Source: Chaudry 2003.

Cost includes debt service and operations.

people and the environment and that all segments of society have a voice in the process." A goal of total water management is to address the interrelationships between all aspects of the "environment and society on a regional basis rather than dealing with each issue discretely and within limited parameters" (AWWA, 1994). Furthermore, the philosophy of total water management recognizes a shift from considering water as an unlimited resource to a limited resource that has variability in quantity and quality.

Water utilities generally have tended to manage water-related issues from a local perspective and with limited interaction with other municipal departments. In the past, there was minimal need for such interactions, but this is changing. As population growth and water supply reliability issues intensify, managing water resources and growth to optimize supply, leverage economies of scale, and manage contaminant loading in watersheds will likely become more important. For these efforts to succeed, greater cooperation is needed between water and wastewater utilities and the development community, regulators, and public institutions. Building trust and fostering a willingness to cede some authority for the greater good will be significant challenges, but they must be achieved in order to reap the benefits of greater regional cooperation.

CONCLUSION

During the next 25 years, population trends will shift the number, ethnicity, and perspectives of water utility customers and will affect the nature and delivery of information to the utility customer base. In addition, these trends will necessitate development of new water

supply sources and press water utilities to plan "bigger" than they have traditionally planned.

Climate change and weather cycles that do not follow predictable patterns will demand that utilities carefully and thoroughly assess their water resources and current use. Models may help water providers prepare for the uncertainties that will accompany severe climate change.

Successfully managing population growth and contaminant loading in watersheds, achieving economies of scale, developing new water supplies, and managing water demands will require utilities to work cooperatively with external agencies, regulators, the environmental community, and the business community to a much greater degree than has been the case historically. Because of these developments, utility professionals of the future must be not only skilled managers but also superb communicators and diplomats.

ACKNOWLEDGMENT

The authors thank the Awwa Research Foundation for funding and project manager Linda Reekie for making this work possible. The authors also acknowledge the guidance of Andrew DeGraca, John Huber, and Rosemary Menard of the Project Advisory Committee. The authors thank the utility managers who gave generously of their time and expertise in the Futures Workshop as well as the experts interviewed at the start of the project. This work reflects their collective wisdom and active engagement. Important logistical support provided by Gloria Rivera, Ryan Reeves, and Lorena Ospina of McGuire/Malcolm Pirnie is acknowledged and greatly appreciated.

ABOUT THE AUTHORS

Edward G. Means III is vice-president of McGuire/Malcolm Pirnie, Irvine, CA. He has 26 years of experience in water utility management and water quality. Nicole West is an engineer with McGuire/Malcolm Pirnie and an expert in natural and engineered aquatic ecosystems. Roger Patrick, president of Competitive Advantage Consulting Ltd., is a specialist in improving the management of water and wastewater utilities.

REFERENCES

AWWA, 1994. AWWA Government Affairs: Total Water Management. MainStream, June 23.

AWWA, 1995. AWWA Government Affairs: Water Conservation and Water Utility Programs. www.awwa.org/Advocacy/govtaff/ watcopap.cfm (accessed 2004).

AWWA, 2004. Water Industry Trends. www.awwa.org/careercenter/ resources/docs/WaterIndustryTrends.cfm (accessed 2004).

Bureau of Reclamation, 2003. Water 2025: Preventing Crises and Conflict in the West. US Dept. of the Interior, Washington.

CDC (Centers for Disease Control and Prevention), 2004. National Center for Health Statistics. www.cdc.gov/ nchs/faststats/lifexpec.htm (accessed 2004).

Chaudry, S. 2003. Unit Cost of Desalination. Final Paper. California Desalination Task Force. www.owue.water.ca.gov/ recycle/desal/ Docs/UnitCostofDesalination.doc (accessed 2004).

Dettinger, M. et al, 2004. Global Warming and Water Supplies: How Much Do We Know? AwwaRF Update to the Strategic Assessment of the Future of Water Utilities. Huntington Beach, Calif.

Hobbs, F. & Stoops, N., 2002. Demographic Trends in the 20th Century. US Census Bureau, Census 2000 Special Report, Series CENSR-4, US Govt. Printing Ofce., Washington. www.census.gov/prod/ 2002pubs/censr-4.pdf (accessed 2004).

Hutson, S.S. et al, 2004. United States Geological Survey (USGS). Estimated Use of Water in the United States in 2000. USGS Circular 1268. http://water.usgs.gov/pubs/circ/2004/circ1268/.

Intergovernmental Panel on Climate Change (IPCC), 2000. IPCC Special Report: Emissions Scenarios.

Johnson, K. & Murphy, D.E., 2004. Drought Settles In, Lake Shrinks and West's Worries Grow. *The New York Times,* May 2.

Maddaus, W.O. & Maddaus, L.A., 2001. Water Demand Management Within the Integrated Resource Planning Process. www.isf.uts. edu.au/whatsnew/Demand_Mgmt_IRP.pdf (accessed 2004).

Mehan, G.T. III, 2003. Everyone Undervalues the True Worth of Water. *The Detroit News.* June 19. www.epa.gov/water/infrastructure/ pricing/About.htm (accessed 2004).

US Census Bureau, 2004. www.census.gov/servlet/QTTable (accessed 2004).

This page intentionally blank.

Drinking Water Contaminant Regulation—Where Are We Heading?

BY FREDERICK W. PONTIUS
(*JAWWA* March 2004)

The current body of drinking water regulations has been pieced together in an ever-more-complex series of rulemakings since the US Environmental Protection Agency (USEPA) first enacted the Safe Drinking Water Act (SDWA) in 1974. With each new amendment of the SDWA, the paradigm applied to regulating drinking water contaminants has changed.

The overall goals of the SDWA and its implementation—safe drinking water for people served by US public water systems—are imperative and worthy of pursuit. But over the 30-year history of the SDWA, an immense regulatory maze and administrative superstructure has developed that may now be at risk of collapsing under its own weight. The sheer complexity of the regulations and the increasing financial resources required of USEPA, state agencies, and water utilities for implementation and compliance with existing and future rules will eventually threaten the long-term effectiveness and viability of the US drinking water regulatory program.

US DRINKING WATER REGULATIONS ARE AT A CROSSROADS

Current regulations have been developed, implemented, and, in some cases, revised through a series of 10 major rulemakings under the SDWA from 1974 to 1993. The number of contaminants regulated rose steadily from 1986 until 1993 (Roberson, 2003). At the same time, the 1986 and 1996 amendments to the SDWA have shifted the regulatory paradigm governing contaminant regulation from the original paradigm created in 1974 (Table 1).

The paradigm required by the version of the SDWA in effect at any given time governs how contaminants are regulated at that time. For example, under the 1986 SDWA, USEPA was required to regulate 83 contaminants and 25 additional contaminants every three years regardless of the need for regulation. It became apparent that such an intense level of regulatory activity could not be sustained by the agency, nor could it be justified in the absence of data indicating that contaminant regulation was needed to protect public health.

From 1992 to 1996, attention was directed primarily to regulatory negotiation of the rules known then as the microbial/disinfection by-product (M/DBP) cluster: the Stage 1 Disinfectants/Disinfection Byproducts Rule (D/DBPR), the Enhanced Surface Water Treatment Rule (ESWTR), and the Information Collection Rule (ICR). Subsequent stakeholder discussions resulted in the Interim ESWTR (IESWTR), the Long Term 1 ESWTR (LT1ESWTR), the Stage 2 D/DBPR, and the Long Term 2 ESWTR (LT2ESWTR).

The 1996 SDWA amendments made substantial changes to the law and mandated the establishment of a series of new drinking water regulations. Since 1996, USEPA has developed, proposed, and finalized regulatory actions and has promulgated eight major rules that are currently being implemented. In 2002, the US Congress amended the SDWA by enacting the Public Health Security and Bioterrorism Preparedness and Response Act, which added several important sections to the SDWA to address water system security.

Tables listing current drinking water standards and their applicability have been presented in a previous article (Pontius, 2003). Currently, National Primary Drinking Water Regulations (NPDWRs) are set for 92 contaminants. These include turbidity, 7 microbials or indicator organisms, 4 radionuclides, 11 DBPs, 16 inorganic contaminants, and 53 organic contaminants. Maximum contaminant levels (MCLs) have been set for 83 contaminants, and 9 contaminants have treatment technique requirements. MCLs and treatment technique requirements are enforceable by USEPA, whereas MCL goals (MCLGs) are health-based goals and are not enforceable. Maximum residual disinfectant levels have been set for three disinfectants. Secondary standards are recommended for

15 contaminants to ensure the aesthetic quality of drinking water, although a few states have adopted them as enforceable standards. On the basis of odor and/or taste thresholds, advisories have been issued for four contaminants.

ARE CURRENT REGULATIONS ADEQUATE?

SDWA specified review process. Under the SDWA, USEPA must periodically review existing NPDWRs and, if appropriate, revise them. Section 1412(b)(9) of the SDWA requires the administrator to review and revise as appropriate each NPDWR not less often than every six years. Revision of regulations must maintain or provide for greater protection of public health.

USEPA recently completed its first six-year review of existing NPDWRs (i.e., currently regulated contaminants). Unregulated contaminants, such as those being evaluated by the Contaminant Candidate List (CCL), are not covered by the six-year review. The current 1996–2002 review addresses NPDWRs promulgated prior to 1997 (referred to as pre-1997 NPDWRs), with the exception of those regulations that are the subject of recent or ongoing rulemaking activity (e.g., arsenic, radionuclides, DBPs, and most microbiological NPDWRs).

The principal goal of the SDWA six-year review is to identify, prioritize, and target candidates for regulatory revision that are most likely to result in a meaningful opportunity for health risk reduction and/or meaningful cost savings to public water systems and their customers while maintaining or providing for greater levels of public health protection. A protocol was developed in consultation with the National Drinking Water Advisory Council and other stakeholders (Figure 1) to identify NPDWRs for which there was a health or technological basis for revision. New information was examined in these key areas: health effects, analytical methods improvements, treatment technology effectiveness, other potential regulatory changes, occurrence and exposure data, and potential economic consequences. Preliminary decisions to revise or not revise were published for public comment Apr. 17, 2002 (USEPA, 2002b).

Numerous factors influence "not revise" decisions. On July 18, 2003, USEPA announced completion of its review of 69 NPDWRs that were established prior to 1997 (USEPA, 2003b). These 69 NPDWRs included 68 chemical NPDWRs and the Total Coliform Rule (TCR). On the basis of the agency's initial review, plus public comments received and other new information, USEPA concluded that it was appropriate to revise only the TCR. Table 2 summarizes the "not revise" decisions, which were based on one of the following reasons.

A health risk assessment is in process. As of Dec. 31, 2002, a detailed review of current health effects information was being conducted or was scheduled. Because the results of the assessment were not available in time for consideration under the 1996–2002 review

cycle, revision of these NPDWRs was not appropriate at this time.

Current regulation remains appropriate after data/ information review. For some NPDWRs, the outcome of the six-year review indicated that current regulatory requirements remained appropriate. New information available either supported the current regulatory requirements or did not justify a revision.

New information is available, but no revision is appropriate at this time because of one or more of the following reasons.

The revision in question is a low priority. In USEPA's judgment, any resulting revisions to the NPDWR would not provide a meaningful opportunity for health risk reduction or cost savings to public water systems and their customers. These revisions are a low-priority activity and therefore are not appropriate for revision at this time for one or more of the following reasons: competing workload priorities, administrative costs associated with rulemaking, and burden on states and the regulated community to implement any regulatory change that resulted.

Information gaps exist. Although results of the review support consideration of a possible revision, the available data are insufficient to support a definitive regulatory decision at this time.

For contaminants for which health risk assessments are in process (as of Dec. 31, 2002) and contaminants for which there are information gaps, results of the updated health risk assessment and the results of any research and information-gathering will be considered during the 2002–2008 review cycle. However, if the results of the health risk assessment or research and information-gathering indicate a compelling reason to revisit the "not revise" decision, the review/revision schedule for that NPDWR may be accelerated.

STATUS OF REGULATIONS DETAILED

Array of regulations issued from 1998 on. Not considered in the six-year review were the Stage 1 D/DBPR and the IESWTR, promulgated Dec. 16, 1998 (USEPA, 2001a; USEPA, 1998b; USEPA, 1998c), the Filter Backwash Recycle Rule, the radionuclides rule, and the arsenic rule. These rules are now being implemented. The final radionuclides rule was promulgated Dec. 7, 2000 (USEPA, 2000c) and applies only to community water systems. It updated the MCLs for radium 226+228, alpha emitters, and gross beta and photon emitters and set a new MCL for uranium (Huber, 2003). The final arsenic rule was issued Jan. 22, 2001, setting a new MCL for arsenic at 10 μg/L (USEPA, 2001b), with a compliance date of February 2006.

Large system compliance with the Stage 1 D/DBPR and IESWTR began Jan. 1, 2002, and small system compliance with the Stage 1 D/DBPR began Jan. 1, 2004. The IESWTR applies only to systems serving

TABLE 1	Comparison of SDWA* paradigms for contaminant regulation		
Provision	**1974 SDWA**	**1986 SDWA**	**1996 SDWA**
Authority to regulate	General authority was provided to regulate contaminants.	USEPA† was to regulate contaminants known or likely to occur and that may have an adverse health effect	USEPA is to regulate contaminants that are known or likely to occur, that may have an adverse health effect, and that have a meaningful opportunity for risk reduction.
Use of science in decision-making	Not specifically addressed	Not specifically addressed	To the degree that an agency action is based on science, USEPA must use the best available, peer-reviewed science and supporting studies conducted in accordance with sound and objective scientific practices as well as data collected by accepted methods or best available methods (if the reliability of the method and the nature of the decision justifies use of the data).
Selection of contaminants for regulation	Contaminants regulated in the NIPDWRs‡ were selected on the basis of recommendations by the National Academy of Sciences and the 1962 US Public Health Service Standards.	A list of 83 contaminants was to be regulated regardless of the need for regulation. An additional 25 contaminants, taken from a Drinking Water Priority List, were to be regulated every three years.	The CCL§ is to be published and updated every five years. USEPA must make a determination to regulate for at least five contaminants from each CCL.
Establishment of health goals	RMCL** to be set at a level at which no known or anticipated adverse health effects occur with an adequate margin of safety.	MCLG†† to be set at a level at which no known or anticipated adverse health effects occur with an adequate margin of safety.	1986 SDWA provision retained
Establishment of enforceable limits	NIPDWRs were to protect health to the extent feasible, using technology, treatment techniques, and other means that are generally available (taking costs into consideration).	MCL‡‡ is set as close to MCLG as feasible. "Feasible" means with the use of best available technology, treatment techniques, or other means available (taking cost into consideration), based on efficacy under field conditions and not solely under laboratory conditions.	1986 SDWA provision retained
Competing risks	Not specifically addressed	Not specifically addressed	An MCL may be set at other than the feasible level if an overall increase in health risk would result at the feasible level. The overall health risk must be minimized by balancing competing risks.
Analysis of costs and benefits	Not specifically addressed except as noted above	Not specifically addressed except as noted above. Costs and benefits were analyzed as directed by Executive Order.	An HRRCA§§ is required. USEPA must determine if benefits justify the cost and may adjust the regulatory limit so that it does.

*SDWA—Safe Drinking Water Act
†USEPA—US Environmental Protection Agency
‡NIPDWRs—National Interim Primary Drinking Water Regulations
§CCL—Contaminant Candidate List
**RMCL—revised maximum contaminant level
††MCLG—maximum contaminant level goal
‡‡MCL—maximum contaminant level
§§HRRCA—Health Risk Reduction and Cost Analysis
***NPDWRs—National Primary Drinking Water Regulations
†††M/DBP—microbial/disinfection by-product
‡‡‡CCRs—Consumer Confidence Reports

10,000 or more people. A long-term ESWTR to extend the IESWTR to systems serving 10,000 or fewer people—the LT1ESWTR—was proposed Apr. 10, 2000 (USEPA, 2000a). The final LT1ESWTR was published Jan. 14, 2002 (USEPA, 2002a), and its provisions reflect the IESWTR. The MCLG for *Cryptosporidium* is set at zero. Filtered systems must remove 99% (2 logs) of *Cryptosporidium*. Critical deadlines and requirements have been reviewed previously (Pontius, 2003). Simultaneous compliance with the Stage 1 D/DBPR, IESWTR, and LT1ESWTR necessitates careful optimization of

treatment processes, which may also provide treatment barriers against unregulated contaminants regardless of source. Control of DBPs to meet the Stage 1 D/DBPR must not result in an increase in microbial risk.

Proper backwash practice is necessary to ensure optimal treatment so that contaminants captured via filtration do not subsequently penetrate the treatment barrier. The SDWA required USEPA to regulate recycling of filter backwash water within the treatment process. A rule was proposed Apr. 10, 2000 (USEPA, 2000a), and a final rule was issued June 8, 2001 (USEPA, 2001c). It

---

TABLE 1   Comparison of SDWA* paradigms for contaminant regulation, continued

Provision	1974 SDWA	1986 SDWA	1996 SDWA
Compliance time frame	Not addressed	18 months allowed for compliance	Three years are allowed for compliance, with two additional years allowed if capital improvements are necessary.
Revision of NPDWRs***	Not addressed	NPDWRs must be reviewed and revised at least every three years	NPDWRs must be reviewed and revised at least every six years.
Antibacksliding provisions	Not addressed	Not addressed	An NPDWR revision must maintain or provide for greater health protection.
Mandated regulations	NIPDWRs and revised NPDWRs were to be issued.	Set NPDWRs for 83 contaminants regardless of need; specify criteria for requiring filtration of surface waters; require disinfection of all public water systems.	Retained the list of 83 contaminants to be regulated; regulations are required for arsenic, radon, filter backwash recycling, the M/DBP††† cluster of rules, and disinfection of public water systems (all surface water systems and, as necessary, groundwater systems).
Right-to-know provisions	Public notification provisions were included so that water systems notify customers when a violation occurs.	Public notification provisions were updated.	CCRs‡‡‡ are required annually. Public notification provisions were updated.
Funding for water system improvements	Not addressed	Not addressed	State Revolving Loan Fund created for water system improvements
Variance and exemptions	Time extensions for compliance allowed in certain cases	Same basic provisions updated	Same basic provisions updated; small system variances allowed, considering affordability
Affordability	Not specifically addressed	Not specifically addressed	Affordability only considered when determining if small-system variances will be allowed
Role of the states	States may assume primary enforcement authority by adopting regulations no less stringent than those of USEPA. States may have more strict limits and regulate contaminants that USEPA does not regulate. In states without primacy, USEPA directly implements the regulations.	Same basic provisions as the 1974 SDWA	Same basic provisions as 1986 SDWA; states with primacy must adopt USEPA rules within two years of promulgation, with an extension of up to two additional years possible.

*SDWA—Safe Drinking Water Act
†USEPA—US Environmental Protection Agency
‡NIPDWRs—National Interim Primary Drinking Water Regulations
§CCL—Contaminant Candidate List
**RMCL—revised maximum contaminant level
††MCLG—maximum contaminant level goal
‡‡MCL—maximum contaminant level
§§HRRCA—Health Risk Reduction and Cost Analysis
***NPDWRs—National Primary Drinking Water Regulations
†††M/DBP—microbial/disinfection by-product
‡‡‡CCRs—Consumer Confidence Reports

---

requires systems using surface water or groundwater under direct influence of surface water to recycle spent filter backwash water, thickener supernatant, or liquids from dewatering processes through the processes of a system's existing conventional or direct filtration system or to an alternate recycle location approved by the state, no later than June 8, 2004. Systems needing to make capital improvements to modify their recycle location, must do so by June 8, 2006. Reporting requirements in the rule were to be met no later than Dec. 8, 2003. USEPA has issued a comprehensive technical guidance manual to assist water utilities in compliance with this rule (USEPA, 2002d).

Stage 2 D/DBPR and LT2ESWTR proposed. A negotiative rulemaking process for the Stage 2 D/DBPR and the associated LT2ESWTR came to a conclusion in the form of a Federal Advisory Committee Act committee agreement approved Sept. 29, 2000. Under the proposed Stage 2 D/DBPR (USEPA, 2003d), MCLs for total trihalomethanes (TTHMs) and the sum of five haloacetic acids (HAA5), i.e., monochloro-, dichloro-, trichloro-, monobromo-, and dibromoacetic acid,

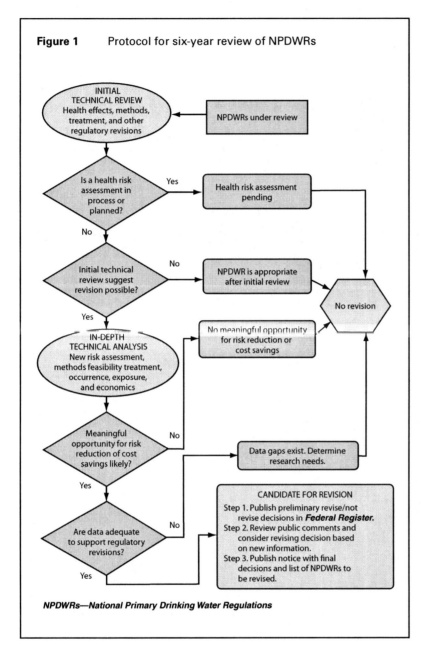

Figure 1   Protocol for six-year review of NPDWRs

NPDWRs—National Primary Drinking Water Regulations

gation for small systems. Stage 2B MCLs of 80 μg/L for TTHMs and 60 μg/L for HAA5 would be determined by LRAAs at the new Stage 2 sampling sites (identified in the IDSE). Compliance with these limits would be required no later than six years after final rule promulgation for large systems and not later than seven and a half years (or eight and a half years in some cases) for small systems.

Under the proposed LT2ESWTR (USEPA, 2003c), water treatment plants would be granted credit toward *Cryptosporidium* removal, depending on the filtration technology currently used (Table 3). Large surface water systems monitor source waters monthly for *Cryptosporidium, Escherichia coli,* and turbidity for 24 months. On the basis of the highest monthly RAA, or two-year mean if 48 samples are collected, the system is assigned to a bin (Table 4) that will require additional removal requirements for *Cryptosporidium.* Small systems sample for *E. coli* biweekly for one year. If the mean *E. coli* level is >10/100 mL for lakes or reservoirs or >50/100 mL for flowing streams, then the system samples bimonthly for *Cryptosporidium.* The average *Cryptosporidium* level is used to determine bin requirements.

Water systems choose from a microbial toolbox (Table 5), a combination of technologies or approaches to meet their additional removal requirements. The proposed LT2ESWTR requires compliance with treatment requirements no later than 72 months after promulgation of the LT2ESWTR for large systems and no later than 102 months for small systems. The proposed LT2ESWTR also includes requirements for disinfection profiling and benchmarking of *Giardia* and virus inactivation.

The final Stage 2 D/DBPR and LT2ESWTR are expected at least 18 months after the close of the public comment period, in mid-2005.

Radon rule delayed. The history and issues associated with USEPA's attempt to set an NPDWR for radon have been reviewed previously (Pontius, 2002; Pontius, 2001). On Nov. 2, 1999, USEPA published a proposed regulation for radon in drinking water (USEPA, 1999). A final rule was required by the SDWA to be issued by

will undergo a phased transition from compliance based on running annual averages (RAAs) across the distribution system to locational RAAs (LRAAs). In the first phase, Stage 2A, water systems will meet TTHM and HAA5 MCLs of 120 and 100 μg/L, respectively, determined by LRAAs at Stage 1 D/DBPR sampling sites. Water systems would continue to meet the Stage 1 RAA limits for TTHMs and HAA5.

An Initial Distribution System Evaluation (IDSE) would be conducted to identify those sampling sites within the distribution system that have the representative highest TTHM and HAA5 levels. An IDSE report is to be completed within two years of final rule promulgation for large systems and within four years of promul-

TABLE 2   Summary of USEPA* six-year review determinations

Determination	Justification	Contaminant	Year Regulated	MCLG† mg/L	MCL‡ mg/L
Not appropriate for revision at this time	Risk assessment in process; chemical currently undergoing a USEPA health risk assessment, including the three initiated as a result of the six-year review (34 NPDWRs§)	Acrylamide	1991	Zero	TT**
		Alachlor	1991	Zero	0.002
		Antimony	1992	0.006	0.006
		Asbestos	1991	7 MFL††	7 MFL
		Atrazine	1991	0.003	0.003
		Benzo(a)pyrene	1992	Zero	0.0002
		Cadmium	1991	0.005	0.005
		Carbofuran	1991	0.04	0.04
		Carbon tetrachloride	1987	Zero	0.005
		Copper	1991	1.3	TT
		Cyanide	1992	0.2	0.2
		2,4-D	1991	0.07	0.07
		1,2-Dichlorobenzene	1991	0.6	0.6
		1,4-Dichlorobenzene	1987	0.075	0.075
		1,2-Dichloroethane	1987	Zero	0.005
		Di(2-ethylhexyl)adipate	1992	0.4	0.4
		Di(2-ethylhexyl)phthalate	1992	Zero	0.006
		Diquat	1992	0.02	0.02
		Endothall	1992	0.1	0.1
		Ethylbenzene	1991	0.7	0.7
		Ethylene dibromide	1991	Zero	0.00005
		Glyphosate	1992	0.7	0.7
		Methoxychlor	1991	0.04	0.04
		Pentachlorophenol	1992	Zero	0.001
		Polychlorinated biphenyls	1991	Zero	0.0005
		Simazine	1992	0.004	0.004
		Styrene	1991	0.1	0.1
		2,3,7,8-TCDD (dioxin)	1992	Zero	5×10^{-8}
		Tetrachloroethylene	1991	Zero	0.005
		Thallium	1992	0.0005	0.002
		Toluene	1991	1	1
		1,1,1-Tichloroethane	1987	0.2	0.2
		Trichloroethylene	1987	Zero	0.005
		Xylenes	1991	10	10
Not appropriate for revision at this time	NPDWR remains appropriate after data/information review (16 NPDWRs)	Barium	1992	2	2
		Dalapon	1992	0.2	0.2
		cis-1,2-dichloroethylene	1991	0.07	0.07
		trans-1,2-sichloroethylene	1991	0.1	0.1
		Dinoseb	1992	0.007	0.007
		Endrin	1992	0.002	0.002
		Epichlorohydrin	1991	Zero	TT
		Hexachlorocyclopentadiene	1992	0.05	0.05
		Mercury	1991	0.002	0.002
		Monochlorobenzene	1991	0.1	0.1
		Nitrate (as N)	1991	10	10
		Nitrite (as N)	1991	1	1
		Selenium	1991	0.05	0.05
		2,4,5-TP (silvex)	1991	0.05	0.05
		1,2,4-Trichlorobenzene	1992	0.07	0.07
		Vinyl chloride	1987	Zero	0.002
Not appropriate for revision at this time	New information available, but no revision appropriate because of low priority (14 NPDWRs)	Benzene	1989	Zero	0.005
		Beryllium	1992	0.004	0.004
		Chlordane	1991	Zero	0.002
		1,2-Dibromo-3-chloropropane	1991	Zero	0.0002
		1,1-Dichloroethylene	1987	0.007	0.007
		1,2-Dichloropropane	1991	Zero	0.005
		Heptachlor	1991	Zero	0.0004
		Heptachlor epoxide	1991	Zero	0.0002
		Hexachlorobenzene	1992	Zero	0.001
		Lindane	1991	0.0002	0.0002
		Oxamyl	1992	0.2	0.2
		Picloram	1992	0.5	0.5
		Toxaphene	1991	Zero	0.005
		1,1,2-Trichloroethane	1992	0.2	0.2
	New information available, but no revision appropriate because of information gaps priority (4 NPDWRs)	Chromium (total)	1991	0.1	0.1
		Dichloromethane	1992	Zero	0.005
		Fluoride	1986	4	4
		Lead	1991	Zero	TT
Candidate for revision	Based on review of "other regulatory revisions" (1 NPDWR)	Total Coliform Rule	1989	zero	TT‡‡

*USEPA—US Environmental Protection Agency
†MCLG—maximum contaminant level goal
‡MCL—maximum contaminant level
§NPDWRs—National Primary Drinking Water Regulations
**TT—treatment technique
††MFL—million fibers per litre
‡‡No more than 5% of the samples per month may be positive for total coliforms. For systems collecting fewer than 40 samples per month, no more than 1 sample per month may be positive.

August 2000, is now scheduled for late 2004, and may be delayed until 2005 or even 2006.

The proposed MCL for radon is 300 pCi/L. An alternative MCL (AMCL) is proposed at 4,000 pCi/L, which would apply instead of the MCL if a state or utility has a multimedia mitigation (MMM) program to lower indoor-air radon. The AMCL is based on the national average outdoor radon level of 0.04 pCi/L. Radon is naturally occurring (primarily from soil gas) and has little potential for deliberate introduction into a water supply.

Under the USEPA proposal, MMM programs must meet four criteria. First, the public must be involved in MMM program development. Second, quantitative goals must be set for existing homes that have been remediated and new homes that were constructed as radon resistant. Third, strategies for achieving goals must be identified. Last, results must be tracked and reported.

Groundwater regulation is pending. The Ground Water Rule (GWR) was proposed May 10, 2000 (USEPA, 2000b). A final rule has been delayed, and the agency has been working to address stakeholder concerns. The SDWA requires that the final rule be issued no later that the final Stage 2 D/DBPR, which is currently expected in mid-2005. However, USEPA hopes to finalize the GWR sometime in 2004.

The proposed GWR establishes a risk-based regulatory strategy for all groundwater systems to address risks through a multiple-barrier approach. The proposed rule relies on five major components: (1) periodic sanitary surveys of groundwater systems requiring evaluation of eight elements and identification of significant deficiencies, (2) hydrogeologic assessments to identify wells sensitive to fecal contamination, (3) source water monitoring for systems drawing from sensitive wells without treatment or with other indications of risk, (4) correction of significant deficiencies and fecal contamination, and (5) compliance monitoring to ensure disinfection treatment is reliably operated where it is used. Correction of significant deficiencies could be accomplished by eliminating the source of contamination, correcting the significant deficiency, providing an alternative source water, or providing a treatment that

TABLE 3	Proposed *Cryptosporidium* treatment credit toward LT2ESWTR* requirements
Plant Type	**Log Credit†**
Conventional treatment (includes softening)	3.0
Direct filtration	2.5
Slow sand or diatomaceous earth filtration	3.0
Alternative filtration technologies	Determined by the state through product or site-specific testing

*LT2ESWTR—Long Term 2 Enhanced Surface Water Treatment Rule
†Applies to plants in full compliance with the Surface Water Treatment Rule (SWTR), Interim SWTR, Interim Enhanced SWTR, and the Long Term 1 ESWTR, as applicable

TABLE 4	Proposed LT2ESWTR* bin requirements†	
Bin Number	***Cryptosporidium* Concentration‡ oocysts/L**	**Additional Treatment Required§**
1	< 0.075	No additional treatment
2	0.075 to < 1.0	1 log (conventional, softening, slow sand, and diatomaceous earth plants); 1.5 log (direct filter plants)
3	1.0 to < 3.0	2.0 log (conventional, softening, slow sand, and diatomaceous earth plants); 2.5 log (direct filter plants)
4	> 3.0	2.5 log (conventional, softening, slow sand, and diatomaceous earth plants); 3 log (direct filter plants)

*LT2ESWTR—Long Term 2 Enhanced Surface Water Treatment Rule
†Applies to plants in full compliance with the Surface Water Treatment Rule (SWTR), Interim Enhanced SWTR, and Long Term 1 Enhanced SWTR, as applicable
‡Total oocyst count as determined by US Environmental Protection Agency method 1622 or 1623, uncorrected for recovery
§Additional treatment for alternative filtration technologies will be as determined by the state, so long as the total *Cryptosporidium* removal and inactivation is at least 4.0 logs, 5.0 logs, and 5.5 logs for bins 2, 3, and 4, respectively.

achieves at least 99.99% (4-log) inactivation or removal of viruses.

Contaminants to be regulated are selected from CCL. The SDWA requires USEPA to publish a list of contaminants for possible regulation, referred to as the CCL. Five or more contaminants from each CCL are to be selected and evaluated to determine whether these contaminants should be regulated by an NPDWR. The first CCL (CCL1) was published in March 1998 (USEPA, 1998a) and listed 60 contaminants (50 chemicals and 10 microbes). In 1998, 20 of the 60 contaminants were classified as priorities for regulatory determination because it was believed at that time that there were sufficient data to evaluate both exposure and risk to public health and to support a determination of whether to proceed to promulgation of an NPDWR. However, c12 of the 20 priority contaminants were found to have insufficient information to support a regulatory determination. In addition, sodium was added to the list of regulatory determination priorities.

TABLE 5   Proposed LT2ESWTR* microbial toolbox log credits, design, and implementation criteria

Toolbox Option	Proposed *Cryptosporidium* Log Credit With Design and Implementation Criteria
Watershed Control Program	0.5-log credit for state-approved program comprising USEPA†-specified elements; does not apply to unfiltered systems
Alternative source/intake management	No presumptive credit; systems may conduct simultaneous monitoring for LT2ESWTR bin classification at alternative intake locations or under alternative intake management strategies.
Offstream raw water storage	No presumptive credit; systems using offstream storage must conduct LT2ESWTR sampling after raw water reservoir to determine bin classification.
Presedimentation basin with coagulation	0.5-log credit with continuous operation and coagulant addition; basins must achieve 0.5-log turbidity reduction based on the monthly mean of daily measurements in 11 of the 12 previous months. All flow must pass through basins; systems using existing presedimentation basins must sample after basins to determine bin classification and are not eligible for presumptive credit.
Lime softening	0.5-log credit for two-stage softening (single-stage softening is credited as equivalent to conventional treatment); coagulant must be present in both stages—includes metal salts, polymers, lime, or magnesium precipitation. Both stages must treat 100% of flow.
Bank filtration (as pretreatment)	0.5-log credit for 25 ft (7.6 m) setback; 1.0-log credit for 50 ft (15 m) setback; aquifer must be unconsolidated sand containing at least 10% fines. Average turbidity in wells must be <1 ntu. Systems using existing wells followed by filtration must monitor well effluent to determine bin classification and are not eligible for presumptive credit.
Combined filter performance	0.5-log credit for combined filter effluent turbidity # 0.15 ntu in 95% of samples each month
Roughing filters	No presumptive credit proposed
Slow sand filters	2.5-log credit as a secondary filtration step. 3.0-log credit as a primary filtration process; no prior chlorination
Second-stage filtration	0.5-log credit for second separate filtration stage; treatment train must include coagulation prior to first filter; no presumptive credit for roughing filters.
Membranes (microfiltration, ultrafiltration, nanofiltration, and reverse osmosis)	Log credit equivalent to removal efficiency demonstrated in challenge test for device if supported by direct integrity testing
Bag filters	1-log credit with demonstration of at least 2-log removal efficiency in challenge test
Cartridge filters	2-log credit with demonstration of at least 3-log removal efficiency in challenge test
Chlorine dioxide	Log credit based on demonstration of log inactivation with contact time table
Ozone	Log credit based on demonstration of log inactivation with contact time table
UV‡	Log credit based on demonstration of inactivation with UV dose table; reactor testing required to establish validated operating conditions
Individual filter performance	1.0-log credit for demonstration of filtered water turbidity of <0.1 ntu in 95% of daily maximum values from individual filters (excluding 15-min period following backwashes) and no individual filter >0.3 ntu in two consecutive measurements taken 15 min apart
Demonstration of performance	Credit awarded to unit process or treatment train based on demonstration to the state, through use of a state-approved protocol

*LT2ESWTR—Long Term 2 Enhanced Surface Water Treatment Rule
†USEPA—US Environmental Protection Agency
‡UV—ultraviolet

Preliminary regulatory determinations were published June 3, 2002 (USEPA, 2002c). Final determinations were announced July 18, 2003 (USEPA, 2003a) and are summarized in Table 6. There is currently particular interest in the timing of future regulatory determinations for other contaminants on the CCL, especially perchlorate and methyl tertiary butyl ether (MTBE). USEPA may take action on CCL contaminants if information becomes available and may proceed with regulatory determinations prior to the end of the next regulatory determination cycle (i.e., August 2006). Alternatively, the agency may monitor, conduct research, develop guidance, or regulate contaminants not included on the CCL to address an urgent threat to public health.

Biological and chemical warfare agents regulated. The SDWA regulatory paradigm presumes that contaminants are either naturally occurring or could occur as a result of natural, human-caused, or industrial pollution. It does not directly address chemical or biological contaminants that could be purposefully introduced into a water supply with intent to kill or harm large numbers of people.

The Public Health Security and Bioterrorism Preparedness and Response Act of 2002 (PL 107-188) was enacted into law June 12, 2002, adding several important sections to the SDWA. The 8,000 water systems serving 3,300 or more people are conducting vulnerability assessments and are updating or revising emergency response plans based on the results from their vulnerability assessments. Regulatory compliance planning must now include emergency planning so that should a natural disaster or terrorism incident occur, the utility will be ready to respond quickly and in a manner that will protect customers, prevent or minimize noncompliance, and avoid legal liabilities.

TABLE 6   Summary of CCL1* determinations to regulate

Contaminant	Description and Potential Health Effects	Determination
Acanthamoeba	A free-living protozoa associated with human infections affecting the eye, lung, brain, and skin	Do not regulate. Regulation would not present a meaningful opportunity for risk reduction. No monitoring data exist indicating occurrence of Acanthamoeba cysts in drinking water, and it is removed by existing filtration practices. Guidance is being provided for contact lens wearers.
Aldrin and dieldrin	Structurally similar insecticides that are probable human carcinogens. They were used from 1950 to 1970 on corn and cotton, banned in 1974 (except for termite control), and completely banned in 1987.	Do not regulate. Regulation would not present a meaningful opportunity for health risk reduction. Aldrin and dieldrin have a low frequency and low level of occurrence in drinking water.
Hexachlorobutadiene	A volatile organic contaminant mainly used to make rubber compounds.	Do not regulate. Regulation would not present a meaningful opportunity for health risk reduction. It does occur in some public water systems but not at a frequency or level of public health concern.
Manganese	Naturally occurring element that is essential for humans and all animal species. Manganese is generally considered to have low toxicity when ingested orally. The major source of manganese intake in humans is dietary ingestion (with the exception of occupational exposure to manganese dusts via inhalation). Drinking water accounts for a small proportion of manganese intake.	Do not regulate. Regulation would not present a meaningful opportunity for health risk reduction. USEPA† is developing a Drinking Water Advisory for manganese.
Metribuzin	A pesticide that is persistent in the environment. It is used as an herbicide on soybeans, potatoes, alfalfa, and other crops. Metribuzin is not classifiable as a human carcinogen, but there may be effects on the liver and body weight from chronic exposure to high doses.	Do not regulate. Regulation would not present a meaningful opportunity for health risk reduction. Metribuzin is infrequently detected in public water supplies and is not known to occur at levels of public health concern.
Naphthalene	A volatile organic contaminant naturally present in fossil fuels and formed when wood or tobacco are burned. The major human exposure is through the use of mothballs containing naphthalene. Usually not found in water because it evaporates or biodegrades quickly.	Do not regulate. Regulation would not present a meaningful opportunity for health risk reduction. Naphthalene is infrequently detected in public water supplies and is not known to occur at levels of public health concern.
Sodium	Naturally occurring element essential for human health. The contribution of sodium in drinking water to the total dietary intake is very small.	Do not regulate. Regulation would not present a meaningful opportunity for health risk reduction. Water systems currently must monitor for sodium at the entry point to the distribution system and report results to public health officials.
Sulfate	Naturally occurring in soil, sediments, and rocks, and also present in the human diet. Ingesting high levels of sulfate in drinking water can cause increased water in fecal matter (diarrhea).	Do not regulate. Regulation would not present a meaningful opportunity for health risk reduction. The adverse effects of sulfate are generally mild and of short duration; they generally occur at concentrations greater than the SMCL‡ of 500 mg/L.

*CCL1—first Candidate Contaminant List
†USEPA—US Environmental Protection Agency
‡SMCL—secondary maximum contaminant level

WHERE ARE WE HEADING?

USEPA contemplates TCR revisions and associated regulations. USEPA plans to consider revisions to the TCR with new requirements for ensuring the integrity of distribution systems. Traditional issues associated with the TCR have been reviewed previously (Pontius, 2000). The TCR and TCR monitoring are intended to address potential unintentional fecal contamination and not the possibility of deliberate biological contamination of source waters and distribution systems. In the summer of 2003, USEPA identified several TCR issues and is currently developing white papers, which are expected to be issued in late 2004.

Related to the TCR is consideration of regulations targeted at water distribution systems. Under certain conditions, contaminants can enter water distribution systems. Cross-connections and negative distribution system pressures can allow contamination, and state programs to address these concerns vary in their effectiveness. USEPA has worked with stakeholders to develop and issue a series of white papers on distribution system issues. (USEPA's distribution white papers are available at http://www.epa.gov/ogwdw/tcr/ tcr.html; summaries of each white paper are available from the USEPA Drinking Water Hotline, 800-426-4791.) Stakeholder meetings have been held to discuss distribution system issues, and distribution system components that could be included in revision of the TCR are under consideration. Proposed TCR revisions are expected June 2006, and final action is due June 2008.

Contamination by a currently regulated chemical or microorganism or a common unregulated chemical or

FIGURE 2   Regulatory framework for a simple surface water system

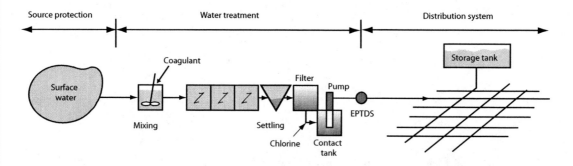

SOURCE WATER PROTECTION REGULATORY PROGRAMS
 SWAP
 SWPP
 TMDL limits
 NPDES permitting

UTILITY SOURCE WATER QUALITY CONTROL PRACTICES
 Algae control
 Chemcal addition
 Hypolimnetic
  aeration

PUBLIC HEALTH SECURITY AND BIOTERRORISM PREPAREDNESS AND RESPONSE ACT OF 2002
 Vulnerability
  assessment
 Emergency
  response plan

WATER QUALITY LIMITS FOLLOWING TREATMENT
 Meet 16 inorganic contaminant MCLs.
 Meet 4 radionuclide MCLs.
 Meet 53 organic contaminant MCLs.
 Meet 15 secondary MCLs.
 If using chorine dioxide, meet chlorine dioxide MRDL
  at EPTDS.
 If using ozone, meet bromate MCL at EPTDS.

TREATMENT REQUIREMENTS
 Achieve at least 2-log *Cryptosporidium* removal.
 Achieve at least 3-log *Giardia* removal/inactivation.
 Achieve at least 4-log virus removal/inactivation.
 Achieve *Legionella* removal/inactivation.
 Meet CFE turbidity performance criteria.
 Meet IFE turbidity requirements.
 Meet FBRR requirements.
 Maintain disinfectant residual at EPTDS
 Maintain overall level of disinfection by meeting
  benchmark if making major changes to
  disinfection practice
 Apply optimal corrosion control treatment.

PROPOSED RADON RULE
 Meet radon MCL, or meet AMCL if state has an
  MMM program.

PROPOSED LT2ESWTR REQUIREMENTS
 Determine log-removal credit for existing treatment.
 Monitor source water for *Cryptosporidium*.
 Determine if additional treatment require based
  on the results of source water *Cryptosporidium*
  monitoring.
 Install additional treatment from microbial toolbox.
 Conduct disinfection benchmarking for *Giardia*
  and viruses, if required.

WATER QUALITY LIMITS IN THE DISTRIBUTION SYSTEM
 Meet TCR limits.
 Maintain minimum disinfectant residuals at
  ends of the distribution system.
 Meet lead and copper action levels at the tap.
 Meet TTHM and HAA5 MCLs determined by
  RAAs across the distribution system.
 Meet MRDLs for chlorine/chloramines.
 If using chlorine dioxide, meet chlorite MCL.

PROPOSED STAGE 2 D/DBPR REQUIREMENTS
 Conduct IDSE.
 Meet Stage 1 D/DBPR MCLs determined by
  RAAs at Stage 1 sampling points.
 Meet Stage 2A MCLs determined by LRAAs
  at Stage 1 sampling points.
 Select Stage 2 sampling points on the basis
  of IDSE results.
 Meet Stage 2B MCLs determined by LRAAs
  at new Stage 2 sampling points.
 Control peak TTHM/HAA5 excursions.

POSSIBLE FUTURE REGULATIONS
 Distribution System Rule
  Possible cross-connection control
   requirements

AMCL—alternative maximum contaminant level, CFE—combined filter effluent, D/DBPR—Disinfectants/Disinfection Byproduct Rule, EPTDS—entry point to the distribution system, FBRR—Filter Backwash Recycling Rule, HAA5—sum of five haloacetic acids, IDSE—Initial Distribution System Evaluation, IFE—individual filter effluent, LRAAs—locational running annual averages, MCL—maximum contaminant level, MMM—multimedia mitigation, MRDL—maximum residual disinfectant level, NPDES—National Pollutant Discharge Elimination System, RAAs—running annual averages, SWAP—Source Water Assessment Program, SWPP—Source Water Protection Program, TCR—Total Coliform Rule, TMDL—total maximum daily load, TTHM—total trihalomethane

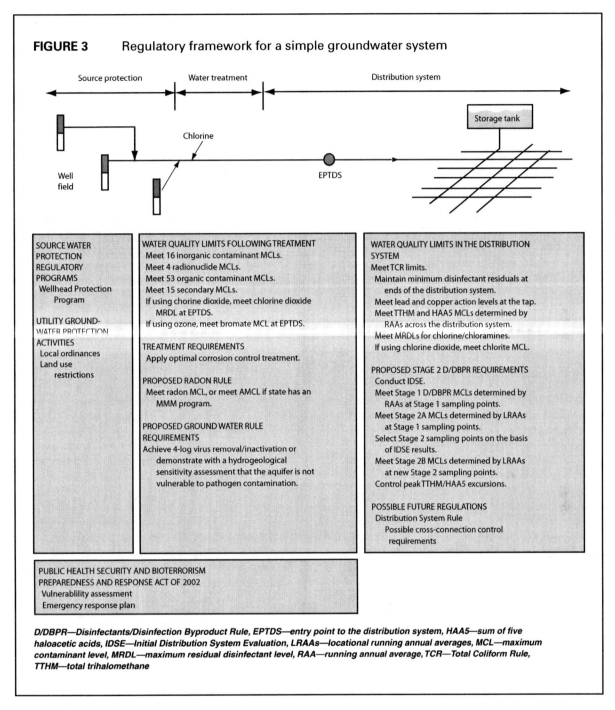

FIGURE 3   Regulatory framework for a simple groundwater system

Source protection   Water treatment   Distribution system

Storage tank

Chlorine

Well field

EPTDS

SOURCE WATER
PROTECTION
REGULATORY
PROGRAMS
  Wellhead Protection
    Program

UTILITY GROUND-
WATER PROTECTION
ACTIVITIES
  Local ordinances
  Land use
    restrictions

WATER QUALITY LIMITS FOLLOWING TREATMENT
  Meet 16 inorganic contaminant MCLs.
  Meet 4 radionuclide MCLs.
  Meet 53 organic contaminant MCLs.
  Meet 15 secondary MCLs.
  If using chorine dioxide, meet chlorine dioxide
    MRDL at EPTDS.
  If using ozone, meet bromate MCL at EPTDS.

TREATMENT REQUIREMENTS
  Apply optimal corrosion control treatment.

PROPOSED RADON RULE
  Meet radon MCL, or meet AMCL if state has an
    MMM program.

PROPOSED GROUND WATER RULE
REQUIREMENTS
Achieve 4-log virus removal/inactivation or
    demonstrate with a hydrogeological
    sensitivity assessment that the aquifer is not
    vulnerable to pathogen contamination.

WATER QUALITY LIMITS IN THE DISTRIBUTION
SYSTEM
  Meet TCR limits.
  Maintain minimum disinfectant residuals at
    ends of the distribution system.
  Meet lead and copper action levels at the tap.
  Meet TTHM and HAA5 MCLs determined by
    RAAs across the distribution system.
  Meet MRDLs for chlorine/chloramines.
  If using chlorine dioxide, meet chlorite MCL.

PROPOSED STAGE 2 D/DBPR REQUIREMENTS
  Conduct IDSE.
  Meet Stage 1 D/DBPR MCLs determined by
    RAAs at Stage 1 sampling points.
  Meet Stage 2A MCLs determined by LRAAs
    at Stage 1 sampling points.
  Select Stage 2 sampling points on the basis
    of IDSE results.
  Meet Stage 2B MCLs determined by LRAAs
    at new Stage 2 sampling points.
  Control peak TTHM/HAA5 excursions.

POSSIBLE FUTURE REGULATIONS
  Distribution System Rule
    Possible cross-connection control
      requirements

PUBLIC HEALTH SECURITY AND BIOTERRORISM
PREPAREDNESS AND RESPONSE ACT OF 2002
  Vulnerablility assessment
  Emergency response plan

*D/DBPR—Disinfectants/Disinfection Byproduct Rule, EPTDS—entry point to the distribution system, HAA5—sum of five
haloacetic acids, IDSE—Initial Distribution System Evaluation, LRAAs—locational running annual averages, MCL—maximum
contaminant level, MRDL—maximum residual disinfectant level, RAA—running annual average, TCR—Total Coliform Rule,
TTHM—total trihalomethane*

microorganism that a water treatment system is not designed to remove or inactivate because that agent would normally not be expected to occur poses a significant threat. Treatment failures, inadequate treatment, or lack of treatment of poor quality or contaminated source waters are well-documented causes of outbreaks of waterborne disease (Craun et al, 2003) and may pose an equal or greater threat than the introduction of an exotic toxin, chemical, or biological agent. This consideration makes it even more important that water utilities focus their

efforts on providing sufficient treatment barriers and not be distracted or hampered by overly burdensome drinking water rules that impose such an administrative burden on water utilities, states agencies, and even USEPA that efforts to protect public health are hindered. Figures 2 and 3 show the regulatory framework for a simple surface water system and simple groundwater system, respectively. Record-keeping and reporting requirements also add to the overall regulatory administrative burden.

Unconventional delivery approaches considered. Rather than treat all water piped into the distribution system to strict USEPA regulatory standards, unconventional approaches to water delivery have been suggested as providing a more economical means of ensuring that the water customers drink meets USEPA regulatory requirements while lowering the cost of treating water for nonpotable uses (Cotruvo & Cotruvo, 2003). Public water systems are considering the use of unconventional approaches for potable water delivery for any number of reasons. Specific driving forces include consumer preference, the need to achieve regulatory compliance with economic efficiency, the need to solve site-specific drinking water contamination problems from unregulated contaminants, the need to protect sensitive subpopulations, the need to provide water during periods of water shortages and drought conditions, and current concerns regarding water system security.

Use of a dual distribution system is an unconventional approach to water delivery that may have application in some water systems as a means of increasing or supplementing the available supply of water and preserving potable water for drinking water uses. Reclaimed wastewater can then be used for irrigation, turf, construction, certain industrial uses, and other outdoor uses, and only potable-use water must be treated to meet SDWA requirements.

Federal and state regulations impose certain requirements that affect the technical feasibility and/or regulatory acceptability of unconventional approaches to drinking water treatment in public water systems. Point-of-use (POU) devices and point-of-entry (POE) treatment are allowed under the SDWA but with certain restrictions. POU devices may be economical compliance technology for small systems for arsenic and radionuclides. Compliance with future USEPA regulations will afford some opportunity for water utilities to consider alternative potable water delivery. Bottled water is not currently allowed by USEPA as a compliance option but may be required as a condition of a variance or exemption.

The presence of drinking water contaminants that have not been regulated by USEPA may motivate some water systems to consider POU devices or POE treatment. The presence of unregulated contaminants such as perchlorate, MTBE, hexavalent chromium, N-nitrosodimethylamine, newly discovered DBPs, and endocrine-disrupting compounds at concentrations of customer concern may cause some water utilities to consider unconventional approaches. The effectiveness of POU devices and POE treatment for removal of these compounds will vary with the technology applied, and POU treatment would not be appropriate for compounds that present a health risk from dermal absorption or inhalation exposures.

Delivering water of two qualities triggers ethical and legal issues. Over time, concerns regarding drinking water–sensitive subpopulations are likely to increase in importance to water suppliers. As a result, some water utilities are considering whether they should provide drinking water of differing qualities to meet the differing needs of their customers. A general quality potable water meeting USEPA regulations would be provided via central treatment. In addition, a higher-quality drinking water would be produced for purchase by certain customers desiring a higher-quality product. Separate provision of a higher-quality water only to certain customers may raise environmental justice objections. It may also give the impression that the potable water provided to other customers is unsafe. More important, provision of a separate higher-quality water may increase the liability of the water system and the potential for toxic tort litigation.

The principal public health advantage to central water treatment and piped distribution of potable water is that all water must meet minimum USEPA regulations. This ensures that customers are protected regardless of how they are exposed (e.g., tap water, outdoor faucet).

Unconventional delivery systems could lead to even more regulation. Although allowances are made in the SDWA and USEPA regulations for a water system to use POU and POE as a compliance option, the possibility exists for consumers to be exposed to harmful concentrations of contaminants if water is consumed from an untreated tap. The possibility of less-than-complete consumer protection associated with unconventional approaches poses the primary regulatory barrier; unconventional approaches will not be acceptable to state and federal regulators without assurances and safeguards that consumers will be protected. In addition, it remains to be demonstrated whether unconventional approaches can be implemented in a way that reduces the overall regulatory burden and complexities faced by water utilities and state agencies or whether these approaches will simply add another layer of complexity.

The next six-year review of NPDWRs is expected to be completed in the August 2008 time frame. That review will include the 68 chemical NPDWRs just reviewed as well as NPDWRs for which new or revised regulations were promulgated between 1996 and 2002 (e.g., arsenic, radionuclides). The current protocol (Figure 1) will be applied to the 2002–2008 review, modified where appropriate to incorporate lessons learned from the current review. As discussed previously, the review may be accelerated for an individual NPDWR, and contaminants from the CCL may be regulated "off cycle" if new information presents a compelling reason to do so. An unregulated contaminant monitoring rule is due for proposal in September 2004, with the final rule due in September 2005. A preliminary notice of the second CCL is expected in early 2004. Preliminary determinations from the second CCL are expected in August 2005 and final determinations in August 2006. A preliminary notice for the third CCL is expected in February 2007 and a final notice in February 2008. However, consideration of ways to streamline drinking water regulations and their implementation during the next six-year review cycle will be critical to the long-term effectiveness and viability of the US drinking water regulatory program.

A report by the Association of State Drinking Water Administrators (ASDWA, 2003) estimated that by 2006, funding available to the states will meet only 62% of state drinking water program needs. Primacy grants are insufficient to carry out all of the provisions of the SDWA for state primacy, and state budgets continue to be strapped. The regulatory tangle and administrative superstructure has strained the financial and personnel resources required to maintain them, especially at a time when state agencies and water utilities face shrinking budgets. In fact, traditionally low water rates might need to be raised to cover the actual costs of state

primacy. Everyone agrees that drinking water regulation and regulatory oversight are necessary, but if the US regulatory program is to survive, we must regulate smartly, efficiently, and economically.

ABOUT THE AUTHOR

Frederick W. Pontius is president of Pontius Water Consultants Inc., Lakewood, CO. Pontius is currently a PhD candidate in environmental engineering at the University of Colorado (CU) at Boulder. He also holds a master's degree in environmental engineering and a bachelor's degree in chemical engineering, both from CU. For more than 25 years, he has worked in the areas of regulatory compliance, water quality, and water treatment and is the recipient of several AWWA Best Paper awards.

REFERENCES

ASDWA (Association of State Drinking Water Agencies), 2003. Public Health Protection Threatened by Inadequate Resources for State Drinking Water Programs; An Analysis of State Drinking Water Program Resources, Needs, and Barriers. ASDWA, Washington.

Cotruvo, J.A. & Cotruvo, J.A. Jr., 2003. Nontraditional Approaches for Providing Potable Water in Small Systems: Part 1. *Jour. AWWA*, 95:3:69.

Craun, G.F.; Calderon, R.L.; & Craun, M.F., 2003. Waterborne Outbreaks in the United States, 1971–2000. *Drinking Water Regulation and Health* (F.W. Pontius, editor). John Wiley and Sons, New York.

Huber, D., 2003. Regulating Radionuclides. *Drinking Water Regulation and Health* (F.W. Pontius, editor). John Wiley and Sons, New York.

Pontius, F.W., 2003. Update on USEPA's Drinking Water Regulations. *Jour. AWWA*, 95:3:57.

Pontius, F.W., 2002. Regulatory Compliance Planning to Ensure Water Supply Safety. *Jour. AWWA*, 94:3:52.

Pontius, F.W., 2001. Regulatory Update for 2001 and Beyond. *Jour. AWWA*, 93:2:64.

Pontius, F.W., 2000. Reconsidering the Total Coliform Rule. *Jour. AWWA*, 92:2:14.

Roberson, J.A., 2003. Complexities of the New Drinking Water Regulations—Everything You Wanted to Know But Were Afraid to Ask. *Jour. AWWA*, 95:3:48.

USEPA (US Environmental Protection Agency), 2003a. Announcement of Regulatory Determinations for Priority Contaminants on the Drinking Water Contaminant Candidate List; Notice. *Fed. Reg.*, 68:138:42898.

USEPA, 2003b. National Primary Drinking Water Regulations; Announcement of Completion of EPA's Review of Existing Drinking Water Standards; Notice. *Fed. Reg.*, 68:138:42908.

USEPA, 2003c. Long-term 2 Enhanced Surface Water Treatment Rule; Proposed Rule. *Fed. Reg.*, 68:154:47640.

USEPA, 2003d. Stage 2 Disinfectants and Disinfection Byproducts Rule; Proposed Rule. *Fed. Reg.*, 68:159:49548.

USEPA, 2002a. National Primary Drinking Water Regulations; LT1ESWTR. Final Rule. *Fed. Reg.*, 67:9:1812.

USEPA, 2002b. National Primary Drinking Water Regulations; Announcement of the Results of EPA's Review of Existing Drinking Water Standards and Request for Public Comment; Proposed Rule. *Fed. Reg.*, 67:74:19030.

USEPA, 2002c. Announcement of Preliminary Regulatory Determinations for Priority Contaminants on the Drinking Water Contaminant Candidate List. *Fed. Reg.*, 67:106:38222.

USEPA, 2002d. Filter Backwash Recycling Rule Technical Guidance Manual. Office of Ground Water and Drinking Water. EPA 816-R-02-014.

USEPA, 2001a. Revisions to the IESWTR and Stage 1 D/DBPR. *Fed. Reg.*, 66:10:3770.

USEPA, 2001b. Arsenic. Final Rule. *Fed. Reg.*, 66:14:6976.

USEPA, 2001c. Filter Backwash Recycling Rule. *Fed. Reg.*, 66:111:31086.

USEPA, 2000a. Long Term 1 Enhanced Surface Water Treatment and Filter Backwash Rule. Proposed Rule. *Fed. Reg.*, 65:69:19046.

USEPA, 2000b. Ground Water Rule. Proposed. *Fed. Reg.*, 65:91:30194.

USEPA, 2000c. Radionuclides. Final Rule. *Fed. Reg.*, 65:236:76708.

USEPA, 1999. Radon. Proposed Rule. *Fed. Reg.*, 64:211:59246.

USEPA, 1998a. Drinking Water Candidate Contaminant List. Notice. *Fed. Reg.*, 63:40:10274.

USEPA, 1998b. Disinfectants and Disinfection Byproducts. Final Rule. *Fed. Reg.*, 63:241:69390.

USEPA, 1998c. Interim Enhanced Surface Water Treatment. Final Rule. *Fed. Reg.*, 63:241:69478.

This page intentionally blank.

The Regulatory Horizon: Implications for Water Utilities

BY EDWARD G. MEANS III, NICOLE WEST, AND ROGER PATRICK
(*JAWWA* November 2005)

A variety of trends are affecting the water utility community. In order to prepare for and help manage the future utility environment, the AWWA Research Foundation (AwwaRF) funded the project "Update of the Strategic Assessment of the Future of Water Utilities."

Increasingly sophisticated analytical technology is detecting contaminants at concentrations that are often beyond our understanding of environmental risk. Protecting our aquatic systems, accommodating and managing population growth, and providing reliable supplies of freshwater represent a significant challenge. Regulations under the Safe Drinking Water Act (SDWA), Clean Water Act (CWA), and Endangered Species Act (ESA) will continue to drive treatment complexity, costs, and decision-making. This article discusses current and upcoming regulations, some of their implications, and potential management strategies identified through the AwwaRF project.

REGULATORY PRESSURES

SDWA. The SDWA, promulgated in 1974, is the federal law that ensures the quality of Americans' drinking water. Under the SDWA, the US Environmental Protection Agency (USEPA) sets standards for drinking water quality and oversees the implementation of those standards. As a result of the SDWA, the number of drinking water regulations has increased over the past 25 years (Figure 1). Although the information in this figure suggests a slowing of USEPA regulatory activity, several of the rules in recent years are far more complex (i.e., involving treatment techniques). An example of this is the Stage 1 Disinfectants/Disinfection Byproducts Rule (D/DBPR). Even though Figure 1 only shows activity through 2001, significant regulatory activity has occurred since that time, including promulgation of the Long-Term 1 Enhanced Surface Water Treatment Rule (LT1ESWTR) in August 2003 and proposal of the Long-Term 2 ESWTR (LT2ESWTR) and the Stage 2 D/DBPR in August 2003.

In order to protect public health, the USEPA has set a goal that 95% of community water systems be in compliance with health-based standards and that 95% of the population served by community water systems receive drinking water that meets health-based standards by 2008.

Meeting the USEPA's goal of 95% of community water systems in compliance with health-based standards by 2008 will require steady progress. In order to reach this strategic target, compliance with regulatory requirements by small and very small water systems will be critical. Figure 2 shows that very small and small systems in the United States accounted for most SDWA health-based violations in 2003. The USEPA goal can either drive those systems to improve their facilities and operations to gain compliance (through assistance from Drinking Water State Revolving Funds) or force more consolidation, privatization, or contract operations to meet the compliance goals.

Compliance with the SDWA has resulted in more people being served by community water systems with no health-based violations even though the goal that 95% of the population served by community water systems receive drinking water that meets health-based standards has not yet been met (Figure 3). Figure 4 suggests that the USEPA and the states may have to focus their compliance efforts on the large and very large systems to attain this particular goal.

Drinking water regulations on the horizon. Under the SDWA, the USEPA has proposed several new drinking water regulations. These include the Radon Rule, the Ground Water Rule (GWR), the Stage 2 D/DBPR, and the LT2ESWTR. The Radon Rule proposes a maximum contaminant level (MCL) for radon-222 in public water supplies. The GWR will require groundwater systems to perform sanitary surveys and hydrologic sensitivity assessments (for systems not providing a 4-log reduction in viruses). The Stage 2 D/DBPR will supplement existing DBP rules by requiring systems to meet DBP MCLs at each monitoring site in the distribution system.

The LT2ESWTR will apply to all systems and will mandate additional *Cryptosporidium* treatment requirements in systems at higher risk for this protozoan. It also contains provisions to ensure that systems maintain microbial barriers as they manage DBP formation.

In addition to proposed regulations, there are

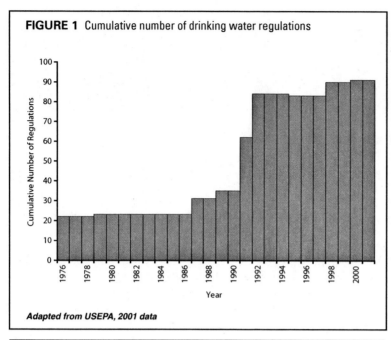

FIGURE 1 Cumulative number of drinking water regulations

Adapted from USEPA, 2001 data

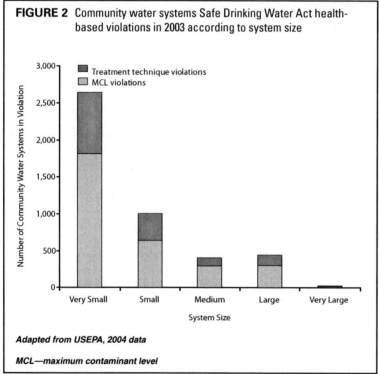

FIGURE 2 Community water systems Safe Drinking Water Act health-based violations in 2003 according to system size

Adapted from USEPA, 2004 data

MCL—maximum contaminant level

years the USEPA must select at least five contaminants from this list for which adequate data are available and determine whether these contaminants should be regulated (the next selection is scheduled for 2006). Given current information on occurrence and health effects, affected utilities should pay close attention to potential MTBE and perchlorate regulations.

Detection of trace levels of endocrine disruptors and pharmaceuticals and personal care products in natural waters has been largely driven by significant advances in analytical technology that allow detection of compounds down to concentrations that were not previously possible. Accordingly, scientists are detecting trace levels of contaminants in water that were not previously detectable. These contaminants are receiving special scrutiny because of publicized evidence of sexual and developmental abnormalities in fish, mollusks, alligators, and frogs exposed to endocrine-disrupting chemicals (e.g., Desbrow et al, 1998). Over the next few years, research on occurrence in drinking water, treatment/removal/destruction technologies, and toxicology of these compounds will help decide whether these chemicals will need to be regulated in drinking water.

The focus of DBP regulation since 1979 has been on chlorinated and brominated DBPs. There are, however, numerous other DBPs for which minimal occurrence and health effects information is available. A 2002 USEPA occurrence study focused on 50 high-priority unregulated DBPs and found elevated levels of DBPs at plants treating with chloramines (Weinberg et al, 2002). Control of the four regulated trihalomethanes by alternative disinfectants does not necessarily guarantee control of other halogenated DBPs and may increase the concentration of some DBPs when compared with chlorination. For example, iodinated DBPs are associated with source water containing bromide and iodide that is disinfected with chloramines (Plewa et al, 2004). The body of technical information on DBP occurrence and risk continues to expand. Notwithstanding the substantial compliance hurdles of the Stage 1 and 2 D/DBPRs, new information will continue to alter the regulatory agenda.

several unregulated contaminants on the SDWA Contaminant Candidate List that may adversely affect public health and occur in drinking water with a frequency and at levels that may pose a threat to public health. These currently unregulated contaminants include methyl tertiary butyl ether (MTBE), perchlorate, other industrial chemicals, chemical herbicides and insecticides, and microbiological contaminants and microorganisms including several types of viruses. Every five

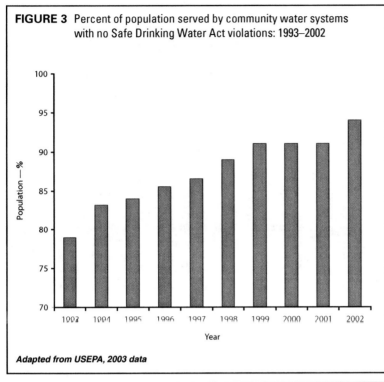

FIGURE 3 Percent of population served by community water systems with no Safe Drinking Water Act violations: 1993–2002

Adapted from USEPA, 2003 data

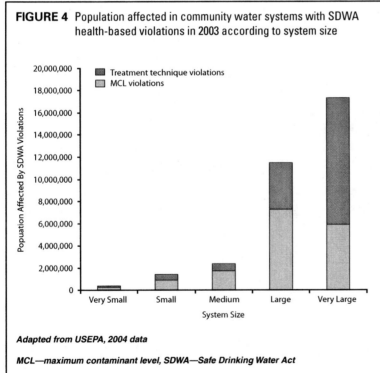

FIGURE 4 Population affected in community water systems with SDWA health-based violations in 2003 according to system size

Adapted from USEPA, 2004 data

MCL—maximum contaminant level, SDWA—Safe Drinking Water Act

N-nitrosodimethylamine (NDMA) and other nonhalogenated compounds are known to be formed under certain disinfection treatment conditions. The 10^{-6} cancer risk level is in the low nanogram-per-litre range. Their formation mechanisms are not completely understood.

NDMA is a potent carcinogen in laboratory animals and can be formed in small amounts during chloramination of drinking water. Barrett et al (2003) found a 90th percentile NDMA concentration of 9.0 ng/L in four quarters of distribution system sampling from 121 North American treatment plants. NDMA was observed in greater than 60% of distribution system samples. NDMA undergoes photolysis in the presence of ultraviolet (UV) light. Accordingly, to the extent that controlling NDMA becomes more important in the future, use of UV light treatment may become more desirable.

Many water utilities will need to raise water rates to fund the increased treatment and monitoring that will be necessary in order for them to comply with new regulations. Small systems will have the most challenges coping with increased operation costs, and many may be forced to consolidate in order to cope with these rising costs. Improved technology is continuing to reveal more contaminants at lower detection levels, which may, fairly or unfairly, negatively affect consumer confidence in tap water.

New technology penetration. In order to meet increasing regulatory standards, many water utilities will have to implement advanced treatment technologies. The use of advanced treatment technologies will also be driven by the increasing use of marginal quality water, such as seawater, recycled water, and brackish groundwater. Greater emphasis may be placed on treatment technologies that can reduce or eliminate a wide array of contaminants at the same time (e.g, membrane filtration). The use of these technologies will also increase as they become cheaper. For example, the cost of desalination has decreased in recent years because of advances in treatment technology and is expected to continue decreasing (Figure 5). Partly as a result, approximately 20 seawater reverse osmosis plants are in various stages of planning along the California coast. Increasing use of these marginal water supplies will present their own regulatory and operational challenges (e.g., boron penetration of seawater reverse-osmosis membranes).

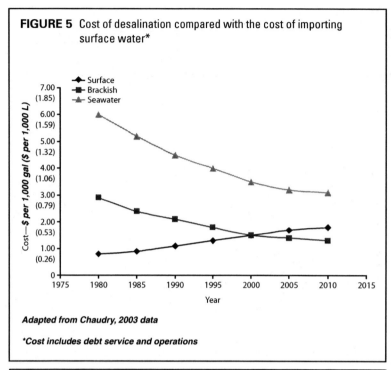

FIGURE 5 Cost of desalination compared with the cost of importing surface water*

Adapted from Chaudry, 2003 data

*Cost includes debt service and operations

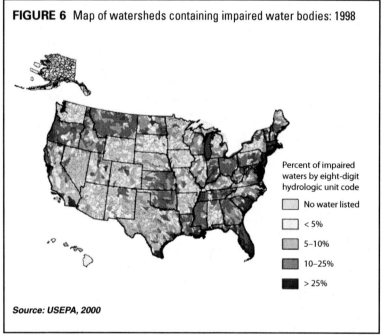

FIGURE 6 Map of watersheds containing impaired water bodies: 1998

Percent of impaired waters by eight-digit hydrologic unit code

No water listed

< 5%

5–10%

10–25%

> 25%

Source: USEPA, 2000

Residuals disposal. Virtually all drinking water treatment processes produce liquid and/or solid residuals. As new and proposed regulations require greater contaminant removal rates, the quantity and quality of residuals may change. Regulation of radionuclides and arsenic will significantly affect the nature and risk of the residuals produced by some utilities and drive higher disposal costs. More stringent sanitary sewer restrictions (driven in part by greater recycled water development)

may preclude disposing of concentrated brines via this method.

CWA. The 1972 CWA established the basic structure for regulating discharges of pollutants into US waters. The CWA grants the USEPA authority for driving the establishment of water quality standards for surface waters and the authority to implement pollution-control programs. Recently, efforts have shifted focus from regulation of discharges from traditional "point source" facilities, such as municipal sewage plants and industrial facilities, to regulation of runoff from streets, construction sites, farms, and other "wet-weather" sources.

As authorized by the CWA, the National Pollutant Discharge Elimination System permit program controls water pollution by regulating point sources that discharge pollutants into US waters. As part of the shift to more holistic watershed-based strategies, states are beginning to establish total maximum daily loads (TMDLs) for impaired waterbodies. A TMDL is the maximum amount of pollutant a waterbody can receive and still meet water quality standards. A TMDL allocates the amount of pollutant allowed in the river between the different sources of the pollutant, including point, nonpoint, and natural background sources.

Effluent guidelines are national standards for wastewater discharges to surface waters and publicly owned treatment works. In September 2004, the USEPA announced a three-year effort to develop technology-based regulations for water treatment plant waste discharges in its 2004 Final Effluent Guidelines Program Plan. This announcement was in response to increasing concern that the drinking water supply and treatment plants have the potential to discharge "nontrivial" amounts of nonconventional and toxic pollutants. In particular, there is a concern that many drinking water facilities have the potential to discharge significant quantities of conventional and toxic pollutants; the sources of these pollutants can include drinking water treatment sludge and reverse-osmosis reject wastewaters. As noted earlier, treatment of drinking water residuals to comply with future discharge regulations could increase overall water production costs.

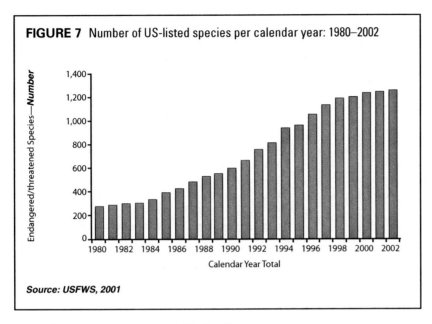

FIGURE 7 Number of US-listed species per calendar year: 1980–2002

Endangered/threatened Species—*Number*

Calendar Year Total

Source: USFWS, 2001

regulations on the timing and flow allowed for pumping drinking and agricultural water.

In 1983 the United Nations established the World Commission on Environment and Development. The commission suggests that the solution is sustainable development—the concept that human activity should "meet the needs of the present without compromising the ability of future generations to meet their own needs." To ensure that water resources are available and of good quality for future generations, water utilities will have to engage and manage their water resources in a sustainable fashion that protects environmental resources and adopt practices that improve the quality of the water resource and its ecosystems.

Discharge regulations may positively affect water utilities by cleaning up source waters, which would, in turn, require less treatment. However, current development patterns may lead to more runoff and possibly an increase in contaminant load. As urban development replaces agricultural land uses, the type and concentration of contaminants entering source waters will likely change.

ESA. Although progress has been made in surface water quality, more than 300,000 acres (121,410 ha) of lakes and 20,000 mi (32,180 km) of rivers and streams remain polluted (Figure 6). As result of habitat destruction, more species are becoming endangered or threatened (Figure 7).

The ESA of 1973 instructs federal agencies to carry out programs to conserve endangered and threatened species and to conserve the ecosystems on which these species depend.

Since 1995, there have been noticeable declines in many amphibian species in the United States. Although habitat destruction is the major reason for this decline, there may be additional factors such as chemical pollution, increased UV light radiation, parasites, and predation (USGS, 2005). As more aquatic organisms become threatened or endangered, water treatment plants may have to reevaluate placement or replacement of water intakes in order to avoid further population declines. For example, loss of juvenile fish at commercial and agricultural diversions in the Sacramento-San Joaquin Delta has been identified as a major contributor to population declines in many native species, such as the endangered winter-run chinook salmon and the threatened delta smelt (CDFG, 2005; WFCB, 2002). Research on fish loss at the diversions may lead to stricter

STRATEGIES

Several strategies were identified during the workshop to help utilities anticipate, shape, and implement future regulations.

Utilities can combine their efforts to advocate for funding to meet regulations. They can lead in research and development (e.g., new treatment technologies, residuals treatment, contaminant occurrence surveys). This leadership can help ensure appropriate regulation and enable the best future technology choices. This may become more difficult in the future should federal funding for water utility research become constrained and utility spending priorities shift. Cutting research activities and engagement in technical forums may be an attractive temptation for utilities as water rate increases draw greater public scrutiny, but few industries became great by cutting research and development.

Utilities can coordinate and cooperate to create a cohesive voice on Capitol Hill, to educate decision-makers, and to advocate for public health. Utilities can strategically engage regulators and actively participate in the legislative process, which will ultimately have a direct effect on regulations, their utilities, and the water community as a whole. For example, by participating in the TMDL process, utilities can leverage the TMDL outcome and potentially reduce costs for treatment, improve water quality, and reduce health risks to customers. Utilities should monitor the regulatory process and ensure that regulations are based on the best available technical information and not a product of political or public pressure.

CONCLUSION

Regulations will continue to pose complex compliance challenges to water utilities. Utilities can take specific steps to manage their future, including:
- providing specific organizational responsibility to track and implement regulations,
  - engaging in the legislative and regulatory process,
  - anticipating regulations,
  - preparing formal compliance plans,
  - articulating the costs of compliance to regulators and the public, and, perhaps most important,
    - conducting periodic internal compliance audits.

These six steps will help ensure that adequate water supplies will continue to be developed, that public health and the environment are protected, and that the costs of doing so are minimized.

ACKNOWLEDGMENT

The authors acknowledge AwwaRF funding and Project Manager Linda Reekie. The authors also acknowledge the guidance of Andrew DeGraca, John Huber, and Rosemary Menard (Project Advisory Committee). The authors thank the utility managers who gave generously of their time and expertise in the Futures Workshop as well as the experts interviewed at the start of the project. This work reflects their collective wisdom and active engagement. Important logistical support by Gloria Rivera, Ryan Reeves, and Lorena Ospina of the authors' staffs

ABOUT THE AUTHORS

Edward G. Means III is vice-president of McGuire/ Malcolm Pirnie, Irvine, CA. Means has 27 years of experience in water utility management, water resources, and water quality. Nicole West is an engineer with McGuire/Malcolm Pirnie in the Irvine, Calif., office, who has experience in natural and engineered aquatic ecosystems. Roger Patrick is president of Competitive Advantage Consulting Ltd. and specializes in improving the management of water and wastewater utilities.

REFERENCES

Barrett, S. et al, 2003. Occurrence of NDMA in Drinking Water: A North American Survey, 2001–2002. Proc. 2003 AWWA Annual Conference, Anaheim, Calif.

CDFG (California Department of Fish and Game), 2005. Central Valley Bay-Delta Branch. Delta Smelt. www.delta.dfg.ca.gov/gallery/dsmelt.asp (accessed Sept. 15, 2005).

Chaudry, S. 2003. Unit Cost of Desalination. Final Paper. California Desalination Task Force. www.owue.water.ca.gov (accessed Sept. 16, 2005).

Desbrow, C. et al, 1998. Identification of Estrogenic Chemicals in STW Effluent. Chemical Fractionation and in Vitro Biological Screening. *Envir. Sci. & Technol.*, 32:11:1549.

Plewa, M.J. et al, 2004. Chemical and Biological Characterization of Newly Discovered Iodoacid Drinking Water Disinfection Byproducts. *Envir. Sci. & Technol.*, 38:18:4713.

USEPA (US Environmental Protection Agency), 2004. Factoids: Drinking Water and Groundwater Statistics for 2003. www.epa.gov (accessed Sept. 16, 2005).

USEPA, 2003. Safe Drinking Water Information Systems/Federal Version (SDWIS/FED). www.epa.gov (accessed Sept. 15, 2005).

USEPA, 2001. Contaminants Regulated Under the Safe Drinking Water Act. www.epa.gov/safewater/creg.html (accessed Sept. 16, 2005).

USEPA, 2000. Atlas of America's Polluted Waters, EPA 840-B-00-002.

USFWS (US Fish and Wildlife Service), 2005. US Species Listed per Calendar Year. www.fws.gov/endangered/ (accessed Sept. 16, 2005).

USGS (US Geological Survey), 2005. Amphibian Research. http://biology.usgs.gov/frog.html (accessed Sept. 16, 2005).

Weinburg, H.S. et al, 2002. The Occurrence of Disinfection By-Products (DBPs) of Health Concern in Drinking Water: Results of a Nationwide DBP Occurrence Study. National Exposure Research Laboratory. Office of Research and Development. US Environmental Protection Agency, EPA/600/R-02/068, Washington.

WFCB (Wildlife, Fisheries, and Conservation Biology), 2002. Fish Treadmill Project: Applied Research to Improve Protection of California's Fishes. http://wfcb.ucdavis.edu/www/faculty/joe/treadmill (accessed Sept. 15, 2005).

The Infrastructure Crisis

BY JOHN E. CROMWELL III, ELISA SPERANZA, AND HAYDN REYNOLDS
(*JAWWA* April 2007)

In 2005, AWWA's Water Utility Council commissioned the authors to develop a report (AWWA, 2006) that would clarify the infrastructure issue and motivate local actions to address the issue. The primary objectives were to (1) articulate the true nature of the need for infrastructure stewardship via asset management and (2) help water utilities convey this need, both internally and externally, to governing bodies, customers, and other stakeholders. This article is derived, in part, from that report.

Much has been written about the infrastructure "crisis." Coverage of the issue began with publication of *America in Ruins* (Choate & Walter, 1981), which used images of ancient Roman ruins to frame the infrastructure issue as an imminent crisis. Many other studies of the nation's infrastructure crisis followed. In addition, the media continues to use catastrophic images and provocative headlines when covering infrastructure issues. These presentations convey the message that the damage is done and US cities are already in ruins. This characterization exaggerates reality—and, ironically, is not consistent with the proper definition of the word "crisis."

According to *Merriam-Webster's Collegiate Dictionary* (2006), a crisis is a "turning point," a key juncture at which decisive action must be taken to avoid damage or harm. Used correctly, crisis is actually a good word for characterizing the nation's existing infrastructure situation.

The two charts shown in Figure 1 appeared in *Dawn of the Replacement Era* (AWWA, 2001). These charts present data from 20 medium- and large-sized US water utilities. Figure 1, part A shows the replacement cost value (in 2001 dollars) of the pipe assets of those 20 utilities, plotted in the historical year in which they were installed. Part B shows a projected ramping up of pipe replacement investment needs in the twenty-first century. This investment generally echoes the historical pattern of pipe installation, with allowances for differences in asset lives resulting from manufacturing eras and operating environments.

Although the graphs in Figure 1 do define a crisis in the real sense of the word, they also convey considerable good news.

• The pattern of pipe installation produced by historical demographic trends results in an echo wave of reinvestment needs. This pattern does not indicate a sudden increase in expenditures, but rather a ramping up of expenditures that extends over several decades—the period we call "the dawn of the replacement era." Consequently, utilities are able to gradually transition to a new steady state that is at a higher level of spending needed to sustain a mature pipe network.

• Experience with the best asset management practices indicates that there are many ways to flatten and stretch the ramping-up phase without compromising the level of service provided to customers. This is especially important as knowledge of asset wear-out processes and application of renewal technologies continue to improve.

There is also considerable risk conveyed in Figure 1. Communities that do not act at this crucial point to implement asset management and ramp up reinvestment will fall behind the curve, leaving the next generation with a mountainous funding problem that will constitute a true threat to sustainability.

FRAMING THE ISSUE

Today's utility managers and governing board members are at a turning point. There is an urgent need to mobilize and sustain proactive investments to prevent catastrophic or disastrous conditions from developing in our infrastructure. Mobilizing action to prevent catastrophe is, in many ways, more challenging than mobilizing action in response to catastrophe. There are broadly held beliefs that an acute crisis is required to motivate significant new spending. However, framing the issue in this manner suggests that a one-time expenditure is an effective response. In contrast, proactive asset management, coupled with a sustained ramping-up of expenditures, is the appropriate response for probable infrastructure needs—a point that must be communicated to stakeholders. Water infrastructure renewal requires a long-term commitment that cannot be sustained if perception of the problem and the reality are not brought closer together.

The sense of urgency that gave rise to the crisis rhetoric is not misplaced. Professionals in the public works arena have known that when urban infrastructure needs to be replaced because of wear and obsolescence, it is difficult to draw adequate public attention and monetary resources to these needs. This is particularly true

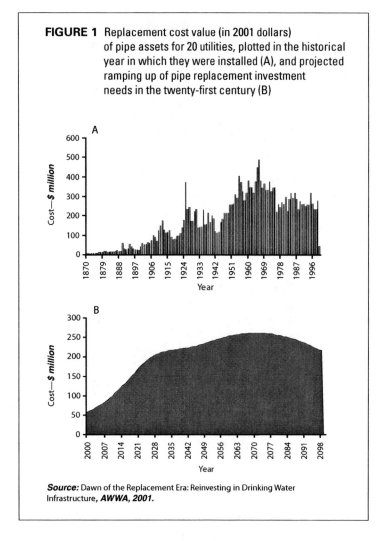

FIGURE 1 Replacement cost value (in 2001 dollars) of pipe assets for 20 utilities, plotted in the historical year in which they were installed (A), and projected ramping up of pipe replacement investment needs in the twenty-first century (B)

Source: Dawn of the Replacement Era: Reinvesting in Drinking Water Infrastructure, *AWWA, 2001.*

more attention than a potential one. Some utility managers regard such news coverage as a victory until a journalist shows up at the door to ask why a crisis has been allowed to develop in the first place and who is to blame—with cameras rolling.

Even if it appears to work once, the crisis approach is not the sustainable funding mechanism needed for water infrastructure renewal. If the issue is framed in this way, utilities may be backed into a corner, and sustainable solutions may be out of reach. To say there's a crisis is like saying there's a fire. Once it's doused, it's expected that the crisis is over.

While developing the most recent AWWA Water Utility Council report (AWWA, 2006), the authors polled utility managers and public affairs professionals to gain a sense of current opinions (see sidebar). Broadly summarized, utility managers believe they are in the difficult position of trying to convey a need for sustained action to prevent catastrophic conditions from developing, while other community needs are made to seem more pressing. Managers are very concerned that the true value of water and wastewater infrastructure is underappreciated. They believe that a long-term commitment to funding the renewal process, rather than a short-term response, is needed to ensure sustainability.

Public officials were also polled. They generally believe that water utilities provide good, reliable service. Their impressions of water infrastructure issues are, for the most part, locally based and not distracted by national media. Although they believe they should "do the right things" to support utility managers in funding needs and educating the community, they note many stark realities that enter into the picture.

First and foremost, locally there are competing demands for funding. In this environment, visible, tangible, and immediate public benefits matter most. The relative lack of visibility of buried pipes and the long-term nature of the renewal process are noted as impediments to mobilizing local action. The effectiveness of crisis conditions, court orders, US Environmental Protection Agency (USEPA) standards, and security concerns are acknowledged as motivating factors. Although the value and importance of water are acknowledged as motivators, the need for a long-term commitment to sustainability as a focus of intergenerational stewardship is regarded as a tough sell in some communities.

Part of the solution to this problem involves framing the issue in a way that is closer to reality. The issue

in the wider context of reinvestment needs in other sectors (e.g., transportation and health). This worry stems from the fact that infrastructure wears out over time and is invisible to most people.

Over the past several decades, major national studies of water infrastructure have been conducted (NCPWI, 1987a, 1987b; WIN, 2000; USEPA, 2002; CBO, 2002). These studies predicted a growing need for reinvestment in infrastructure. They also emphasized meeting those needs to prevent a funding gap from which it would be difficult to recover. The authors of these studies were preoccupied with quantifying the national scale of the problem in terms of billions of dollars. Although the numbers vary between studies, the consistent result was the need for sustained increases in investment.

Although these infrastructure studies documented the prospect that a funding gap could arise, media coverage required more newsworthy headlines, i.e., that a huge gap already exists and our infrastructure is falling apart. The media knows that a full-blown crisis draws should be framed proactively in terms of the need for a sustained ramping up of expenditures rather than as an expenditure gap, which frames the issue as a crisis that requires sudden outlays of cash. As confirmed by our research, every city has a long queue of crisis petitioners, and being relegated to waiting in that line is not the desired outcome. Communities cannot afford to have their utilities backed into that corner.

By framing the issue in terms of the need to ramp up, utilities are able to draw a focus on asset management as the best practice approach to keeping the investment slope gradual, thus avoiding rate shocks (AWWA, 2004). Communities that practice asset management are able to mobilize around finding a sustainable path rather than seeking a limited crisis response.

ASSET LIFE LESSONS

Pipes generally constitute the bulk of a utility's infrastructure and have the longest useful lives. Consequently, they are major drivers in fundamentally altering investment needs for rehabilitation and replacement and are the impetus for ramping up expenditures. This does not diminish the importance of nonpipe assets, which also represent a significant investment and typically require much shorter replacement cycles.

Pumps, treatment facilities, and extensive instrumentation and control systems are also "invisible" to the public. However, these components have shorter life cycles and require earlier replacements, creating a view of water-related infrastructure that requires only periodic injections of capital followed by years of trouble-free operations. In contrast, pipe rehabilitation and replacement will require a sustained flow of expenditures.

In simple terms, pipes are replaced based on the date the pipe was installed and the length of its life. As in any population, age is not the only determinant of the length of a life. All the pipes that are "born" in a given year will not "die" in the same year in the future. For example, pipes placed in corrosive soils or in locations where they are subject to adverse stresses have a decreased life expectancy. The number of years of service for a group of pipes installed in the same year will take the form of a statistical distribution, such as a bell-shaped curve. Within this distribution, the average life of most pipes makes up the middle of the distribution and the

extremes—represented by the "tails" of the distribution curve—are either shorter or longer than the average.

This bell-shaped curve of life expectancy stretches out replacement investment needs over a longer period of time than would be the case if all pipes had exactly the same life expectancy. Stretching out reinvestment needs sounds good, but the total picture is a bit more complicated.

Most water systems installed additional pipes as the utility expanded. Every installment of pipe assets will have its own bell-shaped curve of replacement needs. The succession of bell-shaped curves will be staggered over time, and the distributions will overlap. To get an accurate picture of total replacement needs, the overlapping curves must be added together.

The result is a replacement investment forecast, or echo wave, that reflects the original waves of construction and expansion in a water system. In the past, the expansion rate of pipe networks was a result of population growth, which tended to ebb and flow as waves of growth swept over US cities. Combining the demographic patterns of original construction work with the bell-shaped curves of expected pipe life produces a graph like the one in Figure 2. This graph illustrates the relationship between the original waves of construction

Motivating Local Action on Water Infrastructure

Views from members of AWWA's Water Utility Council and Public Affairs Council on what it takes

"Unfortunately, it seems to require a disaster or catastrophe."

"Sadly, the greatest impact would be an example of catastrophic failure somewhere."

"The tyranny of the immediate has taken over public decision-making."

"It will probably take a crisis."

"Unfortunately, many times it is a disaster or major problem."

"For most communities, it will probably take a significant critical event where it is perceived that the public's health or safety has been put in jeopardy. Then elected officials can champion the solution and position themselves as protecting the public from imminent disaster."

Water Infrastructure at a Turning Point: The Road to Sustainable Asset Management, AWWA 2006.

and the echo waves of replacement for each class of pipe assets. The resulting ramplike shape of reinvestment results because most system expansion occurred during the latter part of the previous century (i.e., most pipes are less than 50 years old) and because even the oldest pipes have long lives (life expectancy for many of the oldest pipes is more than 100 years because of the thick-walled design of the early iron pipes).

Two key variables determine the echo wave of pipe reinvestment needs; one is known, and the other is unknown. The original pipe installation pattern is fixed and known; it can be inferred reasonably well from data on historic population growth trends even if there are no utility records. The second variable—the bell-shaped distribution of pipe life for each annual cohort group of pipe assets—is not known with precision. Moreover, it is not fixed but rather evolving as individual pipes continue to wear out, and it is changeable because of maintenance interventions. Asset management focuses on this "pipe life" variable in an attempt to optimize the rate of reinvestment in terms of total life-cycle costs (flattening the ramp) while maintaining the level of asset performance required to deliver the desired level of service to customers.

In addition to physical failures such as main breaks, pipe performance is affected by line pressure, water chemistry, and microbiology. The wear-out/service deterioration process can be actively managed to extend pipe life by efficiently responding to main breaks and by intervening with maintenance actions to arrest structural decay. In some instances, rehabilitation investment can be substituted for replacement investment to extend pipe life by remediating nonstructural performance issues such as adequate pressure and acceptable taste, odor, color, and microbial safety.

THE VALUE OF ASSET LIFE

Pipes are not a "simple" asset because of the complex way in which they wear out at varying rates over time. This complexity is coupled with the inherent difficulty of assessing and tracking this wear-out process when the assets are buried from view. The importance of these complexities becomes magnified because pipes constitute such a high proportion of the replacement value of a water system's assets.

A utility that falls behind in pipe reinvestment will find it very difficult to catch up. On the other hand, a utility that replaces pipes too early may lose an enormous amount of the remaining value of these expensive assets. There is a lot at stake in determining the right rate of reinvestment. Consequently, it is critical to have an asset management process that seeks to understand and proactively manage the rate at which pipes wear out in concert with a logical, targeted reinvestment program.

Originally, water utility pipe networks were financed through the repeated waves of population and resultant economic growth, reflecting a tremendous store of value. In a city with half a million people, the cost to replace the water pipes might amount to $1 billion. Today, some cities have lost as much as a third of the population they had at their peak. A pile-up of replacement needs, as implied in Figure 2, can present significant financial risk to a utility if these previously staggered investments must be repeated simultaneously, without correlation to the current economic conditions for that community.

There is a potential catastrophe in the making if this large wave of investment needs is not recognized and acted on in advance. A community that allows this wave to overtake its water utility without a proactive strategy will be inundated by it.

Another essential message regarding the ramp-up phase is that with few exceptions ramping up can extend over decades. This allows time to build knowledge of asset life expectancy and an ability to increase asset life through maintenance, targeted rehabilitation, and development of new and/or less expensive technologies. Fortunately, there is time to reduce the uncertainty, flatten the ramp, and ensure that communities derive maximum value from these essential public assets.

RESOLVING THE CRISIS

Utilities face a real infrastructure crisis—a turning point that requires decisive action. The crisis is the result of subtle phenomena and involves many invisible assets. Consequently, the opportunity to take timely action could be missed. This risk is amplified by the fact that a staggering amount of public equity is at stake, not to mention public health. Asset management provides a necessary risk management framework in which uncertainties can be systematically investigated and weighed. This enables decision-makers to make better-informed, timelier decisions about the rate of reinvestment in water infrastructure and, in turn, to better serve their communities.

Utility managers and board members are counted on to apply careful and balanced judgment to issues involving utility policies, operations, capital investment, and customer service. Governing board members often see their main role as that of controlling costs and quality of service. Asset management enables governing board members to also act as stewards of invaluable water infrastructure.

If viewed as unrelated line items in operating and capital budgets, the activities needed to develop, refine, and implement asset management programs are considered in the context of short-term benefits against other spending needs. It is in this context that true catastrophe will ultimately be needed to support changes in expenditures.

When asset management initiatives are grouped into a coherent risk management program, it is possible to see the connections between a given level of capital replacement and parallel activities, such as repair, rehabilitation, and condition assessment. In turn, board members can better appreciate and make the case for ramping up key expenditures. Moreover asset management is a continuous improvement process that facilitates knowledge transfer from one generation of managers and board members to the next. It also provides direction for research and development of repair and replacement technologies and a better understanding of sustainable service levels.

The dynamics of echo replacement waves are such that transition to a steady state in the replacement era will extend decades into the future. Today's utility managers and governing board members stand at a turning point. They cannot solve the infrastructure problem by writing one big check—which is an incorrect perception of this generation's responsibility. This generation is responsible for inaugurating a process of asset management that will facilitate ramping up of reinvestment and guarantee a sustainable flow of valuable public benefits from these valuable assets.

ACKNOWLEDGMENT

This article is derived, in part, from the AWWA report, "Water Infrastructure at a Turning Point: The Road to Sustainable Asset Management" (2006). It was developed for the AWWA Water Utility Council with sponsorship from the Water Industry Technical Action Fund. The authors are indebted to a number of collaborators, including Gary Breaux, Tom Curtis, Paul Demit, Mike Hooker, Gary Lynch, Sue McCormick, John Sullivan, Kurt Vause, Gina Wammock, and Al Warburton.

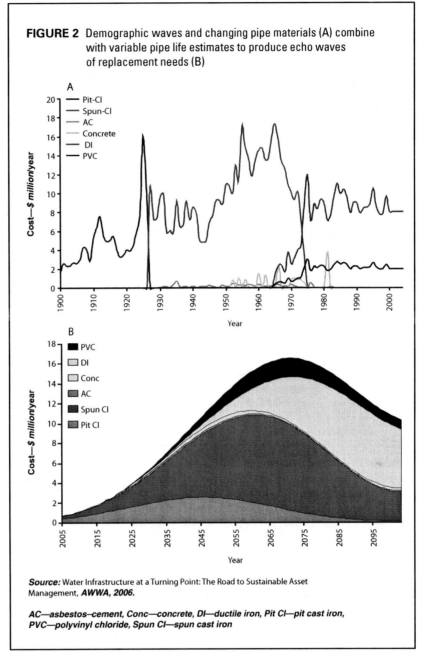

FIGURE 2 Demographic waves and changing pipe materials (A) combine with variable pipe life estimates to produce echo waves of replacement needs (B)

Source: Water Infrastructure at a Turning Point: The Road to Sustainable Asset Management, **AWWA, 2006.**

AC—asbestos–cement, Conc—concrete, DI—ductile iron, Pit CI—pit cast iron, PVC—polyvinyl chloride, Spun CI—spun cast iron

REFERENCES

Association of Local Government Engineering New Zealand, 2006. International Infrastructure Management Manual. National Asset Management Steering Group, Thames, New Zealand.

AMSA (Association of Metropolitan Sewerage Agencies), 2001. Managing Public Infrastructure Assets to Minimize Cost and Maximize Performance. AMSA, Washington.

AWWA, 2006. *Water Infrastructure at a Turning Point: The Road to Sustainable Asset Management.* AWWA, Denver.

AWWA, 2004. *Avoiding Rate Shock: Making the Case for Water Rates.* AWWA, Denver.

AWWA, 2001. *Dawn of the Replacement Era: Reinvesting in Drinking Water Infrastructure.* AWWA, Denver.

Choate, P. & Walter, S., 1981. America in Ruins. Council of State Planning Agencies, Washington.

CBO (Congressional Budget Office), 2002. Future Investment in Drinking Water and Wastewater Infrastructure. CBO, Washington.

Merriam-Webster's Collegiate Dictionary. 2006. Merriam-Webster Inc., Springfield, Mass.

NCPWI (National Council on Public Works Improvements), 1987a. The Nation's Public Works: Report on Water Supply. NCPWI, Washington.

NCPWI, 1987b. The Nation's Public Works: Report on Wastewater Management. NCPWI, Washington.

USEPA (US Environmental Protection Agency), 2002. The Clean Water and Drinking Water Gap Analysis. USEPA, Washington.

WIN (Water Infrastructure Network), 2000. Clean & Safe Water for the 21st Century. WIN, Washington.

ABOUT THE AUTHORS

John Cromwell III consults for utilities and performs policy research for Stratus Consulting Inc., in Washington, D.C. He has bachelor's degrees in biology and economics as well as a master's degree in public policy analysis, all from the University of Maryland. Elisa Speranza is vice-president of CH2M HILL and global team leader for Utility Management Solutions. Haydn Reynolds is a leading authority in the Australian water industry on improving the strategic integration of asset management and capital investment, including the assessment of asset lives based on life-cycle cost concepts.

Workforce Development and Knowledge Management in Water Utilities

BY NEIL S. GRIGG
(*JAWWA* September 2006)

A distribution system manager of a large water utility recently witnessed a living example of critical knowledge and skills loss when two of his three supervisors retired on the same day. Both supervisors had worked for the city for about 30 years each—a combined experience of more than half a century. Their retirements constituted a serious loss of knowledge for the utility's distribution section, a capital- and maintenance-intensive area that generally represents two thirds of a utility's asset value and usually employs 40% of a utility's workforce. Technical knowledge is also held by water utility employees in treatment, source water systems, and staff support. Taken together, the "brain trust" of the nation's water providers is concentrated in approximately 200,000 employees nationwide, and two thirds of these hold technical positions requiring specialized knowledge, skills, and abilities (AWWA, 1996). Significant water utility knowledge also resides in the consulting firms, contractors, and suppliers who constitute a "shadow" brain trust.

In the coming years, US water providers will need to not only replace employees who retire but also add technical personnel. On the basis of such factors as demand from new housing and tighter regulations, it is predicted that water and wastewater systems will be the only growing segment within the utility sector. It is also forecasted that job prospects will be promising for qualified individuals because the number of applicants in this field is normally low (DeNileon & Stubbert, 2005). The positions most likely to be affected by retirement are in the areas of plant operations, distribution and collection field maintenance, administration (nonmanagement), customer service, line supervision, meter reading, engineering, and plant maintenance (DeNileon & Stubbert, 2005).

How can a water utility undergoing such rapid change attract and use an effective workforce? How can it stem the loss of institutional knowledge and transform itself into a learning organization that retains knowledge? The answer lies in a two-pronged strategy that integrates the organizational areas of leadership, training, and human resources with emerging information and knowledge management systems that can be used to better manage assets and improve workforce performance.

This article reviews the problem of knowledge loss in utilities, assesses the responses available to them, and recommends an integrated framework to adapt emerging information and knowledge management systems within existing management systems. It includes a synthesis of the workforce and knowledge management literature, a review of recent Awwa Research Foundation (AwwaRF) projects on knowledge management and workforce initiatives, and a summary of a focus group session on workforce and knowledge management at an AWWA section meeting.

This time of challenge for the water community offers benefits as well. Utilities can use these workforce and knowledge transitions to improve their management. They can right-size without layoffs, access new ideas, optimize management and training, and improve their documented knowledge base (Olstein et al, 2005). Individual employees can also learn about organizational needs with an eye toward improving their own job performances.

UTILITY WORKFORCE AND WORK ENVIRONMENT ARE EVOLVING

As it has emerged during the past two decades, today's workforce "crisis" stems from a convergence of demographics, complexity of work, and new technologies. AWWA surveys and government statistics estimate the size of the water utility workforce at about 200,000 employees working at some 50,000 utilities (USEPA, 2002). If two thirds of these employees are in technical positions, then approximately 132,000 positions are at risk for knowledge loss.

Baby Boomers trigger demographic change. As in other industries, the demography of the water industry is driven by aging Baby Boomers who were born soon after 1946 and are now reaching their sixties. Many of these workers joined utilities some 30 years ago and are retiring at rapid rates. Most of the new workers replacing them were born after 1980.

The typical boomer technical worker in a water utility was hired in the 1970s and likely is male and white. With water utilities representing one of the last "lifetime employment" opportunities, this worker may have enjoyed a stable 30-year career. As a result, the water utility workforce is older and does not reflect the gender diversity found in industries that experienced higher employee turnover. Furthermore, water utilities have found it hard to attract women to engineering and technical jobs, positions in which they are underrepresented in other sectors as well. Although water utility workforces have about the same ethnic and racial diversity profile as similar organizations, white males still hold the key jobs (Herman et al, 2003).

Work environment is changing too. Today's employment environment is more turbulent than that of the 1970s. It is characterized by more global competition, outsourcing, mobility, and uncertainty (Judy & D'Amico, 1997). Advances in technology have made work tasks more complex at almost every level and in nearly every industry. Within water utilities, the core processes of providing safe water—source, treatment, and distribution—have become "high-tech" operations, and today's water utility workers must understand and adapt to scientific discoveries, ever-changing information technologies, and new systems applications.

These complexities are evident in the concerns of water utility managers. In a 2005 survey, some 1,700 of them reported that their top five concerns were regulatory issues, business factors, source water supply, security, and water storage/infrastructure. Workforce factors cited were aging workforce and loss of brain trust, rising skill requirements, and inadequate industry incentives (Runge & Mann, 2005).

To meet these challenges, utilities require specialized skills that incorporate new technologies and meet tighter regulations. Although some knowledge of engineering, science, operations, and maintenance can be acquired in formal programs, much of this expertise comes from on-the-job training and experience in the water industry and does not transfer readily from other industries. Within this environment of rapid change, the mostly publicly owned utilities face cost pressures that put them at a disadvantage and make them less likely to be considered an "employer of choice." Faced with similar workforce problems, private sector firms can adapt by merging, introducing new products, or outsourcing work, but water utilities require a stable technical workforce and must carefully determine the right combina-

tion of in-house capacity and outsourcing.

As water suppliers adapt to address these challenges, they may find organizational change difficult to implement (Westerhoff et al, 1998). Organizationally, utilities are often slow to change, rely on paper operations and maintenance (O&M) documents, suffer from "organizational amnesia," and have limited succession plans (Smigiel et al, 2006; Olstein et al, 2005). Because of cost pressures and downsizing, training is often inadequate. Furthermore, the social contract with water utility workers is changing, and utilities may not be able to offer the same job security as in the past. Factors working against water utilities include defined benefit retirement programs, lack of advancement potential in technical positions, and pressure to hold costs down (Herman et al, 2003).

IMPLEMENTING AN INTEGRATED FRAMEWORK PROMOTES KNOWLEDGE RETENTION

To respond to these challenges, utilities are overhauling their management systems and exploring new concepts. As shown in Figure 1, four management systems (organizational development, human resources, training, and leadership development) focus on the workforce component of improvement, and two (information and knowledge management) focus on the technology component. When managers nod at the statement "Knowledge retention is not just about technology," they are implicitly acknowledging the workforce element.

By taking into account both the workforce and the technology components, utilities can create integrated strategies to deal with knowledge loss. By adopting integrated strategies, utilities are basically admitting that there is no technological "magic bullet" to solve the problem and that improvement requires hard work on management systems as well as innovative methods and technologies. The following sections discuss the six management systems in Figure 1 and the roles they play in knowledge retention.

Organizational development builds on strengths, overcomes weakness. Technology may be alluring, but the key factor in knowledge retention is how well an organization functions. The field of organizational development spans topics ranging from technical questions of structure to employee behavior. From a practical standpoint, the issue is how to create organizational systems that provide the kind of incentives and work environments that encourage employees to do their best, succeed, and create value. An organizational development program should "enable" individual workers, while facilitating workforce improvement and knowledge management.

Much has been learned about organizations in the information age. Older, hierarchical organizations in which information does not move well are out of date. To respond to the need for organizational improvement,

FIGURE 1 Integrated approach to workforce and knowledge management

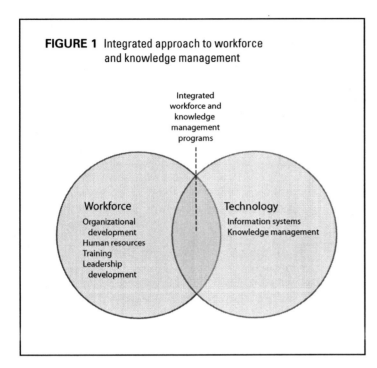

Integrated workforce and knowledge management programs

Workforce

Organizational development
Human resources
Training
Leadership development

Technology

Information systems
Knowledge management

utilities need to become "flatter," right-sized, leaner, and more responsive. The question is how to make this happen? The first step toward improvement is an assessment of organizational strengths and weaknesses.

Auditing an organization for effectiveness in communication and knowledge management requires standards to measure against. AWWA's QualServe program offers three tools to help: self-assessment, peer review, and benchmarking (AWWA, 2005). Organizational development is one of QualServe's four business systems related to water supply; (the others are business operations, customer relations, and water operations).

Under the benchmarking program, four categories of performance indicators were developed. The first is the organizational best practices index that measures seven management practices: strategic planning, long-term financial planning, risk management planning, optimized asset management, performance measurement, customer involvement, and continuous improvement. The second—the employee health and safety severity rate—measures workdays lost per employee per year. The third indicator tracks training hours per employee. The fourth indicator focuses on productivity and measures customer accounts per employee and water delivered per employee. Other public service organizations have similar programs. For example, the American Public Works Association has an agency accreditation program that measures organizational effectiveness.

Organizational assessment programs require more in-depth approaches to measure knowledge retention. Self-assessment or peer review programs might ask these types of questions:

- Does the organization's leadership understand workforce capacity and knowledge programs?
- Are succession plans in place?
- Do information management systems capture knowledge and make it accessible, or does "organizational amnesia" occur?
- Are training and workforce capacity-building programs in place and functioning?
- Does the organization have a working knowledge management system?

Human resources management contributes to building workforce capacity. The human resources staff should be integrally involved in a knowledge retention strategy. Workforce capacity-building is a significant role for the human resources department, which deals with work design, employee recruitment, training, compensation, evaluation, retention, and other activities.

As a field, human resources has expanded its reach and now includes knowledge management in its portfolio, shared with other organizational functions. "In addition to managing traditional human resource functions, [human resources] is expected to . . . continuously improve the company's return on its greatest asset . . . its people" (NHRA, 2005). By working on the team to audit the organization's management systems, the human resources department will gain an awareness of the problems faced and will be able to better respond to improve workforce capacity-building and knowledge management.

Following an organizational audit such as QualServe, it may be necessary to redesign jobs and create new communication patterns that will facilitate knowledge capture in the organization. For example, if it is proving difficult to attract trained workers into complex jobs (such as distribution system renewal), the human resources department might recommend a special training program.

Training is fundamental to capacity development. As a resource for capacity development and knowledge retention, training programs should involve not only human resources staff but also managers across the utility. The training responsibility of every supervisor should be recognized and evaluated. Training is a broad topic, and although everyone generally agrees that training is necessary, it is also easy to defer and ignore.

One way to look at training tasks and status is through the lens of the national organization of "trainers," i.e., the American Society for Training and Development (ASTD), whose vision is to be a worldwide leader in workplace learning and performance (ASTD, 2005). The goals of training embrace capacity development and knowledge retention to bring out the best in individuals and organizations, as evidenced by ASTD's "manifesto": "The ability to learn, and of those who know how to convert that learning into practice (performance), creates extraordinary value for individuals, teams, and organizations. Smart organizations recognize that a learning and performance plan is as much a strategic tool as a marketing or finance plan and that it should get the same kind of tough love from the top: insistence on results and full support if it can deliver" (ASTD, 2002).

Given the knowledge economy and the emergence of new concepts that link people, learning, and performance, ASTD sees the focus of trainers as developing people. It believes that these changes are reinventing whole industries, making old business models obsolete, and turning antiquated thinking upside-down. According to ASTD, the measure of success is not how much you invest in learning and performance improvement, but how you do it and how quickly you move a workforce to demonstrated competence. The society also acknowledges that the focus is not on technology but on the ways in which technology serves learning and performance needs. As appealing as these visionary training concepts may be, the main challenge to water utilities is how to implement them in the real world of utility operations.

Leadership participation encourages employee buy-in. The attributes of leadership that may most enhance utility workforce capacity and knowledge management systems are genuine interest and involvement in the issues and programs. It is natural for leaders and supervisors to focus on daily crises. However, leadership's real interest and concern about the longer-term issues of capacity development and knowledge retention are critical determinants of a utility's success and emergence as a knowledge-enabled organization. Without genuine interest and involvement on the part of leadership, employees may doubt their supervisors' support for the program, which undermines employee support as well.

Succession planning deals with preparing the next generation of leaders and supervisors to take over from current ones and is especially important in the water industry (Olstein et al, 2005). Rather than viewing succession planning as an issue for top leadership alone, the knowledge-enabled utility will have succession plans in place for key positions throughout the organization.

Information is key to technology component of knowledge retention. A critical part of utility knowledge retention, effective information systems address business processes in administrative areas as well as in technical areas of source, treatment, and distribution. Most large utilities are becoming data-centered and use software packages to manage their information. In addition, utility information management systems include work management software to guide employees in structured tasks, such as scheduled maintenance management or work orders.

A utility's information systems can include tasks of data management, records management, and library services—fields that are merging into an integrated sector of information management. Through the information systems, the utility can identify, capture, and manage its explicit knowledge. It can also use the systems to manage tacit (or informal) knowledge once a method to capture and process it has been developed.

The technical functions of a water utility—source, treatment, distribution, and engineering—require a relatively common information architecture that comprises facility and system maps, inventory data, engineering drawings, system O&M manuals, equipment manuals and shop drawings, preventive maintenance schedules, maintenance histories and inspection data, operating and performance records, case and incident files, technical studies, regulatory and legal documents, and other technical background documents. Even this partial list of the information architecture is sufficient to demonstrate that an effective document management system to address these data will be essential for retaining knowledge in the utility.

Knowledge management structure includes more than information systems. "Knowledge management" is an organizing concept for a range of tools and techniques and comprises both a framework for the tools and the tools themselves (including methods and software). One way to view knowledge management is as a new set of tools that enables organizations to learn and make better use of the knowledge available to them. The development of integrated information and knowledge management systems will lead naturally to smart asset management tools.

As a broad concept, knowledge management emerged from the theory of knowledge (epistemology) and has merged with information science. Management information systems and decision support systems were added, and later systems and cognitive sciences were merged into the field to create an integrated field of "knowledge management." Knowledge management adherents include a wide array of researchers, managers, management consultants, and vendors. Because knowledge management has many participants, opinions on its definition vary (Rosen et al, 2003).

In simple terms, knowledge management can be defined as a process that makes important information and experience available for utility employees to use in their jobs. A recent AwwaRF project report provides a more complex definition, describing knowledge management as "a business strategy by which a water utility consciously identifies, captures, indexes, manages, and stores experiences, data, and information and provides methods for easily accessing and acting upon these collective assets (corporate history) in a collaborative environment (learning culture) optimizing the use of (leveraging) people, processes, and technology in support of: effective decision-making, assuring compliance, improving performance, innovation, and business continuity, all on a timely and sustainable basis" (Smigiel et al, 2006). This definition is reflected in Figure 2, which shows several facets of knowledge management and illustrates how the organization creates knowledge, to be managed within the knowledge management system, which can be used by employees to advance organizational work.

A clear understanding of knowledge management helps avoid confusion. The author evaluated the portfolios of several firms that practice knowledge management and developed a list of systems, methods, and tools associated with the concept. The result was a compilation of seven relatively independent categories of tools and methods, followed by examples.

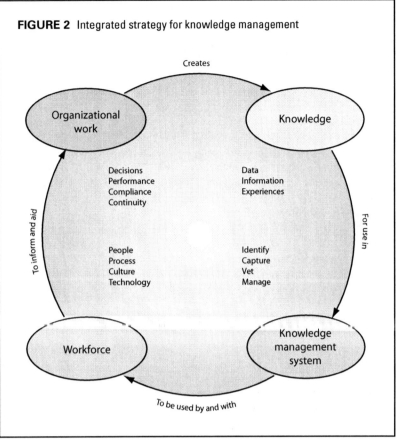

FIGURE 2 Integrated strategy for knowledge management

• Analysis and synthesis tools (e.g., case-based reasoning, meta analysis, scenario planning, social network analysis, knowledge mapping)

• Communications and relationships (e.g., peer mentoring, brainstorming, collaborative technologies, communities of practice, conferencing)

• Information systems (e.g., document management, records management)

• Learning systems (e.g., distance learning, web-based learning)

• Management systems (e.g., intellectual property systems, best practices, project management, workflow management, digital asset management)

• Software (e.g., artificial intelligence, expert systems, knowledge-based systems, knowledge-based decision support, creativity software, data analysis and management, groupware systems)

• Web knowledge portals and systems (e.g., web systems, intranets, knowledge portals)

Within these seven areas and management systems, communication occurs among people and technologies. Information, learning, and analysis systems provide support for the knowledge systems.

To illustrate how these systems can work, imagine an online book publisher who learns from customer orders to market books directly to customers according to their reading preferences. Similarly, a utility's

knowledge management system would take into account the information that is important to its success in customer service, regulatory compliance, and asset management. In the utility, the knowledge required comprises data on source of supply, treatment history and outcomes, customer demand, infrastructure, finances, and administrative systems, among others. This information is found in statistics, records, communications, documents, drawings, reports, and manuals, as well as the tacit knowledge and experience of employees.

Much of this knowledge can be included in the formal information systems, although another approach is needed to capture the tacit knowledge and know-how. The management system required must document tacit knowledge (locations, histories, cases), vet and organize it, make the tacit knowledge available, and create a system to update and improve tacit knowledge.

The industry's understanding of these techniques and systems continues to evolve, and AwwaRF is supporting projects to document utility knowledge management systems (Rosen et al, 2003). One tool that offers special promise is the community of practice. When utilities band together to offer solidarity and training to groups of workers facing the same challenges, they foster rapport, mutual aid, and joint learning. AWWA sections are an effective way to create such communities of practice, and managers should encourage and support employee involvement in them.

Focus groups work to ask and answer the right questions. In an AwwaRF project on knowledge management, a workshop group emphasized the need for practical approaches. It found that research was needed to pinpoint strategies for knowledge retention, integrate them with core management processes, pilot a community-of-practice website, advance organizational development, and develop supports such as resources and infrastructure, better information systems, and standard definitions (Smigiel et al, 2006).

Consistent with the need for a practical approach, the October 2005 meeting of the Rocky Mountain Section AWWA Distribution Systems Committee solicited participants' ideas on workforce capacity and knowledge management by posing this challenge: "As they lose employees, water utilities lose institutional knowledge and need workforce and knowledge retention strategies to perform well in the future. What should those strategies be?"

Some 90 supervisors and line employees were divided into six focus groups and asked to respond to specific questions:

• How does the individual focus group topic (e.g., human resources policies) relate to the problem (i.e., strategies for workforce and knowledge management to stem loss of institutional knowledge when key employees leave)?

• What should the mission statement be for this area so that the utility responds to the problem of knowledge loss in the area? An example mission statement might be "to develop better training programs to capture knowledge and prepare employees to use technology."

• What should the utility's plan of action be to respond in this area? Examples include interviewing and recording knowledge of key employees as they leave or posting knowledge resources on the Internet.

• What are the roles and responsibilities to implement this plan of action? In other words, who in the utility should take on the assignment, what approvals are needed, and who provides the resources?

• What are the main barriers expected? Examples include complacency and resistance from the budget office.

• What is the plan to overcome the barriers?

• What benefits will accrue to the utility and its customers if the action plan is implemented?

Although the focus groups were not able to address all of the questions, they reported a number of insights. Most of the discussion in the groups pertained to leadership and the leadership support needed to set priorities, implement budget and training, and overcome obstacles. Good practices are a necessity. Leaders must help employees take pride in their work to ensure safe drinking water. They must be willing to educate those next in line and not fear training the individuals who will replace them. Supervisors should plan and offer vision. They must emphasize listening, respect, and recognition of their employees. Dissemination of information and positive reinforcement are important. Leaders must make it their job to convince managers and boards of the importance of these issues.

Organizational dynamics also attracted considerable discussion. The groups emphasized that utilities must be well organized and that there should be up-and-down understanding as well as cross-functional understanding throughout the organization. To encourage knowledge transfer, organizational charts need to be flattened and emphasis placed on healthy relationships among coworkers. Obstacles such as resistance to change, fear of failure, and fear of reprimand need to be recognized and overcome. Any strategies to address these and other issues must take into account the differences between large and small utilities.

In discussions of individual responsibilities and responses, incentives attracted significant attention across the board as a way to move employees to perform their required work, share information, and participate in cross-training. Salaries and benefits must be set at attractive levels and jobs structured to give employees more responsibility and ownership of their work. Every position and contribution needs to be valued, not just the "glamour jobs." Incentives can also take the form of opportunities that advance employee capacity, such as training on new equipment and technologies. Certifications are important. Employees like to feel good about their work and think that utilities can do a better job on recruitment by such strategies as interviews with satisfied customers and broadcasts on community-access television showcasing water utility work.

Participants agreed there is a need for training and methods, equipment, and standards but identified lack of time and money as obstacles. The groups recognized that attitude is important and saw value in such employee incentives as getting off work early. Participants also cited such training methods as cross-training, vertical training, and use of apprenticeships and mentoring to disseminate the knowledge of more experienced employees.

Predictably, knowledge management tools and techniques attracted the least attention because many participants were not familiar with them, at least when these were termed "knowledge management." Participants emphasized the importance of the utility's information systems, software, mapping, and data systems, as well as sufficient funds and management systems (including training) to utilize these technologies.

CONCLUSIONS

Loss of critical knowledge, skills, and abilities in water utilities is caused by key employees leaving at the same time that systems and infrastructure are aging, regulatory requirements are increasing, and budgets are under stress. With knowledge and experience walking out the door, utilities are suffering organizational amnesia at a time when they most need that knowledge and experience.

Other organizations have similar concerns, but water utilities face unique obstacles because of the knowledge-intensive requirements of their work, particularly at the critical level of technical supervisors. For small utilities, workforce and knowledge retention are serious problems and must be tackled along with other issues such as management systems and workforce capacity.

Although the field of knowledge management is still evolving, the state of the art is sufficient to create a workforce and knowledge model for large and small utilities. This workforce and knowledge model includes ongoing management systems (organizational development, human resources, training, and leadership development) and emerging information and knowledge management systems. The model must combine human resource policies, management systems, and knowledge initiatives—all elements that constantly evolve. Therefore, the workforce and knowledge model can provide only a snapshot within a general, unchanging framework.

How well the organization functions is perhaps the key factor in workforce capacity and knowledge retention. Audits can help organizations assess their strengths and weaknesses in these areas. However, more in-depth approaches than are currently available are needed to measure workforce capacity and knowledge retention. Questions that might be asked in assessment programs include whether the organization's leadership understands workforce capacity and knowledge programs, whether succession plans are in place, whether information management systems capture knowledge and make it accessible, whether training and workforce capacity–building programs are in place and functioning, and whether the organization has a working knowledge management system.

The field of human resources now sees knowledge management as part of its responsibilities, and human resources staffs represent an integral force in knowledge retention programs and workforce capacity-building. After an organizational audit is completed, it may be necessary for the human resources department to take the lead in recommending special training programs, redesigning jobs, and creating new communication patterns to facilitate knowledge capture within the organization.

Training programs can be a powerful tool for building workforce capacity and knowledge retention. The main challenge for utilities is how to implement effective training programs in the real world of daily pressures. A starting point is for leadership to recognize and commit to training efforts. Programs should involve managers across the utility, and the training responsibility of every supervisor should be emphasized.

One critical determinant of a utility's success as a knowledge-enabled organization is whether utility leaders express genuine interest and concern about capacity development and knowledge retention. Without real interest, employees may doubt whether utility leadership is supportive of their efforts or even interested in their work.

Most large water providers are becoming "data-centered" and use software packages to manage their information systems, which are used more and more to guide structured work and retrieve data and information. The utility's data management, records management, document management, and library systems can be used to identify, capture, and manage its explicit knowledge. Once tacit knowledge is captured, it can be entered into the information systems as well.

The utility's knowledge management structure reaches beyond information management to furnish a new set of tools to enable organizations to learn and make better use of explicit and tacit knowledge. Opinions on the definition of knowledge management vary, and the term serves as a catchall for an array of tools and methods. Simply put, knowledge management is a process to make important information and experience available for utility employees to use in their jobs. This article cited seven categories of knowledge management tools to manage people and technology and facilitate communication among people and between technologies. The knowledge systems that involve people and technology are supported in turn by information, learning, and analysis systems. Much of the utility's knowledge is reflected in statistics, records, communications, documents, drawings, reports, manuals, and the tacit knowledge and experience of employees. Once captured, many of these data can be managed through the utility's information systems, but it may not be practical to transfer all documents to electronic form.

The findings in the article—particularly the findings regarding improving organizational management and leadership—were confirmed during focus group sessions at a workshop on knowledge retention. Participants emphasized the need for flatter, more-responsive organizations; leadership that overcomes fear and instills confidence in employees; effective human resources policies; and the tools and training needed to accomplish goals despite such barriers as ineffective leadership, complacency, and resistance to change. Above all, the focus groups stressed the importance of incentives as a way for utilities to spur individual workers to perform up to their potential.

Large and small water utilities can effectively address their workforce and knowledge retention challenges. The first step is to develop a workforce capacity and knowledge retention plan that includes an organizational development audit, involvement of the human resources staff in review of job design, review of the organization's training programs, and leadership awareness to create genuine interest and involvement in workforce capacity and knowledge management systems. The plan should also provide succession plans for key positions throughout the organization and assessment of information and knowledge management systems to determine their adequacy and areas needing improvement.

Some best practices have emerged to point the way toward knowledge-enabled performance:

• Create a lean, flat, and responsive organization that learns and shares knowledge as it accomplishes the utility mission.

• Develop human resources policies that develop employee capacity and provide the necessary incentives to recruit and retain quality employees.

• Ensure that utility leadership at all levels appreciates the value of the workforce and its need for learning and knowledge retention.

• Acquire and use management systems based on information technology to make the organization's explicit knowledge available to all.

• Develop training programs that promote continuous improvement and knowledge retention.

• Develop knowledge management systems that capture and manage both explicit and undocumented employee knowledge.

In developing knowledge management systems, the concept of community of practice deserves further support among utilities as a means of sharing knowledge, tools, and methods. Managers should take the lead in fostering communities of practice and encouraging employee participation in them.

Although workforce capacity and knowledge retention problems exist at higher management levels, supervisory-level capacity is more at risk because it requires a greater number of employees and specialized technical knowledge, i.e., job-specific knowledge of source, treatment, and distribution. In a sense, this level is similar to the military's noncommissioned officer corps; it retains much of the tacit knowledge that the organization needs to operate.

Problems of workforce capacity and knowledge retention in water utilities can be solved by implementation of existing management and organizational tools, but only if utilities act on what they know and use the tools available to them. Otherwise, water suppliers will relive the story of the county extension agent advising a farmer on how to plant. The farmer replied, "Don't bother me with that—I know how to farm better than I do."

ABOUT THE AUTHOR

Neil S. Grigg is a professor in the Department of Civil Engineering, A205G Engineering, Colorado State University, Fort Collins, CO. He has a BS degree from the US Military Academy at West Point, N.Y., an MS degree from Auburn University, Auburn, Ala., and a PhD from Colorado State University in Fort Collins. Grigg has 20 years of experience in infrastructure management and more than 30 years of experience in public works management.

REFERENCES

ASTD (American Society for Training and Development), 2005. About ASTD. www.astd.org (accessed Nov. 7, 2005).

ASTD, 2002. Leading the Learning Revolution: A Manifesto for the Whole Community of Learning and Performance Professionals. www.astd.org/astd/About_ASTD/manifesto.htm (accessed July 2006).

AWWA, 2005. QualServe. www.awwa.org/Science/qualserve/ (accessed Nov. 7, 2005).

AWWA, 1996. Water:\Stats. Denver.

DeNileon, G.P. & Stubbert, J., 2005. Employment Outlook Good for Operators, Grim for Utilities. *Opflow.* 31:5:1.

Herman, R.; Olivo, T.; & Gioia, J., 2003. *Impending Crisis: Too Many Jobs, Too Few People.* Oakhill Press, Winchester, Va.

Judy, R.W. & D'Amico, C., 1997. *Workforce 2020: Work and Workers in the 21st Century.* Hudson Institute, Indianapolis.

NHRA (National Human Resources Association), 2005. About NHRA. www.humanresources.org/about_main.cfm. (accessed Nov. 2, 2005).

Olstein, M. et al, 2005. Succession Planning for a Vital Workforce in the Information Age. AwwaRF, Denver.

Rosen, J. et al, 2003. Application of Knowledge Management to Utilities. AwwaRF, Denver.

Runge, J. & Mann, J., 2005. State of the Industry Report, 2005. *Jour. AWWA,* 97:10:58.

Smigiel, D.M.; Sulewski, J.G.; & Moss, M.A., 2006. A Knowledge Management Approach to Drinking Water Utility Business. AwwaRF, Denver.

USEPA (US Environmental Protection Agency), 2002. Clean Water and Drinking Water Gap Analysis. www.win-water.org/win_reports/gapreport.pdf. (accessed Nov. 4, 2004).

Westerhoff, G.P. et al, 1998. *The Changing Water Utility: Creative Approaches to Effectiveness and Efficiency.* AWWA, Denver.

2005 Update on Water Industry Consolidation Trends

THE US WATER INDUSTRY CONTINUES TO EXPERIENCE A DRAMATIC REARRANGEMENT OF OWNERSHIP AND AN INCREASING LEVEL OF CONSOLIDATION—A TREND THAT HAS BEEN PERHAPS MOST PRONOUNCED IN THE TREATMENT EQUIPMENT SECTOR.

BY STEVE MAXWELL
(*JAWWA* October 2005)

The ongoing consolidation trend in the water industry has in turn had an effect on the drinking water utility sector of the business—as vendors undergo changes in ownership and evolve their strategic directions. These consolidation trends were reviewed in this column in May 2004, and the evolving growth strategies of the major equipment and service companies in the industry were discussed. In this article, the events and transactions of the past several months will be reviewed, and the earlier discussion of industry consolidation will be updated.

The current game of "musical chairs" in water company ownership really began in earnest in 2003 and early 2004. Many of the big European firms, who had acquired their way into the US water industry during the late 1990s, began to shift direction and started to exit various businesses. Veolia, Suez, and some of the large British water companies led the way in this second phase of water transactions, selling off equipment businesses in order to refocus toward their historical businesses on the operating services side of the industry. This trend has continued, but in addition, many smaller, independent companies are also being snapped up by a new generation of acquirers and consolidators in the water business.

GE VERSUS SIEMENS

As mentioned in the May 2004 column, the US water treatment industry is increasingly being concentrated in the hands of a few large industrial conglomerates. Primary among this group is General Electric (GE), which has acquired a series of major water companies over the past four years. Late in 2004, it acquired one of the largest remaining independent players in the business—Ionics Inc.—a major treatment equipment and services firm. As with many of the companies recently acquired in this industry, Ionics traded hands at a very high multiple—almost three times annual revenues and by some estimates as much as 20 times earnings before interest, taxes, depreciation, and amortization (EBITDA). Many observers cannot understand how acquirers can

afford to pay such steep premiums for water companies. However, high prices are clearly the norm, and industry leaders such as Jeff Immelt, GE's chief executive officer, continue to publicly declare that water represents one of the primary growth opportunities in all of American business. GE now has a water business with more than $2 billion in annual sales (GE Infrastructure, Water & Process Technologies). It also has a big appetite and is in an excellent strategic position to capitalize on future growth opportunities in the industry.

In the earlier review of consolidation trends, the big question was who would step forward to acquire the extensive and far-flung water businesses of USFilter. Many observers assumed GE would acquire this business too, but in the end, it was GE's ubiquitous international competitor—Siemens of Germany—that ultimately bought the USFilter business. With that single strategic move, Siemens became one of the major players in the international water business—and the company has indicated an interest in further expansion. Siemens and GE are both at the top of the list of prospective purchasers in virtually any deal that comes onto the water market.

OTHER PLAYERS SURFACE

Danaher Corp. also emerged as a key player late in 2004 with its acquisition of UV treatment provider Trojan Technologies Inc. Although Danaher has for years been the most aggressive acquirer of businesses in the monitoring and instrumentation side of the water business, this acquisition was the company's first major step into the water treatment and purification business—perhaps signaling a broader appetite in the future. Canada-based Trojan, although a leader in ultraviolet technology, had experienced several years of variable financial performance; nonetheless, this transaction also traded at a very high multiple.

Another new, although not totally surprising, entrant into the water industry this year was 3M Corp., which recently finalized its acquisition of CUNO Inc.

TABLE 1   Key recent water transactions

Seller	Revenue $ millions	Buyer	Date	Valuation $ millions	Multiple of Revenue	Multiple of EBITDA
CUNO Inc.	397	3M Corp.	05/05	1,300	3.3	20*
Ionics Inc.	445	GE Infrastructure, Water & Process Technologies	11/04	1,228	2.8	24.7
Trojan Technologies Inc.	100	Danaher Corp.	09/04	185	1.9	>30*
Culligan Water	682	Clayton Dubilier & Rice Inc.	07/04	610	0.9	7.0*
WTC Industries Inc.	34	CUNO Inc.	05/04	110	3.3	15.0
Veolia—USFilter	1,200	Siemens	05/04	993	0.83	10.0*
Isco Inc.	65.6	Teledyne Technologies Inc.	04/04	80	1.2	12.1
WICOR Industries	750	Pentair Inc.	02/04	874	1.2	10.0

EBITDA—earnings before interest, taxes, depreciation, and amortization

*Estimate

(previously an independent and publicly traded filtration products company). In contrast to many of the other recent sellers in the industry, CUNO had a track record of strong financial performance. Consequently, it also sold at a very high multiple—much more than three times revenues. CUNO's water filter product lines are expected to complement the various existing air-filtration businesses of 3M.

Table 1 lists the buyer and seller in several recent transactions, as well as basic information about the terms of each deal, where available publicly.

Another new name has recently emerged in the water industry—one that few observers expected. Home Depot USA Inc., the retailing giant, has made plans to acquire two businesses in the distribution and infrastructure end of the market—National Waterworks Holdings Inc. and Utility Supply of America Inc. (better known as USABlueBook). Other larger and better-known consolidators in this market, including Pentair Inc. and ITT Industries, were relatively quiet during the past 12 months—perhaps both were focused on the integration of numerous earlier acquisitions that were made in 2003 or early 2004.

It is also worth nothing that private equity investment groups are becoming more significant players in this market as well. Culligan Water was sold in 2004 to the large private equity group Clayton Dubilier & Rice Inc. The giant chemical supplier, Nalco Chemical, was held by a consortium of private equity interests for a few years, although its stock has now been partially returned to the public markets. The National Waterworks Holdings business just acquired by Home Depot had been owned for several years by two major private equity players. Private equity groups are also rumored

to be serious contenders for several other companies currently on the market.

SMALLER DEALS ALSO HAPPENING

Although the bigger companies and the bigger transactions tend to get most of the attention and the press coverage, there are also numerous smaller deals occurring. A couple of these are mentioned in Table 1. Before its acquisition by 3M, CUNO had acquired WTC Industries. In mid-2004, Teledyne Technologies Inc.—another emerging consolidator in this industry—acquired publicly traded Isco Inc., a mid-sized manufacturer of sampling and monitoring equipment, for about $65 million. Several privately held companies in the water sector have also traded hands in the past year or are currently on the market.

With so many major water industry assets changing hands at the same time, the competitive situation in the water treatment equipment industry is a bit up in the air. It will be interesting to watch where various key water treatment assets end up and to see which companies step forward to be major players in the water industry's next generation. Most observers are betting on the various diversified US companies—ITT Industries, GE, Pentair, Danaher—and perhaps several others that have not yet made their first move. The growing participation of equity investors almost surely guarantees that the ownership of many of these assets will continue to change in the future as well. But the big question remains: How will this extensive rearrangement of ownership affect employees, shareholders, competitors, and, finally, customers?

The Consolidation Continues, With GE Leading the Pack

FURTHER CONSOLIDATION WITHIN THE WATER INDUSTRY HAS INCLUDED A FEW WELL-PUBLICIZED ACQUISITIONS AS WELL AS MANY OTHERS THAT HAVE ATTRACTED LITTLE ATTENTION.

BY STEVE MAXWELL
(*JAWWA* May 2006)

The consolidation trend in the water industry continues apace. Since we last looked at things in the October 2005 Market Outlook, there have been several additional large-scale transactions and external investments in the industry. Despite a growing paucity of pure-play water investment opportunities in the United States, general investment interest in the industry continues to boom—and valuations continue to be very high.

The emerging large industrial players (discussed in previous columns) continue to make key moves in their efforts to establish long-term strategic market positions in the industry. Primary among these moves was GE Water & Process Technologies' announcement on March 14 that it would acquire industry-leading membrane manufacturer ZENON Environmental Inc. of Canada. Although not exactly unexpected, this deal shocked the industry, as GE once again stepped up to the plate and paid a very high price to acquire what it perceived as another key asset for its expanding water purification and desalination business. It also marked a strategic shift for GE toward the municipal market. In the past, GE has stated a focus toward the industrial market, but many of ZENON's market opportunities over the long-term future will be in the municipal water and wastewater treatment arena.

GE paid approximately $650 million for ZENON—which was doing about $200 million in revenue and which produced a net operating loss in 2005. This was more than a 50% premium above what ZENON was currently trading for on the Canadian stock exchanges, and many observers felt it was already trading at a very inflated valuation. This sale price translates into more than three times annual revenues and a very high multiple on cash flow. GE has once again demonstrated its huge appetite for and fervent belief in the future of the water treatment and purification business—paying a very steep price (as it earlier did for Ionics and Osmonics) to build a superior competitive position in the membrane filtration business. Company officials have said that they expect to be able to grow their municipal water business by 30% annually in coming years.

OTHER LARGE PLAYERS ALSO MAKE ACQUISITIONS

Another of the industrial players—Danaher Corp.—also continues to make acquisitions in the water treatment and purification arena, although perhaps not on the same scale as GE. Danaher picked up Aquafine Corporation, a relatively small player in the ultraviolet (UV) radiation treatment sector, to complement its Trojan Technologies Inc. unit, which it acquired in late 2004. Danaher indicated that Aquafine would strengthen its product offering to the industrial marketplace and that Trojan would continue to focus on municipal markets.

Other key players in the treatment and purification equipment arena—Pentair Inc. and ITT Industries' fluid technology division—have been relatively quiet on the mergers and acquisitions front for the past two years. ITT has not made any notable acquisitions since its early 2004 purchase of WEDECO AG—the other major competitor to Trojan in the UV treatment market.

As discussed in the January 2006 Market Outlook column, the private investment community continues to scour the water business for attractive opportunities, and there have been two recent and notable transactions in that arena. In January, the well-known investment firm Carlyle Group teamed up with the French company Zodiac S.A. and paid a 30% premium for Water Pik Technologies Inc.—a manufacturer of various health care products as well as pool and spa equipment and systems. Although Water Pik's mix of businesses puts it a little outside the standard company in the water treatment industry, this deal was widely interpreted as further evidence of the private equity community's interest in the broader water services business.

In another significant private equity investment in late 2005, an investment unit of the AIG insurance group (AIG Highstar Capital II LLC) acquired the private water utilities group, Utilities Inc. With some

300,000 customers spread across 17 states, Utilities Inc. is one of the largest privately owned contract operators of water and wastewater utilities in the United States. Although no terms were announced for the purchase (which is still under review by various state public utility commissions), given recent valuations in other deals within the water and wastewater utility sector, it seems likely that this deal also went through at a very steep valuation.

SOME TRANSACTIONS HAVE DRAWN LITTLE ATTENTION

Many other asset-shifting and consolidating transactions have continued to happen at a smaller level—a little below the radar of major media attention in the industry. For example, Teledyne Technologies Inc. quietly continues its efforts to build a significant business in the water treatment and monitoring sector. Following its earlier acquisition of Isco Inc., Teledyne has in the past several months announced the acquisition of three companies: Benthos Inc., a $25 million producer of oceanographic products and security equipment for ports and harbors; MGD Technologies Inc., a small velocity-profiling measurement equipment and service company; and RD Instruments Inc., a $30 million firm focused on the development and manufacture of acoustic doppler water velocity–measurement technology. With these deals, Teledyne has become a major player—and a competitor to Danaher's water quality division, Hach Co.—in the water monitoring and measurement sector of the industry.

Another significant deal that went largely unnoticed in 2005 was the acquisition of Reynolds Inc. by Layne Christensen Co. Reynolds, a major supplier of products and services to the water and wastewater industries, is a 75-year-old, family-owned business producing design–build water and wastewater treatment plants, water supply wells, water intakes, and water and wastewater transmission lines. Reynolds was also one of the largest US providers of cured-in-place pipe services for sewer line rehabilitation. Layne Christensen continues to build a strong position in the water treatment infrastructure market, and the company will now have revenues on the order of $400 million per year.

One result of all of these transactions over the past several years has been a distinct decline in the number of individual water investment opportunities available in the United States. Many of the more sizable, attractive, and pure-play water investment vehicles are gone—having been snatched up by one or another of the big industrial players mentioned earlier. Thus there are very few opportunities left for either a private individual investor or a large strategic buyer. However, a few new opportunities are coming into the public markets. For example, RWE AG just announced that it intends to spin off American Water and turn it back into a publicly traded US company. However, private investors in general must increasingly look to overseas exchanges and markets for investment opportunities. Only 103 of the 359 companies followed by a leading water fund manager, Summit Global Management Inc. of San Diego, Calif., are US companies, and many of those are quite small.

As noted in this column in the past, with so many major water industry assets changing hands at the same time, the long-term competitive situation in the water treatment industry is a bit difficult to predict. It continues to be intriguing to watch the industry's transactional activities—a "chess match" between the various players, with each new move determining the possible future strategic moves of a handful of key consolidators. Also, the growing participation of equity investors almost surely guarantees that many of these assets will change hands again fairly soon.

Economies of Scale in Community Water Systems

BY JHIH-SHYANG SHIH, WINSTON HARRINGTON, WILLIAM A. PIZER, AND KENNETH GILLINGHAM
(*JAWWA* September 2006)

Small water systems face increasingly stringent regulatory requirements under the Safe Drinking Water Act as amended in 1996. According to the National Public Water Systems Compliance Report, more than 54,000 publicly and privately owned community water systems (CWSs) exist in the United States, serving about 252 million people (USEPA, 1997a). Of these, approximately 93% are categorized as "small" or "very small," serving fewer than 10,000 customers (USEPA, 2002). Although these systems serve only 20% of the total population served by all systems, they have received much attention from federal regulators and state and local health officials because they face particular difficulties in complying with federal and state water quality requirements. Because of their size, the technical, managerial, and financial capacities that modern water treatment systems require are often beyond their capabilities. For example, among all the size categories, systems serving 25–500 individuals experience the most violations per 1,000 people served.

Many also believe that supplying customers with water is overly costly for small systems (see Beecher & Cadmus, 2002, for a review of likely mechanisms). At least two distinct kinds of scale economies exist in water supply systems. Capital equipment is the most familiar. There are also scale economies in many ordinary business operations, such as billing, purchasing, and computer systems, as well as in ancillary water treatment and testing operations.

Consolidating water systems—whether through merging smaller systems or through a larger system absorbing one or more small systems—may be a way to reduce the cost of supplying water and to improve the ability of these systems to meet more-stringent regulatory requirements cost-effectively. If consolidation is to achieve scale economies in capital equipment, the systems involved must be geographically close to allow connection to the same water treatment plants. Achieving scale economies in ordinary business operations might not require systems to be physically connected. For example, material costs may show scale economies because large systems may be able to negotiate better

long-term contracts. However, capital and energy costs may be very sensitive to physical connection.

Although the benefits of consolidation are potentially large (Cadmus, 2002; AWWA, 1997), water supply systems remain almost uniquely unconsolidated among all local and municipal services. The reason for this is beyond the scope of this article, but it is time to reexamine the issue, particularly to determine whether policies exist that could enhance consolidation benefits and reduce costs.

In this article the authors use the 2000 and 1995 Community Water Supply surveys (CWSS; USEPA, 2002, 1997b) to examine the potential for achieving reductions in water supply unit costs by increasing system size and consolidating existing systems. To do this, it is key to distinguish system size from other causes of cost variation. System size is only one of many variables that affect water supply cost. Differences in the cost of raw water supply, for example, depend on climate, topography, and geology. The quality of the raw water may affect cost because some raw water supplies will require more expensive treatment than others to meet acceptable health and potability standards. The spatial distribution of the final demand for water will also affect distribution system costs, with higher population densities enabling the fixed costs of the distribution system to be spread over a larger number of accounts. In addition, there may be differences in the efficiency of water supply systems, with some systems obtaining more output than others from the same quantity of inputs. When all else is equal, more-efficient systems will have lower costs. In this work, the authors attempt to separate the cost elements attributable to scale from all other factors to the extent possible, although data limitations make it somewhat difficult.

TWO APPROACHES FOR ESTIMATING PRODUCTION RELATIONS AND SCALE ECONOMIES ARE AVAILABLE

Scale economies postulate that the production cost per unit declines as the volume of production increases, reflecting an increase in input productivity. Different explanations such as topography, indivisibility, organization, and finance can underlie such a phenomenon in other industries (Norman, 1979). Topography reflects the possibility that, for example, the cost of pipes is related to circumference whereas flow is related to cross-sectional area, making it cheaper per gallon to pipe larger quantities. In other instances, a fixed amount of equipment (or knowledge) may be required regardless of volume or in large increments of volume. Organizationally, larger scale allows greater specialization and may result in greater leverage in financial transactions. Scale economies have been observed in industries as diverse as cement manufacture, electricity (Christensen & Greene, 1976), education (Cohn et al, 1989), and banking (Adams et al, 2004).

The production economics literature offers two basic approaches to estimating production relations and scale economies (Coelli et al, 1998). The first approach assumes that all decision-making units (DMUs)—such as firms, plants, and water systems—are technically efficient. Beginning with this assumption, econometric estimation and index methods are then used to study the aggregate technical change, return-to-scale, and optimization rules. This method assumes that the particular DMU is operating on the production possibility frontier, and no further output is technically possible with the given level of inputs. Observed variation along the frontier is assumed to be noise.

A more recently developed approach does not make the assumption of technical efficiency for all DMUs. Here, the goal is to first identify the technology frontier, defined as the maximum output achievable from a given set of inputs. DMUs on this frontier are said to be technically efficient. Then efficiency analysis examines the degree to which other DMUs lie inside the frontier or use a cost-minimizing set of inputs. In the empirical research described in this article, the authors adopt the first approach (although an analysis using the second approach leads to quite similar results). By allowing for flexible scale economies, the efficient scale could be identified (Figure 1).

Researchers studying such relationships have estimated them in a variety of ways. Figure 1 suggests a primal approach—modeling output as a function of inputs. An alternative is characterized by the pioneering work of Christensen & Greene (1976), who instead estimate cost as a function of input prices and scale:

$$\ln C = \alpha_0 + \alpha_Y \ln Y + \frac{1}{2} \gamma_{YY} \ln Y^2$$
$$+ \sum_i \alpha_i \ln P_i + \frac{1}{2} \sum_i \sum_j \gamma_{ij} \ln P_i \ln P_j$$
$$+ \sum_i \gamma_{YI} \ln Y \ln P_i \qquad (1)$$

in which $\gamma_{ij} = \gamma_{ji}$, C is total cost, Y is output, and each P_i is the price of a particular factor input. These researchers apply this approach to understand economies

of scale in US electric power generation. Although Figure 1 shows output per input (the slope of rays through the origin) rising with economies of scale, cost per output should fall with economies of scale. In Eq 1, input prices are used to control for changes both in overall inflation and composition (relative prices). Composition changes are used in the same way that would be necessary in a primal approach through relevant quantities of inputs on the right-hand side.

In this article, the authors adopt a simplified version of the Christensen & Greene (1976) model for empirical analysis. Only the linear terms of the general equation are considered, and the factor prices are ignored because these data are unavailable. The authors do, however, control for the CWS ownership and the type of raw water sources.

USEPA SURVEYS YIELD NECESSARY INFORMATION FOR ESTIMATING SCALE ECONOMIES

Approximately every five years the US Environmental Protection Agency (USEPA) conducts a CWSS, which is designed to obtain data to support the development and evaluation of drinking water regulations. For this project the authors used the CWSS 2000 and CWSS 1995 (USEPA 2002, 1997b). The CWSS database contains detailed operating characteristics and financial information for a large random sample of CWSs. The survey includes questions about annual water production by water source, service population, characteristics of treatment facilities, characteristics of untreated sources, water sales revenues and deliveries by customer category, water-related revenues, water system expenses, and water system assets. Although much information is common to both surveys, each survey also has some unique elements. For example, CWSS 1995 contains more detailed information on the breakdown of water system finances, allowing classification of production inputs into six categories—capital, labor, materials, energy, outside services, and other. For economic analysis, however, this information is not as useful as it might be because expenditures on each of these inputs are recorded, not the quantities and average prices. In any case, this breakdown of cost information is not present in CWSS 2000, and categories of cost are defined differently, making cross-survey comparisons of water system finances somewhat problematic. For example, the past year's capital improvements expenditure is included in the defined value of total water system expenses in the 1995 data set. In the 2000 data set, however, capital improvements expenditures are defined as spending over the past five years and are included as a separate group from total expenses.

CWSS 2000 also contains some questions not present in CWSS 1995, yielding limited information on raw and finished water quality for systems serving more than 500,000 individuals.

FIGURE 1 Productivity and scale economies*

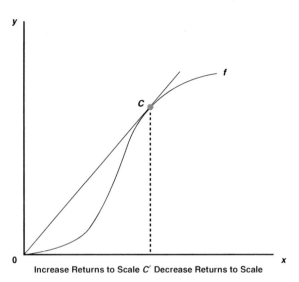

Increase Returns to Scale *C'* Decrease Returns to Scale

*A simple production process in which a single input (x) is used to produce a single output (y). The curve (f) represents the production function, which is the output attainable from each input level (the current state of technology in the industry). A ray through the origin measures productivity at a particular data point. The slope of this ray is y/x (output/input). The figure shows a production relation with initially increasing and later decreasing returns to scale (noticeable by the initially convex and subsequently concave production function). By allowing for flexible scale economies, the efficient scale can be identified (e.g., point C).

TABLE 1   Community water system surveys for systems with population information available

Size Category	Population Served	Number of Systems	
		1995	2000
Very small	25–500	574	336
Small	501–3,300	511	207
Medium	3,301–10,000	270	168
Large	10,001–100,000	361	284
Very large	>100,000	121	251
Total		1,837	1,246

Table 1 defines each category and gives the number of systems in CWSS 1995 and CWSS 2000 that have complete population-served information in each size category.

Sources of raw water also vary among water systems. Table 2 shows the percentage of plants in each size category using groundwater, surface water, and water purchased from other producers. The totals may exceed 100% because some water systems have more than one source type.

CWSS 1995 contains additional detail on the cost of inputs to water water production. Input variables include costs for capital, labor, material, energy, and outside services, along with the "other" category. Depreciation expenses are used as a surrogate for capital costs. Labor costs include direct compensation for salaries and bonuses. Materials costs include expenses for disinfectants, precipitant chemicals, other chemicals, materials and supplies, and purchased water (both raw and treated). Energy costs include expenses for electricity and other energy, such as gas and oil. Outside services costs include expenses for analytical lab services and other outside contractor services. Other costs include other operating expenses, such as general and administrative expenses not reported elsewhere, payments in lieu of taxes, and other outgoing cash transfers. Costs exclude tax payments, which puts the costs of public and private water supply systems on the same footing. The output variable is total water produced in millions of gallons. Total unit cost and factor unit costs are derived using total operating expenses and individual factor costs divided by the total water produced.

CWSS 1995 begins with 1,980 observations. To properly analyze the data set, first the data needed clean-

In the CWSS, water systems are classified primarily according to the number of people served. The full sample sizes for 1995 and 2000 are 1,980 and 1,246, respectively. The population-served information for some systems, however, is missing in the databases.

Article 19

TABLE 2 Percentage of sources using three different types of water

Size Category	Surface Water %	Groundwater %	Purchased Water %
1995 survey			
Very small	30.8	66.2	42.0
Small	23.9	56.9	37.2
Medium	27.8	60.7	31.5
Large	44.0	61.8	33.0
Very large	55.4	55.4	43.8
2000 survey			
Very small	38.7	63.7	21.4
Small	35.7	47.8	26.6
Medium	53.0	46.4	23.2
Large	53.5	45.8	37.0
Very large	67.3	44.2	37.1

ing. The authors eliminated 143 observations that were missing values for the population served; 278 observations that were missing values for total expenses; 86 systems that were either missing a value or recording a zero for total water production (output); 96 observations that were missing a value for ownership class (public or private); 4 systems that were not classified with a public water supply identification number; 1 system that was missing a value for the production source; and 5 outlier systems with impossible costs per unit produced, leaving 1,367 systems for analysis.

For some analyses, the authors also eliminated observations with zero values for the six input factors—553 observations with zero values for capital; 37 for labor; 11 for materials; 41 for energy; 105 for outside services; and 55 for other costs. After the data cleaning, the sample size dropped to 565.

CWSS 2000 contains a subsample of systems also surveyed in 1995. CWSS 2000 is the fifth edition of the survey (USEPA, 2002) and, overall, is much more organized than CWSS 1995. In 2000, investigators asked detailed questions on topics such as treatment technologies, raw water quality, and posttreatment water quality for very large systems and for chemicals such as arsenic and methyl tertiary butyl ether. Similar to CWSS 1995, the 2000 survey collected data on water system operations and finances that are critical to the preparation of regulatory, policy, implementation, and compliance analyses. Unfortunately, CWSS 2000 contains less detailed financial information on drinking water expenses than its predecessor—the detailed

input cost information available in CWSS 1995 is not available in CWSS 2000.

The most interesting aspect of CWSS 2000 is the subsample of systems queried in both surveys. The number of overlapping samples is 132, which is not a huge number but adequate for a simple panel data analysis (panel data means data with multiple observations crossing several years).

To conduct a consistent comparison of the data in the two surveys, information on total expenses, total finished water production, population served, and water source was used.

In the analysis that follows, the authors use three subdata sets of CWSS 1995 and CWSS 2000, depending on the model that's estimated. To summarize these subdata sets, for analysis of costs by type, a sample from CWSS 1995 that contains 565 observations was used. Data are complete for each of the six input factors in these 565 observations. Second, for analysis of total costs, two full data sets for CWSS 1995 and CWSS 2000 with complete data for total operations and maintenance expenses and total water produced were used. The sample sizes are 1,367 and 995, respectively. Finally, for panel data analysis, a data set of 132 system observations that are found in both CWSS 1995 and CWSS 2000 was used.

UNIT COSTS DECLINE AS SYSTEM SIZE INCREASES, BUT THERE'S NO GUARANTEE

The box-and-whisker plot in Figure 2 shows the range of unit production costs (operating costs plus depreciation in dollars per million gallons) for plants of various sizes in 1995 (the graph for 2000 is similar).

Figure 2 gives several interesting insights into production-cost distribution. Most important, unit costs generally decline as system size increases. The 1995 median cost per million gallons for a very small plant is 135% greater than that for a very large plant ($2,653/mil gal versus $1,128/mil gal). However, very little difference exists between the median cost of very small and small plants. The decline in costs with increasing size is also evident in other statistics. The 75th and 95th percentiles also show declines, as does the 25th percentile, at least for small plants and larger. Finally, Figure 2 shows that despite generally falling costs, plenty of overlap remains in costs across all size categories. For example, 20.7% of very small plants and 22.0% of

138

medium plants have a unit cost lower than the median unit cost of very large plants. It can be seen, then, that size does not guarantee lower costs.

To quantify the effect of size and to control for other survey factors that affect costs, we estimate a model that includes data on water source, ownership, and size. Water source variables are designated in the model as "ground," "surface," and "purchased." These categories represent the percentage of raw water that comes from groundwater, surface water, and water purchased from other utilities, respectively. Most utilities in the sample (1,149 of 1,367 in CWSS 1995 and 800 of 995 in CWSS 2000) obtain water exclusively from one type of source. Ownership is a dummy variable indicating whether the system is publicly or privately owned (for comparability, costs exclude taxes paid by private systems). This can be written as

$$\ln C = \alpha + \beta \text{ Public} + \gamma_1 \text{ Surface} + \gamma_2 \text{ Purchased} + \theta \ln W \qquad (2)$$

in which C is annual unit cost in dollars per thousand gallons and W is annual production of finished water (in million gallons). Table 3 shows estimation results for total unit cost. Variables "private" and "ground" are not included in Eq 2. If both the ownership dummies—"private" and "public"—were included in the equation, their sum would be identical to the constant term. To avoid this linear dependency, one of the two must be dropped. Similarly, one of the three ownership variables must be dropped (Studenmund, 2001). The coefficients on the remaining variables represent the differences relative to the base case defined by those two variables. Table 3 indicates that expected costs vary significantly on the basis of water source ("surface," "purchased," and "public" coefficients measure costs relative to groundwater at privately owned systems).

Controlling for size, groundwater systems have the lowest cost; surface water systems are on average 17% more costly than groundwater systems, and use of purchased water is 52% more costly than groundwater. Of course, particular localities may not have unlimited access to different water sources, and it is unlikely that many utilities would be able, by changing water source,

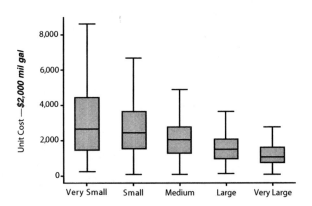

FIGURE 2 Distribution of plant production costs by size*

The line through the box shows the median production costs. The top and bottom edges of the box show the interquartile range (the 75th and 25th percentiles). The horizontal lines (whiskers) above and below the box show the 95th and 5th percentiles.

"Very small" category excludes costs for outside services.

TABLE 3   **Scale economies of total operating expense**

Variable	Number (Standard Error)
Constant	8.34* (0.061)
ln W†	–0.16* (0.009)
Surface	0.17* (0.050)
Purchased	0.52* (0.053)
Public	–0.12* (0.041)
Adjusted R^2	0.46
Observations	565

*Significant to 1% level
†Log change in production

to reduce costs by the amounts suggested by the coefficients in Table 3.

These results also indicate that scale economies exist in the overall operations of these water providers. The coefficients on the size variables (ln W) are actually elasticities. For example, a 1% increase in surface water production reduces unit costs by 0.16%. Although this is statistically significant, the magnitude is fairly modest. On average, for example, a doubling of production volume reduces unit costs by 10%.

The estimate of scale economies was derived by comparing the costs of various-sized systems. These systems may differ in many ways that are not observed, and if those differences are correlated with size, it's possible to attribute their effects on costs to the effect of plant size. Such omitted variables could cause either an over- or underestimate of scale effects. Another way of estimating scale economies is to observe the unit costs of the same water supply system at different production volumes. The advantage of this approach is that it yields direct evidence of the effect of a size change, in contrast to the indirect evidence provided by cross-sectional estimation. In addition, it automatically controls for many system-specific factors that affect costs. The disadvantage is that data are much harder to obtain.

Fortunately, a subsample of 132 water supply systems was surveyed in both CWSS 1995 and CWSS 2000. For these systems, the change in reported unit costs (in constant 2000 dollars) was regressed against the change in production volume. To begin, the authors regress $\Delta \ln C$, the log change in unit costs in constant dollars between 1995 and 2000, against the log change in production $\Delta \ln W$, which gives the results shown in the "Difference between 1995 and 2000" row of Table 4. More generally, Table 4 shows estimates of the elasticity of unit cost with respect to water volume for a variety of model variants of the original equation, including component costs. Source and ownership controls are included and excluded as shown in the table.

Clearly, the largest variation in Table 4 is between the cross-section estimates, using data from a single year, and the difference estimates using differenced data across the two samples from CWSS 1995 and 2000. The

elasticity of unit costs with respect to water volume is much larger in this difference model, more than double any of the nondifferenced estimates. On the basis of the estimated elasticity of roughly 0.47, a doubling of water volume would lower unit costs by almost 30% (e.g., $2^{-0.47} = 0.72 \sim 1 - 30\%$).

Why is the differenced estimate so much larger? Most likely, the short time horizon from 1995 to 2000 implies that capacity and other potentially fixed factors of production are not changing, and it is simply the gains to higher-capacity utilization rather than true scale economies that are being measured. If 0.47 is the true value, it is also hard to imagine what omitted variable would lead the nondifferenced estimates to be smaller—i.e., a variable correlated with size and costs in the same direction. If increased capacity utilization is the real story behind the differenced estimates, that would not occur with a consolidation (or at least not to the degree that is seen with more water being distributed through the same physical system).

In addition to observing total costs, the individual component costs using the more detailed financial data available in CWSS 1995 can also be considered. Figure 3 shows the average unit costs of production for each of these factors by the size (population served) of the system. Figure 3 indicates that the average unit costs of production fall as system size increases for all of the six factors of production, but not necessarily at the same rate. Table 4 quantifies these economies of scale for each component, regressing the component cost per thousand gallons on the log water supplied, with and without source and ownership controls.

It is interesting that the greatest economies of scale exist in the cost categories for capital, outside services, other, and materials. Labor and energy costs exhibit the fewest economies of scale of the six factors. This suggests that larger systems may be relatively better than smaller systems at bargaining for and receiving outside services and materials at lower costs. In addition, it is important to note that these gains do not necessarily depend on water systems becoming physically interconnected.

The smaller elasticity value for labor may imply that larger systems are more economically efficient, as a result of, for example, negotiating contracts and eliminating redundant positions.

TABLE 4	Factor elasticity (with respect to total water produced) estimations* using various models			
Cost Category	Model 1	Model 2	Model 3	Model 4
Capital (1995)	–0.175	–0.176	–0.169	–0.170
Labor (1995)	–0.129	–0.140	–0.122	–0.134
Material (1995)	–0.186	–0.150	–0.188	–0.151
Energy (1995)	–0.120	–0.130	–0.111	–0.122
Outside services (1995)	–0.219	–0.223	–0.204	–0.208
Other costs (1995)	–0.181	–0.181	–0.166	–0.167
Total (1995)	–0.125	–0.116	–0.121	–0.112
Total (2000)	–0.118	–0.130	–0.125	–0.134
Difference between 1995 and 2000	–0.475	–0.468	–0.470	–0.463
Source controls		Yes		Yes
Ownership controls			Yes	Yes

*All estimations are significant to 1% level.

However, the labor economies of scale are not as substantial as those from outside services, other, and materials costs. Similarly, the lower elasticity value for energy costs suggests that larger systems may have to pump water farther per unit delivered, preventing the gains from economies of scale that accrue from reducing other types of costs by increasing water system size.

CONSOLIDATION CAN RESULT IN COST SAVINGS UNDER SEVERAL SCENARIOS

By the use of estimates of the elasticity of unit costs with respect to system output, calculations of the potential cost savings from consolidating all small water systems throughout the United States can be made. From the original model, for each system, the authors used

$$\Delta \ln C = \theta \Delta \ln W, \qquad (3)$$

in which C is unit costs (dollars per thousand gallons) and W is the volume of water supplied. If $\Delta \ln C$ is the change in log unit costs, $WC\exp(\Delta \ln C)$ is the new total cost. Therefore the total cost savings (relative to the observed costs) in the sample is calculated as

$$\sum_i W_i \times C_i \times [1 - \exp(\theta \times \Delta \ln W_i)] \qquad (4)$$

i.e., the old costs minus the new costs. Given θ, the only question is what to assume about $\Delta \ln W_i$ in the experiment on consolidating water systems.

As a starting point, the authors computed the cost savings associated with giving systems below the median supplied water level the scale economy gains associated with moving their output to the median level (304 mil gal). The authors did not assume that the output of these systems would actually rise, but that somehow they would be able to realize those gains.

Carrying out these calculations, the authors found that among the 565 facilities in CWSS 1995 for which detailed cost data are available, such an exercise would realize a $9 million cost savings. Of the total costs of $2.2 billion, this is less than half of 1%. Among the water systems below the median size, however, with total costs of only $67 million, this is a cost savings of 14%. For more than 40 systems, cost savings of more than 50% could be achieved. Table 5 shows total as well as individual component cost savings as a percentage of

total expenditure of water systems below the median size.

The authors also simulated the cost savings for the following three scenarios: (A) Combine small water systems serving <500 people with metropolitan water systems serving >50,000 people; (B) Combine small water systems serving <500 people with medium water systems serving 3,300–10,000 people; and (C) Double the size for the systems serving <500 people. For scenarios A and B, the medians of total water produced in each population group—6,506 and 5,800 mil gal for systems serving >50,000 people and between 3,300–10,000 people, respectively—are used as the targets for scaling up the small systems. For scenario C, two times the current total water produced is used as the target for scaling up. Table 6 gives the total and individual components of cost savings for water systems serving <500 people (using the 565-sample data set) for the three scenarios.

The total cost savings for small systems serving <500 people in the 565 samples are $1.5 million; $700,000; and $280,000 for scenarios A, B, and C, respectively. Generally, the major cost savings are from capital, labor, materials, and other costs. Although costs for outside services have relatively large elasticity because the cost share is small, the cost savings in this category proves to be relatively small. Also, cost savings from energy are small relative to other factors, indicating that on average, an increase in system size saves less money from reduced energy costs than from any of the other types of expenses. One potential explanation for this is that water pumping costs increase as system size increases. On the other hand, the cost categories of materials, outside services, and other do not rely on interconnection to obtain savings. If the

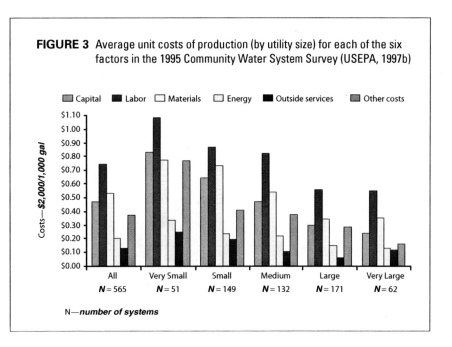

FIGURE 3 Average unit costs of production (by utility size) for each of the six factors in the 1995 Community Water System Survey (USEPA, 1997b)

TABLE 5 Total and individual component cost savings among the water systems below the median size

Cost Category	Cost—%
Capital	2.7
Labor	3.2
Materials	3.3
Energy	0.8
Outside services	0.9
Other costs	2.1
Total costs	13.5

TABLE 6 Total and individual component cost savings percentages for the water systems serving <500 people in the 565-sample data set

Scenario	A	B	C
Capital	12.6	4.3	2.7
Labor	13.7	5.9	2.5
Materials	11.6	5.8	2.2
Energy	3.5	1.9	0.6
Outside services	4.2	2.5	0.8
Other costs	10.9	6.3	2.0
Total costs	56.5	26.3	10.5

Although physical interconnection may be difficult, consolidating administrative tasks and bargaining power may allow water systems to realize some of the potential economies of scale. The variation in unit costs across and within system sizes, however, is large, and size explains only a part of the cost distribution. Viewed another way, a high degree of technical inefficiency in terms of large cost differences exists among systems with the same input structure.

A variety of estimation techniques and cost measures yield broadly consistent estimates. It is apparent that groundwater and surface water are, not surprisingly, cheaper than purchased water. Public systems consistently have lower costs than private ones. Certain cost components also exhibit economies of scale to a greater or lesser extent than others.

By quantifying the economies of scale, it appears that doubling a system's production would lower unit costs between 10% and 30%, depending on the model and the cost component. Consolidating small systems into a large system, however, could double the small system's scale several times over, resulting in gains of 50% or more. The authors' estimates suggest that by giving all systems below the median system size the efficiency gain associated with moving to the median system size, savings of $9 million would be realized in the sample of 565 facilities. Extrapolated to the entire population of roughly 50,000 systems, more than $1 billion could be saved. Similar though slightly smaller estimates exist when considering merging small systems with larger systems.

The majority of cost savings are in the capital, labor, materials, and other cost categories. Although increased efficiencies associated with capital and some labor input (e.g., production versus administration and management) would require interconnection, other efficiency gains would not. To the extent that these gains arise from increased bargaining power, they are all achievable without interconnection.

Compared with overall supply costs, these numbers might be considered small at about 4.2%. To the small and very small systems realizing the gains, though, they are large. Of course, it's not possible to claim that all these benefits could actually be realized. The fact

cost savings are scaled up to the national level using survey sample weights, the cost savings are $1,115 million (all facilities to median size), $794 million (scenario A), $417 million (scenario B), and $140 million (scenario C). To scale calculations to the national level, the savings at each facility is multiplied by the sampling weights used to scale the original sample up to national values. The sampling weights are inflated to reflect the reduced sample size (565) in such a way as to maintain the original distribution of facility size (e.g., across very small, small, medium, large, and very large). Among factors that likely do not require interconnection, such as materials, outside services, and other costs, the savings are $500, $380, $218, and $70 million, respectively.

CONCLUSIONS

In this article, the data sets from CWSS 1995 and CWSS 2000 were evaluated to examine the production costs of water supply systems. The authors found that smaller systems tend to face higher unit production costs across the full range of production inputs.

that so many small water supply systems still exist suggests the possibility of costs or barriers to consolidation that are not readily apparent in the CWSS data. One obvious example is the cost of interconnection when trying to physically bring a rural population into a large system, though some gains may not require this. In any case, as small water systems confront increasing regulatory challenges, consolidation may offer some hope of reducing that regulatory burden. Additional research is needed, however, on the local causes of cost heterogeneity in water supply systems and the localized costs and benefits of consolidation.

ACKNOWLEDGMENT

This article was developed under USEPA Cooperative Agreement No. 82925801-0. USEPA reviewers made comments and suggestions that were intended to improve the article's scientific analysis and technical accuracy. The views expressed in this article, however, are those of the authors, and USEPA does not endorse any products or commercial services mentioned here. The authors are grateful to the three anonymous reviewers for their valuable comments. They thank Ray Kopp, Julie Hewitt, and Kris Wernstedt for valuable suggestions at the start of this research. The authors also appreciate the comments and discussion from John Bennett, Evyonne Harris, Michael Osinski, Carl Reeverts, Brian Rourke, Peter Shanaghan, David Travers, and Kathleen Vokes. The authors also thank Richard A. Krop for providing the CWSS data sets.

ABOUT THE AUTHORS

Jhih-Shyang Shih is a fellow in the Quality of the Environment Division at Resources for the Future, Washington, DC. Shih has 15 years of experience in environmental management and policy. He has BS and MS degrees from the National Cheng Kung University in Tainan City, Taiwan, and a PhD from The Johns Hopkins University in Baltimore, Md. He was also a research associate in the Engineering and Public Policy Department at Carnegie Mellon University in Pittsburgh, Pa., from 1991 to 1995. Shih is a member of the Association of Environmental and Resource Economics, the American Association for the Advancement of Science, and the Institution for Operations Research and Management Science. Winston Harrington and William Pizer are both senior fellows at Resources for the Future. Kenneth Gillingham is a graduate student in the Department of Management Science and Engineering at Stanford University, Stanford, Calif.

REFERENCES

Adams, R.M.; Bauer, P.W.; & Sickles, R.C., 2004. Scale Economies, Scope Economies, and Technical Change in Federal Reserve Payment Processing. *Journal of Money, Credit, and Banking,* 36:5:943.

AWWA, 1997. Feasibility of Small System Restructuring to Facilitate SDWA Compliance. AWWA, Denver.

Beecher Policy Research Inc., & Cadmus Group Inc., 2002. Scale and Scope Economies for Water Systems: Illustrations. Draft Report. Cadmus Group, Boston.

Cadmus Group, 2002. Small Drinking Water System Consolidation: Selected State Program and Consolidation Case Studies. Cadmus Group, Boston.

Christensen, L.R. & Greene, W.H., 1976. Economies of Scale in U.S. Electric Power Generation. *The Journal of Political Economy,* 84:4:655.

Coelli, T.J.; Rao, D.S.P.; & Battese, G.E., 1998. *An Introduction to Efficiency and Productivity Analysis.* Kluwer Academic Publ., Boston.

Cohn, E.; Rhine, S.L.W.; & Santos, M.C., 1989. Institutions of Higher Learning as Multi-product Firms: Economies of Scale and Scope. *Review of Economics and Statistics,* 71:2:284.

Norman, G., 1979. Economies of Scale in the Cement Industry. *Journal of Industrial Economics,* 27:4:317.

Studenmund, A.H., 2001. *Using Econometrics: A Practical Guide* (4th ed.). Addison Wesley, Boston.

USEPA, 1997a. National Public Water Systems Compliance Report. http://water.wku.edu/readingroom/nacr97.pdf (accessed July 23, 2006).

USEPA, 1997b. Community Water System Survey. EPA 815-R-97-001a. www.epa.gov/ safewater/cwsreprt.pdf (accessed July 23, 2006).

USEPA, 2002. Community Water System Survey 2000. EPA 815-R-02-005a. www.epa.gov/safewater/consumer/pdf/cwss_2000_volume_i.pdf (accessed July 23, 2006).

This page intentionally blank.

Envisioning the Future Water Utility

BY GARRETT P. WESTERHOFF, HY POMERANCE, AND DAVID SKLAR
(*JAWWA* November 2005)

The fundamental responsibility of US water utilities is the same as it has always been—to provide safe, reliable, affordable, and secure systems that protect the public health and the environment. Although utilities continue to do an excellent job, many are feeling the powerful influence of unrelenting societal and global business community changes.

According to a national survey of field experts, the most critical issue for these utilities is leadership and management—acquiring, training, maintaining, and assuring a continued stream of strong, smart leaders to guide organizations into the future. Two related issues—proper utility asset management and adequate financial resources gained through public recognition of the value of water—are also paramount in the survey respondents' statements. This article explores the respondents' views on why leadership and management are so critical to the success of the US utility industry, and what utilities can do to position themselves to address such issues.

RESPONDENTS IDENTIFIED KEY ISSUES

In 2004, to help envision what issues water utilities will face in the coming decade, Malcolm Pirnie surveyed an expert panel comprising 71 national leaders from water and wastewater utilities, regulatory agencies, academic institutions, and elected and appointed officials, representing 21 states and the District of Columbia. Using a three-round interactive Delphi polling technique, Malcolm Pirnie Inc. challenged the respondents to predict the future, tapping into their broad perspective on significant issues and future trends. The three rounds of the Delphi technique include:
- identifying significant issues,
- prioritizing significant issues, and then
- rank-ordering high-priority issues.

Interestingly, the issues the survey respondents identified were similar to many current concerns—ensuring water quality and reliability, meeting regulations, responding to customer needs, controlling costs, operating efficiently, retaining a knowledgeable workforce, and maintaining the integrity of utility facilities.

Moreover, all of these issues are expected to intensify because of financial resource constraints. All of the respondents agreed that adequate financial resources are a priority issue.

Another highly prioritized issue was utility leadership and management. In fact, three quarters of the respondents ranked leadership in the top three issues they would confront in the coming decade. This conclusion was echoed in research for the AWWA book, *The Evolving Water Utility: Pathways to Higher Performance* (Westerhoff et al, 2003), and in subsequent interviews and discussions with water utility leaders.

The respondents also indicated that leadership and management would be significantly different in the future. Uwe E. Weindel, executive director of the Williamsport (Pa.) Municipal Water Authority, explained, "Leadership needs to be different in order to effectively manage a more demanding consumer environment. A more businesslike approach needs to be taken."

Because water utility leadership is needed at many levels, maintaining outstanding leadership and management will require both developing existing leaders and planning for leadership recruitment and succession. Enhancing the skills of existing staff and recruiting the best personnel possible are essential to having the right mix of people and skills to address future challenges. Wade Miller, executive director of the WateReuse Association (Alexandria, Va.), believes that a focus on workforce management is essential to success. "For the workforce itself, as well as for the leadership, training and education will also be essential elements of success. Training is sometimes viewed as a luxury; it's not—it is essential."

ENSURING LEADERS FOR TOMORROW IS CRITICAL

Most of those surveyed believed it was critical that utilities begin developing succession plans and enhancing the leadership skills of existing leaders to ensure an outstanding level of management to address future challenges. Singling out midlevel professionals for training "to provide a cadre of experts that can fill future leadership positions" was a suggestion proposed by consultant Jerome Gilbert (J. Gilbert Inc., Orinda, Calif.).

"Within the next decade, many more of us will be retiring," said John Cromwell, "and I think the warnings that have been sounded about the changing demographics of the replacements are significant. This must be high on every utility's radar; it needs action right now."

A QUIET CRISIS EXISTS IN LEADERSHIP DEVELOPMENT

The combination of the shrinking workforce, the changing role of the utility leader, and the new skills utility leaders must possess to move their organizations forward have created a quiet crisis in leader development. These external factors converge to create pressure to find more effective ways to plan and deploy leader development strategies that will ensure the right level of leadership capability across all functions of a utility operation (Westerhoff, 2005).

In addition, current utility leaders will be retiring from the industry in large numbers during the next 5–10 years. This is especially disconcerting because it typically takes a minimum of five years to prepare an individual with several years of experience to assume leadership responsibility. Thus, utilities are encouraged to start the process immediately. However, they must first identify what a leader must be able to do and what leadership should look like.

GREAT LEADERS DON'T JUST HAPPEN

Many organizations start developing leadership learning strategies by defining the profile of an effective leader. In *The 7 Habits of Highly Effective People*, (Covey, 1989), Covey suggests organizations start by identifying their desired specific competencies and levels of business and technical acumen and experience. For example, the San Diego County Water Authority uses a "competency model"—its leadership indexes— to guide the development process and measure results.

Developing leaders is not, however, about creating lists of people or skills. It is about making a serious commitment to the time and resources required to identify, plan, develop, and manage the next generation of leaders. Three leader competency categories seem to reflect the general direction of the industry:

• emotional direction—driving effective interpersonal relationships at individual, team, and public levels;

• strategic thinking—driving development of vision and planning and operational excellence; and

• business and technical leadership—driving complex financial and environmental decision-making.

Leaders must demonstrate proficient skills and abilities in all three categories.

SEATTLE'S PROGRAM PREPARES FOR THE FUTURE

A number of water utilities that are focused on succession planning and leadership development have begun to plan programs and processes to systematically address their needs. Seattle (Wash.) Public Utilities (SPU) created a leadership training and development program for 100 employees representing all job classifications and divisions of the organization. The program included three tracks—one for supervisors and managers, one for members of the SPU leadership team, and one for all other employees.

The program's primary objectives are

• Create a common vision for SPU leaders and an appreciation of the challenges that leaders face.

• Provide formal and informative learning experiences to enhance participants' abilities to develop the leadership competencies necessary to manage amid continuous change.

• Strengthen participants' understanding of their own leadership strengths and development needs and their effect on the organization's accomplishments.

• Increase participants' leadership effectiveness and implement long-term individual leadership development plans.

• Understand individual roles and responsibilities in relationship to overall organizational goals.

UTILITY ASSETS MUST BE EFFECTIVELY MANAGED

One of the key areas in which a powerful leader can make a significant difference to a water utility—identified by survey respondents as one of the top three concerns for the future—is ensuring thoughtful management of the organization's assets. Asset management is "delivering a specific level of service to customers at an optimal life-cycle cost with a strategy that ensures long-term sustainability of public assets." For water utilities, it is a focused, strategic effort to better manage business and physical infrastructure with a long-term service level and financial perspective.

A thoughtful asset management program includes comprehensive plans to maintain assets over time and ensures that they perform according to established service levels and design criteria throughout their useful lives. According to the head of a New England utility, an asset management program will have long-term positive effects on maintaining reliability of service and maximum operations and maintenance (O&M) efficiencies—getting the most out of the infrastructure.

Asset management is a systematic approach to utilizing proven methodologies and techniques. It does not need to be complicated and expensive, but is, rather, a good stewardship of resources. As John E. Cromwell, managing economist, Stratus Consulting Inc. (Washington) put it, "In the next decade we need to focus on finding ways to think smarter about what to do and how best to do it."

One survey expert, Celine Hyer, manager of engineering and environmental services, Hillsborough County Water Department (Tampa, Fla.), proposed a "shift from reactive to proactive maintenance to preserve existing assets," and suggested that utilities "begin an asset management program, including an inventory and condition assessment of all utilities' above- and belowground assets."

A STRATEGIC PLAN SETS THE STAGE

For water utilities, an asset management program can begin with a new or existing strategic plan (Figure 1). An asset management approach can ensure that the strategic plan is fully implemented, goals are achieved, and the plan responds to the needs and demands of key stakeholders, including customers, regulators, and elected officials.

To begin, the utility assesses high-level strategic goals and further refines them into formal and established service levels that cover reliability, customer service, and environmental compliance. The utility can then develop an infrastructure plan, a business plan, and a financial plan that will effectively deliver these service levels in an efficient manner. Using this approach strengthens long-term capital improvement plans and ensures that they are integrated with O&M strategies. It also addresses a utility's organization, business processes, technology, and data and ensures that these elements can support overarching service level objectives with available financial resources.

ASSET MANAGEMENT MUST BE EMBRACED UTILITYWIDE

The overall mission of asset management is to guide the implementation of a strategic plan and strengthen capabilities and decision-making throughout the organization (Figure 2). For an asset management program to be successful and achieve its stated goals and objectives, an overall change in thinking, approach, strategy, and process must be embraced across all departments within the organization. Each department and job function has a role to play in an asset management program and must understand its overall effect.

Asset management is about "businesslike management of assets," so essentially it cannot be considered a stand-alone program that is purely in the realm of one group. Instead, it is a multidisciplinary approach that includes engineering, operations, maintenance, financial, and business functions.

Above all else, asset management requires top-down executive and management support and an enabling organizational structure. The leader of a major utility in the Northeast describes it as an organizational commitment at the top, driven by the top organizational person—a continuous leader who routinely revisits the issues. This is dramatically different from the physical asset–focused approach that many utilities have adopted. Because it provides greater sustainability, it can have a significant effect on bottom-line results and deliver quantifiable performance improvements.

SERVICE LEVELS DEFINE UTILITY GOALS

Service levels are the critical foundation of an asset management plan, reflecting the utility's ultimate goals and objectives. Defined as a utility's commitment to deliver service at a specified level of quality and reliability, service levels are a charter or contract between the utility and its customers. They are also tangible metrics that are routinely reported and tracked against established targets. According to a West Coast utility director, service levels are a key impetus for implementation of an asset management program because they drive cost.

At its core, asset management encompasses everything a utility does to achieve its service level objectives, not just condition assessment and rehabilitation programs. This includes a utility's

- organizational structure,
- culture,
- business processes,
- information technology systems,
- data and information, and
- day-to-day office and field decisions.

FIGURE 1 Strategic approach to water utility asset management

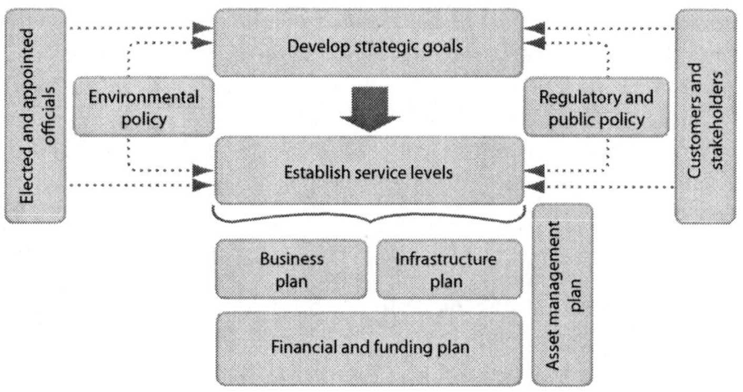

A utility's asset management plan (including business, infrastructure, and financial plans) should be developed from set strategic goals and service levels. These goals and levels should incorporate reliability, customer service, and environmental and regulatory compliance.

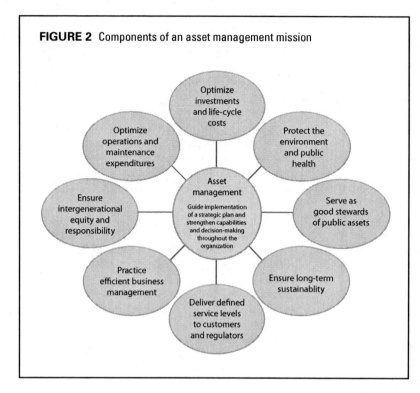

FIGURE 2 Components of an asset management mission

on identifying those physical assets that are likely to fail and ensure that projects initiated will have a direct effect on meeting this target service level. Employees are a key part of this process and must embrace these service levels goals.

AN ASSET MANAGEMENT PLAN INCLUDES THREE ESSENTIAL ELEMENTS

An asset management plan typically consists of three major elements/plans:

Infrastructure plan. To effectively and efficiently deliver target service levels, the utility must have a comprehensive, well-documented infrastructure plan including O&M strategies, capital investments, and a well-managed life-cycle plan for all of its major infrastructure assets (e.g., plants, pumping stations, sewers/mains, valves). Developing such a plan includes an up-front risk and criticality assessment, condition assessment, and an analysis of historic failures and events.

The plan is perhaps the most involved phase of asset management, because it involves integrating many ongoing efforts, including

- asset inventory and valuation,
- modeling and capacity analysis,
- preventive and corrective maintenance strategies,
- renewal and replacement criteria,
- rehabilitation programs, and
- decision support tools.

Criticality, capacity, risk, and condition analyses identify target areas for preventive maintenance budgets. When funds are limited, available dollars must be directed toward infrastructure that has a high probability and/or consequence of failure. Part of this strategy is accepting that many noncritical infrastructure components can be run to failure, because it is cost prohibitive and inefficient to try to prevent all failures.

Survey respondents speculated that utility leaders will need to make tough choices and hard decisions. Alan Roberson, director of regulatory affairs for AWWA, strongly believed that the infrastructure issues will not be easily solved. "Many systems are reluctant to make the significant rate hikes that would be needed to keep their systems up-to-date, and significant federal monies are not likely," he said.

Business plan. A sound business plan ensures that a utility is able to successfully achieve its overall asset management goals through optimized business processes, information technology, and data. Utilities have

To be successful, asset and utility management must be linked to and aligned with the utility's overall strategic goals. This requires consistently defined service levels and detailed performance measures that quantify gaps and determine success.

An important driver for an asset management program is customer expectation, according to one utility expert. The customer wants excellent service at a low cost whereas the utility wants great service and cost efficiency. Utilities must understand what the customer will accept at a reasonable cost.

Well-defined service levels provide clarity of focus for an organization and are the key link between a utility's strategic plan and its asset management plan. In essence, service levels drive strategic decision-making down to the day-to-day activities performed in the field and help to prioritize investment decisions.

Defining service levels. Generally, service levels can be categorized into four major areas:

- reliability,
- quality,
- customer service,
- regulatory measures.

A comprehensive set of service levels should appropriately cover each of these areas. For example, a 60-minute target response time for water main breaks will require that a utility invest in the proper business processes, work standards, and technology to achieve this goal. In addition, if a utility targets no more than 10 water main breaks per month, its capital replacement, rehabilitation, and renewal programs must focus realized that business and financial skills are just as critical as technical and engineering competencies in managing a utility and performing day-to-day operations, as well as in an asset management program.

In order to respond to increasing pressures from stakeholders, elected officials, and the public, utilities must operate in an efficient, businesslike manner with supporting organizational structures, goals, and accountabilities. According to a utility manager in the Southeast, part of implementing an asset management program is dealing with the politics of implementing a rate increase. This includes how to prioritize the work done, especially in a multijurisdictional area in which politics can be a complicating factor that diverts needed resources.

Although many utilities have ongoing business performance improvement initiatives, not all have successfully integrated these activities into their asset management plans and strategies. It is, however, critically important that the principles of asset management filter into all areas of an organization so that everyone is unified.

Financial plan. Finally, a utility needs to fund an asset management program with a long-term view through the most efficient method. This requires detailed and accurate financial forecasts and modeling tools that are updated regularly. It is critical to tie O&M and capital programs into financial models and rate/funding projections on an ongoing basis to ensure that planned rate structures can properly finance an evolving asset management plan. These should be routinely updated as an organization's information and data improve and as regulations and situations change. Financial plans should include

- debt service coverage,
- cash flow requirements,
- reserve requirements,
- capital funding/new debt service, and
- rate effects.

TYING IT ALL TOGETHER

Developing a clear and overarching asset management strategy is the first step in a long journey, but continued and successful implementation of that strategy is the challenging part. It doesn't have to be a daunting task, however, if the asset management plan is implemented using a structured methodology.

Implementation strategies should focus on developing well-defined and specific initiatives carved into sizable chunks that an organization can handle. Asset management is a large effort—and it can't all be done at once. Besides having an overall long-term strategy, one survey respondent recommended a modular approach, saying, "You can get lost if you take on too much at once—it's important to see some quick, early results, so carefully pick out some critical areas, and then try one piece at a time."

Utilities can use a staged and cyclical process through strategy, gap analysis, implementation, and repetition of the cycle on an ongoing basis through specific initiatives and continuous improvement.

Successful implementation efforts will yield lasting benefits over time and transform an organization into an "asset management culture," ensuring long-term success by

- moving from a technical mindset to business-minded leadership built on a solid technical foundation;
- creating greater understanding of the true costs to provide service through life-cycle cost analysis and performance management;
- understanding long-term service level and cost implications of both capital and O&M investment decisions;
- improving capital project and investment prioritization through formal business case analysis;
- understanding long-term financial decision-making that enables more accurate financial projections, ensures adequate funding and rate structures, and helps maintain bond ratings; and
- creating improved trust, understanding, and transparency with the public, elected officials, and other stakeholders.

THE VALUE OF WATER MUST BE COMMUNICATED

One significant responsibility of a water utility leader is ensuring adequate financial resources to meet future challenges. This was considered a major priority by the survey respondents and affected virtually all other issues. Although a combination of federal funding and rate increases was judged to be the likely scenario, in the end most financial resources will need to come directly or indirectly from those who benefit from the system.

One respondent remarked, "We need to stop whining about unfunded mandates and the need for the feds to step up. The step is ours and probably ours alone. Get the local politicians to understand this, give them political covers and rising public expectations to do the right thing, and get moving."

Virtually all respondents agreed that to gain support for increased rate structures that provide financial resources, utility leaders must do a better job of communicating the value of water to the public, customers, rate regulators, and local elected and appointed officials. In fact, tomorrow's successful water utility will have to be respected and trusted and become the source of quality information with a well-defined product value.

Chris Hill, deputy manager of the Metropolitan Domestic Water Improvement District (Tucson, Ariz.), pointed out that leaders will be "increasingly challenged to provide the level of service that customers expect" while balancing the justification of higher rates.

Another respondent suggested, "The utility has to be more visible than through its monthly or quarterly bill. The utility has to be positioned as the expert on drinking water and associated water resource issues in its region. Education on the subject must start early, and an elementary education program is key. It is the duty of the utility to keep the customer informed of the value of the service provided."

Undervalued and underpriced water leads to inadequate support for capital and operating funding requirements. Although some respondents think the public is beginning to appreciate the value of water, many think it is time to create and deliver a "value" message.

COMMUNICATING, BRANDING ARE STRATEGICALLY IMPORTANT

According to Alan Roberson, communicating the need for asset management to the public is a major part of being successful—utilities need to pay more attention to the rate payers. Many forward-thinking water utility leaders realize that creating and effectively communicating a value message to stakeholders and elected and appointed officials has grown in strategic importance. Significant water rate increases are necessary to repair and rehabilitate US water infrastructure, but consumers now have alternatives to tap water (e.g., bottled water and point-of-use devices). Consumer concerns over water safety and aesthetics have pushed bottled water sales in the United States to nearly $10 billion annually. Customers (bill payers) now pay approximately the same amount for bottled water and point-of-use devices as they do for tap water through water rates. In addition, water utilities must compete for the attention of customers, consumers, and the public, penetrating the barrage of print, television, radio, and Internet messages consumers receive.

Against this backdrop, water utilities must improve the public's appreciation of water. If customers understand water's value in the quality of their lives and are willing to pay more for it, the utility will not only build a foundation of public and political support for rate increases, but customers will also use water more wisely.

Defining water's value. In creating a value message, utilities must begin by defining the value of water. In large part, this has been accomplished in a recently published Awwa Research Foundation report, "The Value of Water: Concepts, Estimates, and Applications for Water Managers" (2005). To effectively communicate this message, utilities must stress the values that resonate with them, including cultural attributes, spiritual and emotional values, environmental issues, recreational issues, economic development, water supply and infrastructure reliability, and water quality and aesthetic issues.

Communicating the value. Effective communication is a unique science, particularly in the social/political environment that surrounds a water utility. Developing an effective communications program and tools requires expertise in several areas, including
- behavioral sciences,
- marketing and communication,
- water utility knowledge,
- media communication, and
- branding.

As part of this communication effort, water utility leaders must recognize that the operation and performance of their organization is primarily affected by three groups of people:
- customers, special interest groups, and other stakeholders;
- elected and appointed officials; and
- utility leadership.

Each group affects and is affected in various ways by the others as well as the media, regulators, and other politicians. If a water utility is to function effectively and efficiently, the interests of these disparate groups need to be understood and, as much as practical, aligned to a common utility vision and mission.

This requires understanding and appreciating each group's interests, issues, and needs as well as the most effective communication vehicles to fit those needs. Other building blocks of an effective communication program include a mutual knowledge of the significant influences affecting water utility operations. It's then important to build mutual trust and foster open, effective communication.

UTILITIES NEED TO ESTABLISH A BRAND

One way the utility can communicate the value of water is to brand itself, its products, and its services. Branding drinking water and water utility services will be different from branding a competitive commercial product or service, but it will still create value and provide focus for the water utility's communications program. Most people understand that water is important—but that doesn't translate into support for the utility, even when it has built trust with stakeholders. A powerful brand can cut through noisy clutter and heighten awareness of a product and service. Most critically, a brand strategy must be integrated into a utility's overall business strategy.

One survey respondent said, "As an industry, we need to educate our customers much better. This can be partially accomplished by partnering with the press, educating them on the issues we face."

CONCLUSIONS

In another decade, water utilities will be significantly different than they are today. These differences will involve how the utility is led, how it manages its assets, and how it obtains the necessary financial resources to meet continued challenges to achieving its mission (Westerhoff et al, 2003, 1998).

Forward-looking water utilities must develop integrated approaches to address the many significant influences that will affect them within severe financial constraints. The primary pathways to such an integrated approach will include

- outstanding leadership,
- thoughtful long-term planning and financing strategy,
- businesslike management of assets, and
- stakeholder alignment.

As in the past few decades, long-term strategic planning is mission-critical for utilities. One survey respondent put it this way: "A focus on planning is also a key element of success. As Covey says, 'Plans are worthless, but the planning process is invaluable.' Involve key staff; get their buy-in."

Strategic long-term planning is critical. Survey respondents unanimously agreed that long-term planning is critical if water utilities are to prepare themselves to meet the challenges coming in the next 5–10 years. Many water utilities have a long-term plan. Some call it just that; others call it a strategic plan or a business plan. By whatever name, subject this plan to an annual "strategic conversation," and test its premises against changing situations.

Respondents provided insights that are likely to affect all water utilities in the near future. Water utility leaders can ask these questions as they review their existing plan or develop a new one:

- Have we recognized the changing demands on the leadership of our utility, and have we included specific goals and objectives to ensure the development of the leadership that our utility will need?
- Have we adequately addressed the long-term asset management needs of our utility, and have we included a phased asset management program as a specific objective?
- Have we included a comprehensive communication and branding plan that will enable us to communicate the value of our water and services to stakeholders and align our utility leadership with our elected and appointed officials in a manner that will result in adequate financial resources to achieve our mission?
- Do we have a long-term financial plan that has the support of our elected and appointed officials?

Predicting the future isn't a simple proposition, but the experts in Malcolm Pirnie's Delphi poll provided valuable perspectives and food for thought on the future direction of America's water utilities.

ABOUT THE AUTHORS

Garrett P. Westerhoff is chairman emeritus of Malcolm Pirnie Inc., Fair Lawn, NJ. He is a professional engineer and planner, a member of the National Academy of Engineering, and an honorary member of AWWA. Hy Pomerance is vice-president of and David Sklar is an associate in the Red Oak Consulting Division of Malcolm Pirnie Inc.

REFERENCES

Covey, S.R., 1989. *The 7 Habits of Highly Effective People: Powerful Lessons for Personal Change.* Simon & Schuster Inc., New York.

Raucher, R.S. et al, 2005. The Value of Water: Concepts, Estimates, and Applications for Water Managers. Awwa Res. Fdn., Denver.

Westerhoff, G.P. et al, 2003. *The Evolving Water Utility: Pathways to Higher Performance.* AWWA, Denver.

Westerhoff, G.P. et al, 1998. *The Changing Utility: Creative Approaches to Effectiveness and Efficiency.* AWWA, Denver.

Westerhoff, G.P. et al, 2005. It's all About Leadership? *Underground Infrastructure Management,* January/February.

This page intentionally blank.

The Growing Role of Private Equity in the Water Industry

IN THE PAST FEW YEARS, EXPANDING STRATEGIC AND INVESTMENT INTEREST IN THE WATER INDUSTRY HAS ALMOST BECOME A FEEDING FRENZY. THUS IT IS NOT SURPRISING THAT A RAPIDLY EXPANDING INTEREST—AND OWNERSHIP ROLE—IN THE INDUSTRY ON THE PART OF THE PRIVATE EQUITY COMMUNITY IS ALSO OCCURRING.

BY STEVE MAXWELL
(*JAWWA* January 2006)

During the past decade—as water shortages and water quality issues have attracted greater concern and greater public attention—many water companies have expanded their capabilities, and many new firms have sought to enter the business. Recently, this expanding strategic interest in the water industry has led to hundreds of companies competing to buy assets and to establish a foothold in an industry that is expected to be one of the world's most crucial businesses in coming decades. Thus it is clear why we are also now seeing a rapidly growing interest in the water business from another corner of the economy—the private equity (PE) community.

The PE business, like most industries, tends to exhibit somewhat of a "herd mentality," and at the moment, water is hot. Hundreds of PE firms seem to be tripping over one another to strategically position themselves in the water industry. Typical Wall Street investor conferences on the water industry attract dozens of PE scouts trying to get up to speed on the industry. In many current merger and acquisition transactions, PE groups are serious (even aggressive) buyers, often willing to pay prices that are higher than industry or strategic buyers can justify—a dramatic change from just a few years ago. PE groups have already acquired several major and well-known assets in the US water business—and in fact have already bought and sold several major companies, including Nalco Co. and National Waterworks Inc. Major water companies that are currently held by PE interests include Culligan, Utilities Inc., F.B. Leopold Co. Inc., and Ashbrook Simon-Hartley.

The increasing participation of PE investors may be contributing to the current (and unsustainably high) valuations that currently exist for water companies. It also seems certain that the PE community will have a significant future effect on water business ownership and broader strategic trends within the industry. Some stereotypes about PE firms are starting to emerge, but not all PE firms are the same. In this article, the PE community will be examined in more detail, and the pros and cons of greater PE involvement in the water industry will be investigated.

GROWING AND SELLING COMPANIES

What is a PE firm? PE firms typically raise a pool of capital from wealthy individuals, families, or institutions such as university endowment funds, corporate pension funds, major banks, investment firms, or insurance companies. They then effectively extend this pool of capital with additional borrowed capital and make investments in established companies, usually for a specific period of time. PE firms attempt to grow and enhance the companies in their portfolios in order to eventually sell them at a profit to another owner. PE groups are often confused with or lumped together with venture capital firms, but there is a distinct difference. Venture capitalists invest in small, fast-growing, high-risk opportunities, whereas PE firms generally invest in larger and more established companies.

In a recent special report, *The Economist* referred to PE firms as the "new kings of capitalism." Growing from just a small handful of firms in the 1970s that were generally associated with wealthy families, the PE business has boomed during the past 20 years. In the 1980s, PE firms like KKR and the Carlyle Group developed unsavory reputations as corporate raiders, acquiring several well-known firms through hostile and highly leveraged transactions. Today, however, PE firms are much more established and active in the mainstream of US business. As *The Economist* recently reported, "In the 1980s, PE was a place for mavericks and outsiders; these days it attracts the most talented members of the business, political, and cultural establishment, including many of the world's top managers." Examples include such well-known business and political names as Jack Welch, Lou Gerstner, Paul O'Neill, John Major, and George Bush Sr.

PE firms today have shed their image as swashbuckling wheeler-dealers and instead are involved in trying to offer an economic and strategic environment in which portfolio firms can better pursue well-thought-out and logical growth strategies—free from the constraints of unrealistic parent company expectations and/or the pressures and volatility of the public stock market.

In 1991, investors committed $10 billion to PE funds. Driven by the technology and dot-com boom of the late 1990s, this figure peaked at $160 billion in 2000. In the first six months of 2005, Thomson Venture Economics reported that venture capital funds had raised $12 billion in capital and that PE buyout funds had raised almost $36 billion. It has been estimated that today there are as many as 2,700 major PE firms worldwide—many have just a few hundred million dollars in capital, but several have funds in the multibillion dollar range.

The typical PE fund is obligated to return the profits on its investments to investors at some specific point in time in the future. It is these requirements that have led to many of the typical fears and criticisms leveled at PE groups—that they have a short-term focus and are only interested in buying businesses so that they can quickly turn them at a profit. However, this is an oversimplification.

WHY WATER?

Why are PE firms so interested in the water industry? First, PE firms aren't just interested in water—they are interested in any market or business opportunity in which they see the potential to grow the value of individual companies and thus increase the value they can return to their shareholders.

The total amount of capital sitting in PE funds is now at an all-time peak, and the availability to stretch that capital through borrowing has rarely been as attractive as it has been in the past year. The huge funds that were raised in the late 1990s were typically directed toward investment opportunities in the high-tech and telecommunications industries—opportunities that largely collapsed in 2000 and 2001. As a result, many PE managers find themselves today with huge amounts of capital that they need to invest elsewhere and have thus started investigating a much broader array of industries in which to invest those funds.

Water and other environmental businesses are just one of the industries that have begun receiving more scrutiny from PE funds. Those skeptical of all this PE activity surrounding the water industry point out that the fundamental opportunities in the water industry really haven't changed that much in the past few years; they suggest that the current level of interest in the water industry is mostly a result of PE firms desperately looking for new places to put their huge sums of capital to work. Other environmental industry sectors, such as the engineering/consulting business and the

environmental testing and monitoring business, have also seen a dramatic increase in PE interest in the past few years. Thus there is probably some support for the argument that PE firms have developed more interest in the water industry as other (previously more attractive) investment opportunities have dwindled.

On the positive side, however, it is clear that the water industry is particularly attractive to PE firms for several reasons. First, it is perceived to represent strong and very consistent growth over the long-term—and certain sectors of the water industry offer the allure of high profitability as well. Furthermore, the water industry offers another characteristic that PE firms typically seek—it is a relatively fragmented industry. This aspect offers PE firms the opportunity to consolidate businesses in order to build larger and stronger—and more valuable—companies.

AFTER A COMPANY IS PURCHASED

What do PE firms do after they buy a company? From a broad economic perspective, PE firms can play several important roles in the economy. In brief, PE firms can

• identify undervalued assets and fix them up through such methods as tightening up business controls and management systems, enhancing financial management, splitting apart assets that have no synergy or compatibility, or creating better distribution channels;

• consolidate companies, building larger companies in industries that may be overly fragmented or in situations in which economies of scale can be achieved;

• hold on to stronger companies that are operating in weaker or out-of-favor businesses, waiting for valuation levels to increase; or

• force management—in a variety of ways—to be more responsive to shareholder value.

PE investors use a number of techniques to improve the profitability and size of a company they have acquired so that they can sell it for more than they paid for it. These techniques may include employing greater capital resources for growth, tightening up the company's operating processes and controls, improving distribution and marketing channels, making better use of commercial leverage opportunities, and various other management processes and controls in which a PE firm and its principals may be very experienced.

PE investors are typically attracted to industries that are relatively fragmented—industries in which the acquisition of one platform company can be used as a foundation for adding other companies. The ability to consolidate and build a larger (and probably more valuable) company is a key consideration for many PE firms because they often have both the transactional experience and the capital to make such strategies work.

Sometimes the PE firm may employ a reverse strategy—buying a larger and more diversified company whose businesses may not fit together strategically and then breaking it up into individual pieces so that the sum of the values of the separated pieces is greater than the value of the previously whole company. These sorts of corporate reorganizations and spin-offs may take several years to efficiently and effectively implement, and they also require the transactional and financial engineering skills that most PE firms have.

As with any investor, PE firms prefer to buy low and sell high. Many PE investors try to identify industry sectors or individual companies that are at least temporarily undervalued—perhaps because of poor recent performance despite strong industry fundamentals or maybe because the industry is simply out of favor with investors. (Although some PE firms seem to be implicitly banking on this approach in today's water industry, it seems highly unlikely that this will be a successful course of action given the current stratospheric valuations of many water companies.)

PE firms design their acquisitions in a bewilderingly wide array of structures, but they always try to give incentives to the existing management team so that it will help the PE firms in the process of building and realizing a greater future value. This typically includes the use of stock options or phantom stock, earn-out purchase structures, or some other means of encouraging the existing management team to maximize shareholder value.

EXIT STRATEGIES

Ultimately, of course, almost all PE firms eventually look to sell an acquired company so that they can pay back their investors. There must be some type of logical and eventual exit strategy, although investment time horizons may vary considerably among different types of PE firms. Exit strategies can generally be divided into two approaches:

• selling the portfolio company to a larger industry or a strategic buyer or

• taking the portfolio company to the public stock markets through an initial public offering of stock.

Both of these methods assume that the value of the larger and more "tuned-up" company will allow the PE firm to return a handsome profit to its investors.

Increasingly, however, there is a third exit alternative—selling the portfolio firm to another PE group that is hungry for investment opportunities. This development, which has occurred in a couple of recent water industry transactions, has fueled concern that valuations are getting out of hand. The thinking is that the PE community is helping to bid prices so high that it won't be possible to make a decent return on investment—that some "greater fool" will eventually come along and take a huge loss.

THE EFFECT OF PE FIRMS ON THE WATER INDUSTRY

As mentioned at the outset, there have already been several major PE investments in the water industry, as well as a number of smaller and perhaps less visible purchases. In addition, numerous PE firms are currently and aggressively pursuing water transaction opportunities or are looking at water related firms that are not actively on the market.

Several of the key PE investments in the water industry to date are summarized in Table 1. Virtually all of these transactions have been divestitures to a PE firm from a larger corporate entity that held the water business as a subsidiary unit—as opposed to independent private companies selling directly to PE firms. However, it seems likely this will change in the near future, as many private and family-owned businesses are beginning to seek (or are being pursued by) PE groups. (The valuations for most of the transactions in Table 1, either in an absolute sense or in terms of a multiple on revenue or earnings, have not been published.)

There is one key constraint on the eventual level of PE investment in the water industry. There just aren't that many suitable targets—in terms of business mix and size. In reality, there are not that many pure-play companies out there that are sizable enough to attract the interest of the typical PE firm. For example, in one of the most sought-after sectors of the business—the booming membrane filtration sector of the water treatment and purification business—there really are very few pure-play and desirable companies available. The only firms large enough to attract PE firm interest would be companies like ZENON and Pall, and as might be expected, these are already valued at extremely high levels.

In summary, there are a lot of excited investors from both the industry and the financial community who are chasing relatively few attractive opportunities that are already very highly priced. In this vein, Debra Coy, one of the leading stock analysts in the water industry, authored a recent water industry report titled Investing in Water: Froth, Bubbles, and Sludge (Coy, 2005). It stressed the paucity of exciting investment vehicles in this industry and the discouragingly high valuations at which these relatively few major water companies are already trading.

THE PROS AND CONS OF PE INVOLVEMENT

What should we make of this increasing interest and involvement of PE investors in the water industry? First, it is clearly a plus to sellers in the water industry—there are many more potential buyers (Table 1), and average valuations have been inexorably creeping upward. PE firms will definitely help solve the ownership transition challenges that many water firms will face.

TABLE 1   Private equity investment in the water industry

Seller	Buyer
USFilter/Waterworks*	Thomas Lee, JPMorgan, et al
Suez/Nalco Co.	Blackstone, Apollo, Goldman
Sachs, et al	
US Filter/Culligan	Clayton, Dubilier & Rice
Bridgepoint/Alcontrol†	Candover Partners
East Surrey Water	Terra Firma Capital Partners
Suez/Northumbrian	Aquavit Partners
Infiltrator Systems	American Capital Strategies
Pionetics‡	Ngen
Nuon/Utilities Inc.	AIG Highstar Capital
RWE/Ashbrook Simon-Hartley	Blue Sage Capital
RWE/F.B. Leopold Co. Inc.	PNC Equity
Amtrol§	Cypress Group

*Equity groups later sold business to Home Depot in 2005 at roughly twice its original purchase price.
†Example of one equity group selling to another
‡Early-stage venture-capital type of deal
§Partial equity investment

Second, many PE firms can bring more than just money to the challenging situations faced by small and privately owned water companies. They bring extensive business operations experience, distribution and marketing contacts, and financial management and control system experience, all of which can help such water companies become more profitable. The chief executive officer of National Waterworks Inc., upon the sale of his business to Home Depot, said the following about his PE owners—"Both firms, with their insights into the industrial and distribution sectors, have assisted us in further developing our distribution platform and continuing to execute our business strategy."

In addition, a PE firm can bring a more stable operating environment to the purchased company so that any potential conflicts with a larger corporate owner and the pressures of public ownership are both effectively removed. Finally, from the perspective of selling management, careers are much more secure than they might have been in a sale to a strategic buyer—in fact, a PE firm will probably only do the deal if it has confidence in the capabilities and the commitment of existing management.

There are, of course, also disadvantages and/or potential risks associated with selling to a PE firm. The biggest disadvantage—or the biggest perceived drawback—involves a PE firm's presumably short-term focus on profitability at all costs. Yes, PE firms are certainly going to be scrutinizing the business for ways to operate more cost-effectively, but perhaps this is something the company should be doing anyway (and it doesn't necessarily imply an unwillingness to commit capital for attractive growth opportunities). Second, many sellers worry about the short time commitment or holding period typified by PE investments, which is often about four to six years (or less) before the business is sold. However, this aspect is changing, as some PE firms are now happy to hold on to attractive investments for a longer period, particularly if the investments are growing and profitable. Also—and this may or may not be a disadvantage—most PE firms will expect current ownership, or at least the current management, to stay in place for a few years in order to help the new owners recognize projected growth and profit targets. Contrary to the situation of a sale to a strategic buyer, the owners in this scenario cannot simply sell the business and disappear. They will be expected to retain some ownership—to keep some "skin in the game." They will thus be counted on to stay and help grow the business to the next level, but they will also be eligible to participate in whatever value appreciation the new team can create—to "have another bite at the apple" in the PE parlance.

There are some observers within the water industry who believe that PE ownership is not appropriate in certain sectors of the water business—particularly the utility sector. They argue that it is not a good idea to put public water supply issues and control over scarce water resources in the hands of purely financial buyers who do not have utility operating experience and who have an ownership horizon that is often as short as four or five years. Nick Debenedictis, the chief executive officer of the utility company Aqua America Inc., says that utilities should be governed by independent and transparent boards of directors that are driven by a long-term commitment to customers and not by private investors focused on short-term financial gain—which seems to make pretty good sense. In short, PE investments in this area may not be in the public interest because of short-term ownership, short-term focus on maximizing profitability, lack of transparency in management and governance, and possibly excessive levels of debt.

How will all of this affect the individual seller? If you are thinking about selling your firm or even just part of your firm, there are certainly plenty of PE firms out there who may be interested in your business, and there are many who will claim to be knowledgeable about the water industry. Potential sellers should try to find a firm that brings a set of skills that can help them grow and better manage their businesses, i.e., find a PE firm that can offer a set of strengths and attributes above and beyond simply the amount of money it brings to the table. Contrary to popular opinion, PE firms are not all the same—there may be one out there that can bring new skills and competitive advantages to a company and that fits well with the company's longer-term objectives.

It seems certain that PE firms will play an increasingly significant role in the water industry, at least in the intermediate term, and that they will help to effect a wide range of ownership transition situations and consolidation opportunities in this business. Some of the smarter PE investors will probably reap good financial gains from their participation in the industry, but it seems likely that some will lose their shirts. Like it or not, it appears that PE investment will continue to have a greater effect on the water business, and ways for water equipment and services companies to turn this to their advantage may exist.

REFERENCES

Coy, D., 2005. Investing in Water: Froth, Bubbles, and Sludge. Stanford Group Company, Houston.

The Economist, 2004. The New Kings of Capitalism. *The Economist,* Nov. 25.

This page intentionally blank.

The Role of Water Conservation in a Long-Range Drought Plan

BY WILLIAM B. DEOREO
(*JAWWA* February 2006)

To advocates of water conservation, it seems intuitive that reducing water demands in normal times through increased efficiency and elimination of waste should prove useful in enabling water systems to better ride out the periods of shortages in water supply caused by droughts. However, there are legitimate questions about how water conservation programs should tie into drought-response plans and whether there are situations in which conserving water during normal times may either fail to reduce the effects of droughts or even make matters worse. This article discusses the role of water conservation in drought management planning and examines the performance of a hypothetical community over a 25-year study period in order to identify factors that affect how water conservation improves or hinders drought response.

NUMEROUS FACTORS AFFECT INTERACTION OF WATER CONSERVATION AND DROUGHT

Understanding how water conservation interacts with drought is no simple matter. The relationship between water conservation and drought is complicated by a variety of factors, some unchanging and certain, others evolving and unpredictable.

• Each water system has its own characteristics, community, and customer base.

• The end uses of water in the system under baseline conditions constitute critical information that affects the potential for reducing demands through conservation.

• The amount of growth in the system must be understood because the strategies for dealing with drought in a growing system will differ from those in a system at build-out.

• The mix of water supplies with respect to surface and groundwater supplies must be defined.

• The water rights of the system and the legal context governing diversions, storage, and return flows must be understood.

• The amount of storage in the system plays a significant role in how water supply shortages affect customers.

• The nature of the drought determines the extent of its effects. The system's response depends on both the severity and the duration of the drought as well as how quickly the drought is recognized.

Any analysis of a water supply system must consider each of these factors and take into account a time period long enough that the interaction among all these factors can be observed. This is at its heart an inductive process in which conclusions about the role of conservation in drought planning are drawn from observations of specific performance factors of realistic systems under both normal and drought conditions.

A fundamental concern of many water managers is the phenomenon commonly referred to as demand hardening. In this process, all of the low-value uses are eliminated from the system. Then, when shortages do occur, there are no more low-value uses to curtail, there is less flexibility in managing demand, and cuts must be made in more valuable uses. The concern is that demand hardening could result in more damages to the customers and economy than would be the case if low-value uses were still available to eliminate. This is a valid concern, but it requires careful analysis of the individual system's specific characteristics to determine whether demand hardening is a real problem.

Baseline use patterns must be identified. It is virtually impossible to understand how much water can be saved in a system without first developing a reasonable understanding of the uses to which water is put in the system and the relative amounts devoted to each use. This is important because each end use of water demonstrates a different pattern in three critical areas: the volume of water that can be saved, the time it takes to effect this reduction, and the lost value to the customer (or damage) that is caused by the reduction.

On one extreme is the water system in which a high percentage of the demand is consumed by low-value uses that are easily shed. The classic example is large amounts of irrigation of turf grass with inefficient irrigation systems. In times of shortage, curtailing these uses

159

and reducing their demands on the system are fairly straightforward matters. This approach may work for short periods with little damage, but if extended, the loss in landscaping can be substantial.

At the other extreme is the system in which a high percentage of the demand is for indoor domestic uses and commercial/industrial process uses. With these uses, the lead time for making substantial reductions in demands typically is longer because of the need to install new fixtures, appliances, and control devices.

Growth rates affect system conservation. A system at build-out is adding little or no new net accounts each year. Old customers may be replaced by new ones who have different end uses and demand patterns, but overall the system does not show dramatic increases in treated water demand from new account activity. With all other factors held constant, annual demand remains fairly consistent from year to year. In these systems, demand can be reduced by conservation over time.

In growing systems, however, demand for new treated water continuously rises. If all other factors are held constant, either new water must be brought into the system to meet this increased demand or the increased demand will reduce surpluses. If the system's growth occurs during a wet period, additional yields may somewhat mask increases in demand. However, the system will run into trouble during times of drought unless system managers have anticipated the potential and made proper allowances.

Sufficient water supplies and water rights can offset effects of drought. No water system is immune to drought, but those with more senior rights will be less susceptible to shortages. Understanding the reliability of any system requires a fundamental knowledge of the yield of divertible and storable water supplies over a time period considered to represent the long-term hydrology of the area. This process is fraught with uncertainty; where records are scarce, the historical record may not reflect past droughts. Evidence from indirect sources, such as tree rings, has shown that there have been long periods of low precipitation that preceded the start of record-keeping in many areas. Add to this the uncertainty created by climate change, and the reliability of water supplies from rain and snowmelt becomes more problematic to the person responsible for long-range planning.

In addition to water rights and their variable yields, the legal requirements of the water system must also be understood. All water systems, even those that are in riparian areas, have constraints imposed on them by law. In riparian systems, no water user is allowed to reduce the quantity or quality of the stream to the point where other users are deprived of their rights to use it. In appropriations and permit system areas, each user has specific limits on volumes and flow rates of diversions. In addition, many systems have obligations to return flows to the stream at specific points in order to maintain historic stream flow patterns. These obligations must be understood and included in planning in order to capture the interaction of diversions and return flow obligations, which are often made from wastewater effluent.

Water systems that rely primarily on groundwater are not subject to droughts to the same extent as those that rely on precipitation. However, groundwater systems still must consider the effect of drought in the form of reduced water levels in their alluvial wells or depletion of a nontributary aquifer through groundwater mining. These special cases are not a main focus of this article, but they deserve careful analysis in their own right.

Storage must be adequate and put to use. Reducing demand in times of surplus will have little benefit if it does not result in more water being available in times of shortage. Carryover storage in raw water reservoirs allows surpluses in wet years to be available to make up for shortages in dry years. The amount of storage relative to the annual demand of the water system will affect the best water conservation strategy over the long term. However, the importance of conservation and system efficiency does not diminish as carryover storage drops. The same drought that inflicts serious but manageable shortages in the efficient system may result in a catastrophic shortage for the system with little storage and a lot of waste.

The nature of a drought is unpredictable and often unrecognized. For purposes of this discussion, a drought is defined as a period of decreased precipitation that results in reductions in the amount of water available to the water system in the form of diversions for direct use or storage. Drought severity is characterized by the deficit between the actual annual water supply and the average supply. Drought duration is the interval from when system storage fails to reach at least 67% of capacity to when it refills to this level.

This discussion assumes that the key to recognition of the drought on the part of the system operators is unusually low runoff that fails to fill storage during the spring. Therefore, maximum storage in the system each spring is the primary determinant for drought recognition. Other parameters such as decreased summer precipitation and soil moisture are certainly important. However, because it is impossible to predict snowpack from summer precipitation and because snowpack and winter/spring precipitation are essential to providing annual water supplies, this discussion focuses on spring reservoir storage levels as the key factor for drought recognition.

A "MODEL" COMMUNITY CONFRONTS DROUGHT

A long-range operations model helps account for system variables. The operation of a municipal water system involves the interaction of many elements. Changing one element often affects the others, making it difficult to predict how any set of changes will affect overall operations over time. The most logical approach to account for this level of complexity is an operations model that mimics the system operations and can be run with various combinations of demands, supplies, storage, and short-term drought restrictions.

Such a model, developed by the author over several years, was used to create a simple case study for the purpose of demonstrating how various parameters can be included in the evaluation process. For this exercise, the case study was predicated on a system in which demands come almost exclusively from single-family residential customers. This simplified the analysis without sacrificing the validity of the results.

Model spans 25 years. The model was operated on a monthly time-step over a 25-year study period. The starting year is set at 1990, but most of the data used for the model are not historical. As modeled, the water system experiences only minor growth over the 25 years. At the start of the period, 20,000 accounts are on line, which increases to 23,000 accounts in 2014, the last year of the study (Figure 1).

Household water use changes over time. At the beginning of the study, the water use in the model households is set at a level typical of standard homes built prior to the 1992 National Energy Policy Act (NEPA). Total annual household use averages 151 kgal (0.46 acre-ft), measured at the water meter. Average indoor use is 70 gpcd, and the average home has 2.8 residents; thus, typical indoor use per household amounts to 71 kgal per year (0.22 acre-ft). Average indoor use for the entire system totals 4,367 acre-ft. Outdoor use is 4,890 acre-ft per year, or 79.67 kgal per household. The assumed average irrigated area per home is 4,000 sq ft, equivalent to ~20 gal/sq ft of irrigated area. The assumed theoretical irrigation requirement for the area is 18 gal/sq ft, based on a reference evapotranspiration (ET) of ~30 in. per year, a typical landscape mix of mostly lawns and some shrubs, and the irrigation efficiencies normally found in residential developments.

These data paint a portrait of a community of homes that is fairly typical of established neighborhoods and with considerable potential for saving water through both indoor and outdoor conservation measures. It is assumed that indoor use can be reduced by 35% (from 70 to 45 gpcd) through a comprehensive interior retrofit program. Outdoor use can be reduced by approximately the same percentage through Xeriscape™, improvements in efficiency, and better controls. The model also assumes that the system has the potential of replacing 10% of its irrigation demand with nonpotable reuse programs. A restricting assumption, however, is that these efforts require from five to

seven years to take hold, effectively eliminating their use as short-term drought responses. One exception is that the model assumes it is possible to reduce outdoor use by 25% through restrictions and with little loss in value. However, reductions greater than 25% and for longer than two years are assumed to create significant losses to standard (non-water-wise) landscapes. In addition, it is assumed that short-term reductions in indoor uses of 15% are easily achievable for an indeterminate period of time, but reductions greater than 15% create unacceptable economic losses.

All new customers to the system are assumed to have demands that reflect implementation of NEPA, which mandated 1.6-gpf toilets, 2.5-gpm showerheads, and 2.5-gpm faucet aerators.

Outdoor demands are allowed to rise and fall in proportion to the annual ET as a percentage of the long-term average ET. These fluctuations, combined with indoor demands and water use for new homes, result in the baseline demand pattern shown in Figure 2. These demands include ~10% real (physical) water losses, which typically are caused by distribution system and service line leakage. The study assumes that these losses can be reduced to 5% over a seven-year period by implementation of a leakage management program. Again, because of the lead time that is involved, this water conservation strategy cannot be used as a short-term drought response.

At the beginning of the study, the model water system is assumed to have no special water conservation program other than requiring new customers to meet the NEPA requirements. The system does not have a leak detection program, nonpotable irrigation system, indoor retrofit program, or outdoor efficiency program in place.

Over the course of the study period, the average demand is 11,163 acre-ft. The maximum demand is

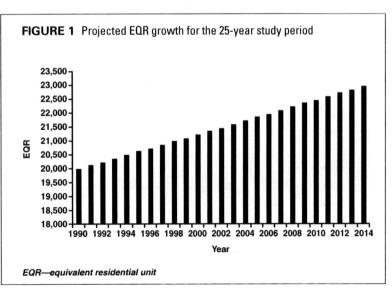

FIGURE 1 Projected EQR growth for the 25-year study period

EQR—equivalent residential unit

161

Article 22

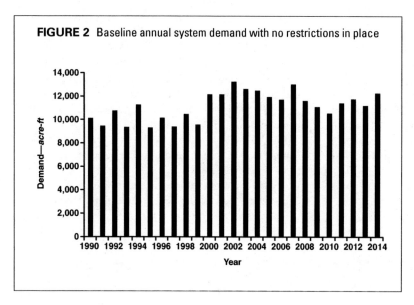

FIGURE 2 Baseline annual system demand with no restrictions in place

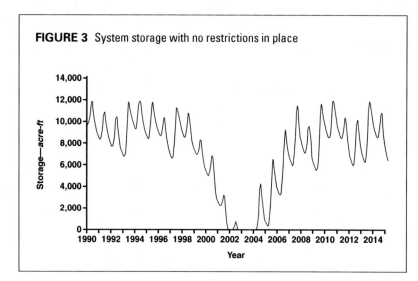

FIGURE 3 System storage with no restrictions in place

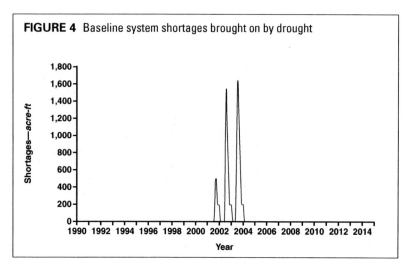

FIGURE 4 Baseline system shortages brought on by drought

13,235 acre-ft in 2002 (the 13th year of the study period). The minimum demand is 9,292 acre-ft in 1995 (the 6th year of the study period).

Model sets up drought that affects system sources and storage. The system has both groundwater and surface water supplies available to it. Groundwater supplies yield a constant 250 acre-ft per month, which is assumed to be stable over time. Surface water supplies comprise both senior and junior water rights. Under normal conditions, the surface supplies yield an average of 10,700 acre-ft per year of divertible and storable yield.

The system has 12,000 acre-ft of storage in which any of its surplus surface diversions may be stored. This capacity is equivalent to approximately one year of annual demand. Losses from evaporation and seepage account for an average of 1% per month of the volume in storage.

Under the constructs of the model, a drought occurs in the system beginning in 1999, the ninth year of the study period. As in the real world, system managers have no prior warning.

In the first year of the drought, the yield of surface water rights is reduced by 28%. In the following years, yields rise to 90% of average (2000), fall to 70% of average over the next two years (2001 and 2002), and dip even further the next year (2003) to 50% of average. In 2004, the system fully recovers, with yields rising to 125% of average.

Figure 3 shows the monthly reservoir storage with no constraints on the system. As the figure indicates, the system experiences severe shortages beginning in 2002 and continuing through 2004. Some recovery occurs in 2004, but there would be no assurance the drought is receding until at least 2006 or 2007. Monthly shortages to the system are shown in Figure 4. The total shortage to the system is 12,280 acre-ft over a four-year period. The maximum shortage of 5,870 acre-ft occurs in 2003 and represents 46% of the unconstrained demands.

As modeled, the drought represents a water supply challenge spanning eight years. Droughts of this magnitude are well-documented in the historical record. Whether such an event constitutes a crisis depends in great part on how it is handled.

There are two basic responses to the drought situation created by the model. The first option is to leave the baseline demands as they exist and rely on short-term restrictions and rationing. The second approach is to implement a comprehensive water conservation program at the beginning of the study period. Although the system managers do not know a drought is coming, the model assumes they begin work on a demand-management program in year one of the study as part of a long-term conservation effort.

How water rationing leads to a $400 million drought. Under the water rationing approach, it is assumed that the drought is not recognized until the spring of 2000 when the reservoir storage fails to exceed 8,000 acre-ft. Only then are restrictions put in place, starting with outdoor uses and then ratcheting up to indoor restrictions as needed to eliminate the shortages. During the first year of the drought, managers cannot know its duration, so they are unlikely to impose draconian measures. The model assumes that in the first year a 25% reduction in outdoor use is requested and that customers respond immediately. This measure eliminates the shortage in 2001, but because yields remain down and the reservoirs again fail to fill, it becomes necessary to increase the outdoor restrictions to 50% and also request a 10% reduction in indoor uses. It is assumed that these restrictions are maintained and increased as long as storage levels fail to reach at least the 8,000 acre-ft mark (Figure 5).

Table 1 shows the decreasing percentage of demands that the rationed system can meet without causing storage levels to continue to decline. As the table shows, reliance on these restrictions for drought management does not just "trim the fat" but cuts to the bone, severely restricting all outdoor water use. In 2001, only 50% of outdoor demands are met, and indoor uses are curtailed by 10%. In 2002 and 2003, it is necessary to cut outdoor deliveries by 75% and reduce indoor uses by 15%. In 2004 and 2005, all outdoor use is eliminated and indoor use is held at a 15% reduction. These levels of rationing for outdoor use eliminate the type of plant material in place at the start of the drought, leaving only those plants that can survive on what little precipitation falls. Basically all nonnative trees, shrubs, and grasses perish. By 2007 when supplies return to "normal" levels, the

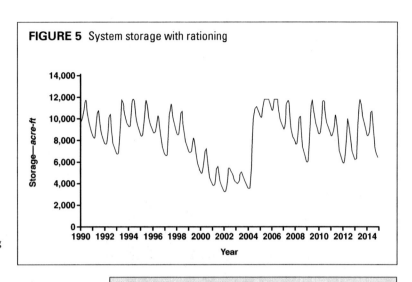

FIGURE 5 System storage with rationing

TABLE 1	Percentage of demand met under restrictions required to stabilize the system	
	Demand Met—%	
Year	**Indoor**	**Outdoor**
1990	100	100
1991	100	100
1992	100	100
1993	100	100
1994	100	100
1995	100	100
1996	100	100
1997	100	100
1998	100	100
1999	100	100
2000	100	75
2001	90	50
2002	90	25
2003	85	25
2004	85	0
2005	85	0
2006	85	90
2007	85	100
2008	100	100
2009	100	100
2010	100	100
2011	100	100
2012	100	100
2013	100	100
2014	100	100

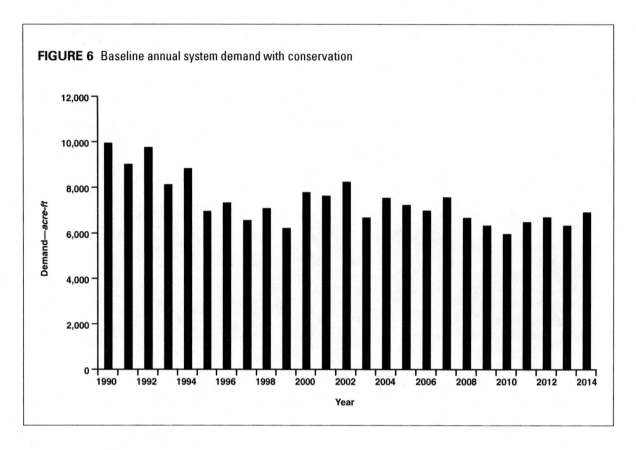

FIGURE 6 Baseline annual system demand with conservation

entire community must be relandscaped. If the cost of relandscaping is projected at $20,000 per house, then the total cost to the community is $400 million.

Launched early enough, water conservation helps reduce drought costs. An alternative scenario assumes that starting in 1990, the community launches a comprehensive water conservation and management program. The conservation/water management program consists of the following elements.

Indoor measures:
• All new customers and 10% per year of existing customers install 1.6-gpf toilets until all toilets in the system meet that standard.
• All new customers and 10% per year of existing customers install high-efficiency showerheads and faucet aerators until all faucets and showers meet that standard.
• Half (50%) of new customers and 10% per year of existing customers install high-efficiency clothes washers until 60% of all clothes washers in the system are high-efficiency appliances.
• Half (50%) of new customers and 10% per year of existing customers install high-efficiency dishwashers until 80% of all dish washers in the system are high efficiency appliances.

Outdoor measures:
• A target of 35% reduction in gross irrigation use is established for homes employing water-wise

landscapes, better irrigation control, and more efficient irrigation systems.
• Half (50%) of all new customers and 5% per year of all existing water utility customers meet this standard until all landscaping demands are reduced by 35% from the baseline rates.
• A nonpotable supply system is constructed that replaces 10% of the utility's outdoor demands with nonpotable supplies.

System measures:
• A leak management program is implemented that reduces the 10% annual water loss to 5% over a seven-year period.

These measures result in a reduction in system demands (Figure 6). Demands actually decrease over the period, even with the same system growth as shown in the baseline case. Figure 7 shows that under this scenario the system's spring storage levels approach the 8,000 acre-ft trigger in only a single year, 2003. The model assumes that system management requests a 25% reduction in outdoor use even though the actual storage is slightly above 8,000 acre-ft. No additional restrictions beyond this single reduction are required.

Under this conservation scenario, the system passes through the same eight-year drought period with virtually no problems. The key to this success is that conservation measures are set in motion early enough that the more difficult and time-consuming retrofit proj-

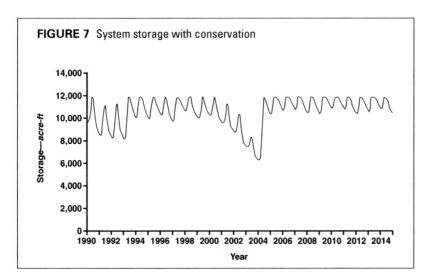

FIGURE 7 System storage with conservation

ect and system programs are largely complete by the time the drought occurs.

The system with conservation measures in place ahead of time avoids the $400 million of real damages suffered by the system on restrictions. Clearly implementation of the conservation program carries its own substantial costs. If full replacement value of the new fixtures and appliances is assigned to each house, the cost is ~$1,500 per home. The cost to implement the landscaping improvements for better controllers, audits, and sod replacement is ~$3,000 per home (on average). The total cost to the community then is $4,500 per home, or $90 million.

Although the costs to implement the leakage management program and nonpotable supply system must be taken into account, these costs are not on the same plane as the devastating monetary damage caused by the restriction model. Furthermore, the conservation measures bring benefits to the community that go beyond the effects of demand reduction. They represent an investment in new and better technology, a more robust and beautiful landscape, and a more efficient water distribution system.

SUMMARY

In the model scenarios described here, it is hard to imagine that the total loss of all landscape in the community would ever be considered a better outcome than the implementation of the comprehensive water conservation program. Even assuming a conservation program cost of $100 million, this figure is only a quarter of the cost incurred by reliance on restrictions alone as the drought response method. Furthermore, not all of these costs are borne by the utility, and they can be managed over many years rather than imposed unexpectedly and without plan by a natural disaster.

The issue of demand hardening can be addressed by looking at how much damage is caused by the 25%

reduction in outdoor water use imposed under the conservation scenario. If the community's landscapes are managed on such a slim margin that a 25% reduction in one year causes as much damage as the seven years of restriction shown in Table 1, then this is a case of demand hardening negating the benefits of conservation. However, the water-wise landscapes envisioned as part of the conservation program obtain most of their savings from better scheduling and reduced waste. They still allow significant amounts of turf, which can easily survive being shorted for a single year. Therefore, it seems unlikely that a 25% reduction in outdoor use on the water-wise landscapes would cause significant long-term damage.

The data presented in this case study suggest that water conservation can play a significant role in helping most water systems manage all but the most serious droughts. The key to the success of the water conservation approach is that it is implemented ahead of time, and increased storage levels in the raw water system are in place before the drought's onset. The water yielded by long-term demand reduction is far greater than that obtained through water rationing and restrictions.

By the time the managers of the system without conservation realize they are entering a drought it is too late to capture significant savings. They are left with no alternative but imposition of strict water-use restrictions, which hit outdoor uses the hardest. Additional scenarios could be run with higher growth rates and different combinations of water rights and storage, but the same general conclusions are likely to emerge—the best drought strategy is to be prepared, take the long view, and have conservation measures well under way before supplies become scarce.

ABOUT THE AUTHOR

William B. DeOreo is president of Aquacraft Inc., Boulder, CO. He has BS and MS degrees from the University of Colorado at Boulder and a bachelor's degree from Boston University, Boston, Mass. A member of AWWA, the American Society of Civil Engineers, and the National Society of Professional Engineers, he has focused much of his professional career on understanding the effects of water conservation on water system demand and reliability. He was principal investigator for the AWWA Research Foundation study on residential end uses of water and has developed water conservation plans for several municipal systems.

This page intentionally blank.

Aquatic Ecosystem Protection and Drinking Water Utilities

BY SANDRA L. POSTEL
(*JAWWA* February 2007)

As one of the most publicly visible stewards of the earth's water sources, drinking water utilities are uniquely positioned to exert a leadership role in the emerging field of ecologically sustainable water management. In important ways, this field is integrating the traditional goals of water management with those of ecosystem conservation in order to sustain a broader spectrum of the valuable goods and services on which human communities depend.

The water strategies of the twentieth century helped provide much of the human population with drinking water, food, electricity, and flood control. Indeed, it is difficult to imagine today's world of 6.5 billion people and $55 trillion in economic output without the vast network of water infrastructure now in place—from dams and reservoirs to wells, pumps, and canals. This infrastructure, however, has disrupted the functioning of aquatic ecosystems on a large scale. If future human needs are to be met without costly and irreparable harm to ecological health, new strategies will be needed that incorporate a broader set of ecological goals into water planning and management.

Drinking water utilities will play an increasingly important role in designing and implementing these new strategies. Globally, municipal water use accounts for less than 10% of total water demands. Because municipal demands are concentrated geographically, however, they can disproportionately affect rivers, lakes, aquifers, and related aquatic ecosystems (FitzHugh & Richter, 2004). Moreover, the world has recently crossed an important demographic threshold: more than half the human population now lives in urban areas, and the ratio of urban to rural dwellers is likely to continue to increase for the foreseeable future. This development will heighten tensions over water between farms and cities, a trend already visible in parts of the world (Postel & Vickers, 2004). Much of the world's urban growth will occur near coasts, heightening competition for water in the downstream reaches of many river basins and threatening the health of estuarine and marine ecosystems.

Trends in the United States mirror these global trends to some degree. Now numbering some 300 million, the US population is projected to reach 420 million by 2050 (US Bureau of the Census, 2004). Estimates compiled by the US Geological Survey (USGS) show that withdrawals for municipal water supplies are the fastest-growing category of withdrawals nationwide, rising 8% between 1995 and 2000, compared with a 2% rise for total withdrawals nationwide (Hutson et al, 2005). The prospect of a changing climate that may include an increase in the number and intensity of regional droughts underscores the importance of water managers getting ahead of the game by proactively planning and managing for this emerging new era of water tensions, worsening shortages, and myriad threats to ecological health.

This article details the nature and importance of aquatic ecosystem health and ecological services as well as the ways in which drinking water systems affect and influence aquatic ecosystems, and it offers some methods, tools, and examples that show how more ecologically sustainable water management can be achieved.

PROTECTING AQUATIC ECOSYSTEM HEALTH AND SERVICES IS ESSENTIAL

Aquatic ecosystems provide numerous irreplaceable benefits. Aquatic ecosystems encompass rivers, streams, floodplains, lakes, wetlands, underground aquifers, springs, and coastal estuaries (where freshwater meets and mixes with saltwater). All of these ecosystems are part of and are supported by the hydrologic cycle—the sun-powered movement of water between the sea, air, and land that sustains life on earth. To the extent human actions alter this cycling of water, they affect the ecosystems that are sustained by it.

Aquatic ecosystems provide numerous goods and services of significant value to humans (Postel, 2005), including

• water supplies for irrigation, industries, cities, and homes;

This article was originally developed as a concept paper by the author at the request of The Nature Conservancy and was modified for publication in JOURNAL AWWA. It is being published with the permission of The Nature Conservancy.

• fish, waterfowl, mussels, and other foods for people and wildlife;
  • water purification and filtration of pollutants;
  • flood mitigation;
  • drought mitigation;
  • groundwater recharge;
  • water storage;
  • wildlife habitat and nursery grounds;
  • soil fertility maintenance;
  • delivery of nutrients to deltas and estuaries;
  • delivery of freshwater flows to maintain estuarine salinity balances;
  • recreational opportunities;
  • aesthetic, cultural, and spiritual values; and
  • conservation of biodiversity, which preserves resilience and options for the future.

The ancient Egyptians thrived for several thousand years on the ecological services provided by the annual flood of the Nile River, which delivered water and nutrients to their fields, carried off harmful salts that had accumulated in the soil, and supported a diversity of fish. Because most ecological services lie outside commercial markets and are not priced in conventional ways, they tend to be undervalued and underappreciated.

In recent years, economists have attempted to place monetary values on some of these ecosystem services, and although these estimates are imperfect, they do provide society with a ballpark sense of ecosystem worth and importance. The recently completed Millennium Ecosystem Assessment reported that the benefits provided by wetlands—including fishing, hunting, recreation, water supply, flood control, water purification, nursery functions, and biodiversity habitat—totaled on average $1,325/acre per year (Millennium Ecosystem Assessment, 2005). The estimated value will vary considerably from one site to another; the ecological services provided by wetlands in Massachusetts, for example, have been valued at up to $6,250/acre per year (Breunig, 2003). Although approximate, such estimates underscore that it is unwise (and often uneconomical) to drain, fill, or convert wetlands to other uses without first assessing the importance of the work they do and the potential value of what will be lost. The same is true for the alteration of rivers and other aquatic ecosystems.

Freshwater ecosystems must be in reasonably good health in order to provide the valuable goods and services that economies depend on. Ecological health refers to the condition or state of an ecosystem's life-support functions and processes, which directly affects its ability to deliver goods and services to human communities. Just as doctors check blood pressure, cholesterol levels, and heart rate to see if these values fall within a range essential for good human health, scientists similarly assess certain ecosystem attributes to determine whether they fall within a range essential for good ecological health.

For water managers, placing the words "ecologically sustainable" alongside their primary responsibilities—whether those entail generating hydropower, delivering irrigation water, or supplying cities with drinking water—requires that they make the protection of ecological health an even more central goal of their planning, operations, and management. To date, attention to ecological effects (e.g., through environmental impact assessments) has typically come after a project is conceived of and designed, with the intent of mitigating the most egregious harm to ecosystems. Protecting ecological health, however, demands a different framework, one that proactively integrates ecological goals into water planning and project design from the start. In this way, effects on aquatic ecosystems can be minimized, not just mitigated. Moreover, by focusing on ecological goals early on, water planners and managers have a much greater opportunity to develop creative solutions to the challenge of meeting human and ecological water needs at the same time.

Ecological goals focus on four areas. In most cases, the ecological goals that water managers will need to integrate into their work fall into four broad (and to some degree, overlapping) areas: environmental flows, source water protection, water quality, and groundwater management.

Environmental flows. The health of rivers and streams depends to a large degree on water flows. Until fairly recently, water laws and management guidelines have concentrated on the maintenance of minimum flows during dry periods, i.e., making sure rivers and streams have at least some water in their channels. During the past decade, however, freshwater ecologists have demonstrated that this minimum-flow paradigm, although perhaps adequate to support recreational boating and fishing, is not sufficient to protect ecological health. Specifically, they have identified four major types of ecological consequences resulting from the alteration of river flows. First, flow alteration can impair or destroy the physical habitats of a river's channel (e.g., riffles and pools) and floodplain. Second, it can prevent aquatic species from receiving the cues or opportunities they need to fulfill their critical life stages, including reproduction. Third, flow alteration can disconnect rivers from their floodplains and fragment them longitudinally, depriving species of important habitats and disrupting ecological services. Fourth, altered flows can allow populations of exotic and introduced species to expand at the expense of native populations, which disrupts food webs and ecological functions (Postel & Richter, 2003).

Consequently, the minimum-flow paradigm is being replaced by a new approach to river management based on the natural flow paradigm. This approach calls for water managers to sustain or replicate a river's natural pattern of variable flows—the pattern of high and low flows, as well as periodic floods and droughts—that the river historically exhibited and to which the myriad life forms in the river have become adapted. (Postel & Richter, 2003; Poff et al, 1997; Richter et al, 1997). The approach does not call for or require a return to the "natural" state, but it does entail maintaining a flow regime that resembles the natural historical one to a sufficient degree to sustain the ecological functions of the aquatic system (Figure 1).

Source water protection. The health of rivers, lakes, and other aquatic systems is closely tied to the condition of the watersheds that feed into them. The quantity, quality, and timing of runoff coming out of a watershed may vary greatly with land-use patterns, vegetative buffers around streams and reservoirs, and other landscape features. It is well accepted, for example, that watersheds with significant areas of protected or well-managed forests offer water purification services that are superior to other landscape

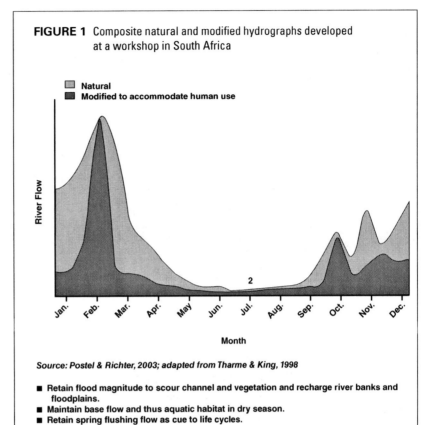

FIGURE 1 Composite natural and modified hydrographs developed at a workshop in South Africa

- Natural
- Modified to accommodate human use

River Flow

Jan. Feb. Mar. Apr. May Jun. Jul. Aug. Sep. Oct. Nov. Dec.

Month

Source: Postel & Richter, 2003; adapted from Tharme & King, 1998

- Retain flood magnitude to scour channel and vegetation and recharge river banks and floodplains.
- Maintain base flow and thus aquatic habitat in dry season.
- Retain spring flushing flow as cue to life cycles.
- Vary base flow in wet season but with removal of some floods.

When modified to accommodate human uses, the river still retains a flow pattern that keeps it healthy and close to the natural flow regime.

types. For example, a survey of US water suppliers by the AWWA Source Water Protection Committee and the Trust for Public Land (TPL) found that water treatment costs are inversely related to the proportion of the watershed protected by forests (Barten & Ernst, 2004; Ernst, 2004). Table 1 shows this relationship between forest cover and predicted treatment costs.

In addition to higher-quality source water, protected watersheds provide other valuable ecological services as well. Land-use activities in aquifer recharge zones can have severe consequences for groundwater, with implications for water supplies and ecosystems sustained by underground water flows. For all these reasons, water managers are increasingly being called on to protect source watersheds. Some hydroelectric producers and drinking water suppliers are even finding that it makes economic sense to pay for the protection of watershed lands they do not own or control because of the improved supply reliability or water quality this protection provides (Postel & Thompson, 2005).

Water quality. Although many countries, including the United States, have made substantial improvements in water quality during the past several decades,

pollution of various kinds continues to jeopardize the health of aquatic ecosystems. Concentrations of nutrients (primarily nitrogen and phosphorus) have increased substantially in rivers throughout the world, which is causing eutrophication, harmful algal blooms, and high levels of nitrate in drinking water sources. In many industrialized regions, riverborne nitrogen has increased up to fivefold from preindustrial levels. As rivers and streams carry these pollution loads to the sea, coastal zones are becoming overenriched with nutrients, leading to a growing number of hypoxic (low-oxygen) dead zones that pose serious threats to coastal fisheries. For example, more than half of the coastal bays and estuaries in the United States are degraded by excessive nutrients (Howarth et al, 2000). In addition, chemical contamination from pesticide use, industrial and military discharges, and other sources is a continuing problem in industrial countries and a growing problem in the developing world (Postel, 2005). The USGS recently established that pesticides are found in virtually all US rivers and streams, often at levels harmful to aquatic life and fish-eating wildlife (Gilliom et al, 2006).

TABLE 1	Forest cover and predicted water treatment costs		
Watershed Forested—%	Treatment Costs per 3,785 m³—$	Average Annual Treatment Costs—$	Cost Increase Over 60% Forest Cover—%
60	37	297,110	–
50	46	369,380	24
40	58	465,740	57
30	73	586,190	97
20	93	746,790	151
10	115	923,450	211

Source: Postel & Thompson 2005; adapted from Ernst 2004.
Table reflects responses from 27 US water supply systems and is based on treatment of 22 mgd (83,270 m³/d), the average production of the water suppliers surveyed.

Groundwater management. Although hidden from view, underground aquifers are inextricably linked to rivers, lakes, wetlands, and other aquatic ecosystems. Natural drainage of water from groundwater aquifers often provides the base flow that keeps rivers running during summer months and dry spells. Shallow groundwater often sustains essential wetland habitat either seasonally or year-round. Water stored underground is also an important reservoir during droughts. Water professionals have increasingly recognized the importance of managing surface water and groundwater as the interconnected systems they are, but conjunctive use and management have often been difficult to achieve.

As rivers, lakes, and streams become less available and more expensive to tap for new supplies, many water providers, farmers, and private enterprises are turning to underground aquifers. Signs of overpumping, i.e., extracting water faster than it is being recharged, are pervasive and spreading. Water tables are dropping in many parts of the world, including important farming and urban areas of China, India, Mexico, the Middle East, north Africa, and the United States. As much as 10% of current global food production (and closer to 25% in India) may rely on the unsustainable use of groundwater (Postel, 1999). Thus, managing groundwater to meet future human needs while also sustaining its important ecological functions constitutes a major challenge to water managers.

UTILITIES SIGNIFICANTLY INFLUENCE ECOSYSTEM HEALTH

The principal role of a drinking water provider is to supply water sufficient to meet the health and safety needs of the residents within its service area. Although many utilities are guided by an environmental ethic as they strive to fulfill this purpose, most will

inevitably affect the river, stream, lake, or aquifer that serves as the source of supply. **Common water supply methods harm ecosystems.** Most utilities draw water in one of four ways: directly from a river or stream, directly from a natural lake, through wells that tap underground aquifers, or from artificial reservoirs created by a dam. Each of these methods can cause significant ecological harm. For instance, direct withdrawals from a river can reduce flows to levels harmful to fish and other aquatic life. Such flow depletion may also concentrate pollutants and push important water quality parameters outside a river's healthy range. Similarly, withdrawals from a natural lake can lower lake levels, harm riparian and aquatic habitats, and increase pollutant concentrations. When groundwater is a primary supply source, pumping from wells may exceed rates of recharge, resulting in long-term depletion of the supply. Even if pumping causes only short-term depletion and aquifer levels recover during wet years, the temporary lowering of water tables may damage ecosystems dependent on that groundwater for base flows.

Dams built to create a reservoir to store water supplies have perhaps the most obvious ecological consequences and rank near the top of the list of causes of riverine habitat loss and impairment (Silk & Ciruna, 2005; Postel & Richter, 2003). Dams transform part of a flowing river into a lake, an entirely different aquatic environment. Dams physically block the migration routes of various fish species, often preventing fish from reaching critical spawning grounds. Because dam operations alter flow patterns and the delivery of sediment, nutrients, and organic matter, they often influence river ecosystems for tens or hundreds of miles downstream of the reservoir. Consequently, even though a utility's supply concerns focus on water in the reservoir, the ecological effects can ripple out considerably above and below the reservoir.

To the extent that utilities draw on water bodies in watersheds other than their own, they can export ecological damage to other regions. This is a common occurrence among cities that have grown considerably in size. As their service areas and customer water demands grow, they reach out to sources farther away to augment supplies. A utility's total effect on ecological health and ecosystem services often increases in scale and severity commensurately.

Most drinking water systems affect the health of their source waters to some degree, and the impacts of large urban systems can be especially serious (Postel, 2005; FitzHugh & Richter, 2004). Here are a few US examples:

• Atlanta, Ga., draws most of its water from Lake Lanier on the Chattahoochee River and Lake Allatoona on the Etowah River. These multipurpose reservoirs, constructed in the 1950s, have greatly altered the natural flow regimes of the two rivers. Many of the native aquatic species in the Etowah are now locally extinct; a nonnative trout fishery has developed in the Chattahoochee because of cold-water releases from Lake Lanier. There is concern that Atlanta's growing water demands will damage the prized fisheries of Florida's Apalachicola Bay (the source of 90% of Florida's oysters) as increased withdrawals from the Chattahoochee system reduce flows into Florida's Apalachicola River, which supports the bay.

• San Antonio, Texas, relies on the Edwards Aquifer as its principal source of drinking water. The aquifer is also a major source of irrigation water in south-central Texas. By the early 1990s, heavy groundwater pumping from the aquifer had reduced flows into San Marcos and Comal Springs, further jeopardizing seven species that were listed under the federal Endangered Species Act.

• Los Angeles, Calif., reached beyond the basin of the Los Angeles River to the Owens Valley and Mono Lake watersheds in order to meet the demands of its exploding population during the twentieth century. This expansion had dramatic consequences. Owens Lake completely dried up. Los Angeles's diversion of water from four of the five tributaries feeding into Mono Lake caused the lake's level to drop more than 40 ft, drained half of the lake's volume of water, and doubled its salinity. Ecological health and aquatic life in these ecosystems suffered accordingly.

• New York, N.Y., reached out to the Delaware and Catskills watersheds when it outgrew the water supplies available from the Croton and other local basins. The string of reservoirs the city built has greatly altered the flow regimes of Schoharie and Esopus creeks in the Catskills and of the Neversink River and east and west branches of the Delaware River. The ecological effects include the apparent disappearance of most mussels from the Delaware River and damage to the river's American shad run, once the largest shad run in the country.

The common theme running through these examples is that the ecological health of the rivers, lakes, and groundwater sources that were tapped to supply drinking water was not sufficiently taken into account at the time of project conception and planning. In some cases, the ecological ramifications of the water projects were simply not fully understood. Even today, however, knowl-edge of ecological effects rarely leads to adequate measures to minimize or avoid them. In addition, once the dams, reservoirs, diversion canals, or groundwater wells are constructed, all too often little consideration is given as to how this infrastructure might be operated to protect the health and functioning of aquatic communities.

There is a visible trend, however, toward policies and management strategies that place greater emphasis on ecological health and would in turn influence water utility operations. At the international level, much attention has been paid to South Africa's 1998 National Water Policy Act, which calls for the establishment of a "Water Reserve." The reserve consists of two parts—one intended to protect the health of people and the other to protect the health of ecosystems. The first part is a non-negotiable water allocation to meet the basic drinking, cooking, and sanitary needs of all South Africans (on the order of 20 L/d per person). The second part of the reserve is an allocation of water to support ecosystem functions so as to secure the valuable services they provide to South Africans. Specifically, the act specifies that "the quantity, quality, and reliability of water required to maintain the ecological functions on which humans depend shall be reserved so that the human use of water does not individually or cumulatively compromise the long term sustainability of aquatic and associated ecosystems" (South African National Water Act, 1998). The water determined to constitute this two-part reserve has priority over licensed uses (such as irrigation), and only this water is guaranteed as a right.

Following South Africa's lead, a number of national and international conferences, commissions, legislative directives, and laws have called for similar approaches. A significant development occurred in 2001 at the International Conference on Freshwater in Bonn, Germany. Delegates from 118 countries included in their recommendations to the following year's World Summit on Sustainable Development that "the value of ecosystems should be recognised in water allocation and river basin management" and that "allocations should at a minimum ensure flows through ecosystems at levels that maintain their integrity" (International Conference on Freshwater, 2001). In effect, water authorities worldwide are being called on to overhaul water policies and management practices in order to safeguard freshwater ecosystems.

Several other international policy initiatives aimed specifically at improving ecological health are reshaping water management strategies as well. In Australia, for example, water withdrawals from the Murray-Darling River Basin, the nation's largest and most economically important river system, have been capped in order to arrest the severe deterioration of the system's health. After a tripling of withdrawals between 1944 and 1994, the Murray's flow dropped to ecologically harmful levels, wetlands and fish populations decreased, and salinity levels and the frequency of algal blooms increased. The capping of withdrawals means that new water demands in the Murray-Darling Basin are met primarily through conservation, efficiency improvements, and water trading. Although scientists point out that the cap alone will not restore the river to health, it at least arrests the decline while other restoration strategies are put in place (Postel, 2005).

In the European Union, the 2000 Water Framework Directive established criteria for classifying the ecological status of water bodies and called on member countries to prevent any deterioration in this status and to bring all water sources to at least "good" status (European Parliament & Council of the European Union, 2000). In December 2005, US governors and Canadian premiers in the Great Lakes Basin approved agreements to implement the 2001 Annex to the Great Lakes Charter, which calls for no significant net degradation of the basin's freshwater ecosystems (Postel & Richter, 2003; Council of Great Lakes Governors, 2001).

Given this growing need and expanding mandate to incorporate ecological health into traditional water supply planning, it is fortunate that water utilities already have a useful mix of tools, methods, and strategies at hand to achieve these missions. Most utilities, however, have a long way to go in realizing the potential of these measures to improve ecological health, which is good news for aquatic ecosystems: there are untapped alternatives ready to be deployed.

TOOLS ARE AVAILABLE TO SAFEGUARD AQUATIC ECOSYSTEMS

Water planning and management have evolved considerably over the past two decades. Many of the concepts and ideas embedded in the emerging framework termed "ecologically sustainable water management" (ESWM) are similar to those in the integrated water resource management (IWRM) framework familiar to many water utilities. Just as IWRM expands on the least-cost planning philosophy of integrated resource planning by giving more attention to environmental assets (Beecher, 1995), so ESWM is a further evolution of planning concepts and goals that incorporates new scientific understanding about ecological health and ecosystem services (Richter et al, 2003). As discussed previously, attention to environmental flows, source water protection, water quality, and groundwater management will be key features in the implementation of ESWM strategies. Because many of the tools and strategies described in the following sections cut across these categories, they are purposefully not grouped under a specific one. This summary of ideas and examples is intended to be illustrative, not exhaustive, and includes only strategies that have demonstrated potential to help meet ecological goals.

Reduce water demand through effective conservation.
Perhaps the most powerful tool available to every water provider—regardless of whether its supply comes from a river, lake, artificial reservoir, or underground aquifer—is demand reduction through water conservation. Long perceived only as an emergency response to drought, water conservation is now a proven strategy to achieve long-term demand reduction (Vickers, 2001). By reducing the volume of water needed to adequately serve its customer base, a utility allows more water to be left in the natural environment to satisfy ecological needs. It may also eliminate the need for a new dam, permit the downsizing of a proposed dam, reduce the size or number of groundwater wells, and decrease the amount of energy and chemicals needed to treat and distribute the water supply.

The most effective conservation strategies to use vary from one city to another but almost always include attention to fixing leaks in the distribution system, reducing per capita household use (indoor and outdoor), and promoting conservation through pricing and rate structures, ordinances, and such incentives as rebates for installation of water-efficient appliances or payments to replace thirsty grasses with native landscaping (Vickers, 2001). The ecological benefits from these measures can be substantial, and they usually save consumers money as well. In Seattle, Wash., a conservation effort is helping keep more water in streams for salmon. In Massachusetts, an aggressive conservation program for the Boston metropolitan area, launched in 1987, brought water use to a 50-year low in 2004 and has indefinitely postponed the need for a diversion of the Connecticut River (Postel, 2005). In addition to aiding the restoration of Atlantic salmon populations and fisheries in the river, the conservation program saved Boston-area residents more than $500 million (in 1986 dollars) in capital expenditures alone (Postel & Vickers, 2004).

An expanding tool to guide conservation efforts is the establishment of benchmarks that help both utilities and consumers distinguish efficient water-use patterns from inefficient ones. For example, many AWWA utility members and some states have established 10% as their benchmark for distribution system water leaks and losses (nonrevenue water). This benchmark lets utilities know that if their percentage of nonrevenue water is 20% or 30% (and such rates are not uncommon), their operations are considerably outside the bounds of efficient water management. Similarly, in January 2006, the state of Massachusetts issued policy guidance for water permitting that sets a performance standard for residential water use of 65 gpcd in basins with medium to high water stress and 80 gpcd in others. This guidance also established nonrevenue water performance standards of 10% and 15% for the two basin categories, respectively. By tying the permit process to such benchmarks, state policymakers are attempting to build efficiency into municipal water systems and water use and thereby augment the volume of water available for aquatic ecosystems.

In a similar vein, water officials in the United Kingdom have identified benchmarks for water use in various types of public facilities, including offices, prisons, schools, and hospitals. Based on the water use levels of the most-efficient facilities in each category, these benchmarks provide useful and achievable targets. For instance, the United Kingdom study found that the benchmark for best water-use practice in offices was 31% lower than the median water use in offices. Similarly, water use in the most-efficient colleges and universities was 35% less than the median level. Overall, the differential among the various categories ranged from 17% to 47%, indicating substantial potential for water savings in these public facilities—water that might then be made available to freshwater ecosystems (UK Environment Agency, 2003).

Manage water within the bounds of an ecological flow prescription. As utilities attempt to incorporate new ecological goals into their management process, many may find themselves asking, "How much water does a river (or lake or aquifer) need to sustain its ecological functions?" Ecological flow prescriptions attempt to answer this question. Some prescriptions can be stated fairly simply: for example, one river's prescription might be that flows always remain within 20% of the natural flow. Other prescriptions may be quite complex, describing a range of flow targets to achieve for different times of the year. In the case of a heavily dammed river, the prescription might set target flows for five different components of the river's flow regime (big floods, small floods, high-flow pulses, low flows, and very-low flows) because each component serves a different ecological purpose and is essential to the river's overall health.

Implementing environmental flow prescriptions usually requires that water managers modify their operations to some degree, such as adjusting the schedule (timing and volumes) of reservoir releases, storing water differently, or withdrawing water at different times or in different places. One strategy, for instance, is to capture more water than needed during times of high flow, store it for later use, and then withdraw less water during periods of low flow. Such an approach can be particularly effective when the water that is skimmed off when flows are high is stored off-site (away from the natural river channel) rather than in an on-channel reservoir. Storing water off-channel typically causes less ecological harm than conventional dam-and-reservoir storage—there is no physical barrier to fish passage, less change in water temperature, and less entrapment of sediment and nutrients. As a result, off-channel storage can make it easier to achieve ecological goals related to water quality, temperature, and fish habitat as well as to meet specific flow recommendations.

Prescriptions can also be established to ensure that groundwater pumping does not cause ecological harm. In the case of the Edwards Aquifer, this prescription took the form of a cap on groundwater withdrawals in order to sustain groundwater flows into surface springs that support the federally endangered Texas blind salamander and the fountain darter. In response to lawsuits, the Texas Legislature established the Edwards Aquifer Authority in 1993 and set a cap of 555.3 million cubic metres (mcm) on annual pumping from the aquifer through 2007 and a more stringent cap of 493.6 mcm by 2008 (Edwards Aquifer Authority, 2005). As in Australia's Murray-Darling Basin, this cap has fostered an active water market (typically irrigators sell water to San Antonio) and has also encouraged more conservation within San Antonio, where per capita domestic use is now considerably lower than in most Texas cities.

Coordinated management of surface water and groundwater offers another tool for some utilities to meet environmental flow prescriptions. Floodwaters can be directed to infiltration zones to recharge aquifers, for instance, and then this underground supply can be tapped during dry periods. This can ensure that enough surface water is available to prevent river flows from dropping below the prescribed level.

Plan for ecosystem allocations during droughts. Ecologically sustainable water management requires that an ecosystem's water needs be met during times of drought as well as abundance. With forethought and innovative planning, utilities can meet both ecological flow requirements and consumer needs during droughts. Without such planning, however, utilities may respond to a drought in ecologically harmful ways, e.g., by completely eliminating outflows from their storage reservoirs in order to save as much water for their customers as possible, an approach that leaves the river downstream with little or no flow at all.

There are various ways utilities can ensure that both their customers and freshwater ecosystems receive the water they need to weather droughts. Perhaps the most important is to have in place an aggressive water conservation strategy specifically for droughts, one that is consistent with, but distinguishable from, the utility's long-term demand-reduction strategy. This drought strategy may begin with voluntary measures but must transition to mandatory measures if the drought persists and if reservoir capacity or stream flows drop to unacceptably low levels. Researchers have found that voluntary restrictions on outdoor water use are of limited value; it is the cities with the most stringent mandatory restrictions that realize the largest water savings (Vickers, 2006).

In response to a record-breaking drought during the summer of 2002, the city of Cheyenne, Wyo., implemented lawn-watering restrictions during the month of July that lowered average demand to 18.1 mgd compared with 34 mgd for July of the previous year—a 47% reduction. As a result, the city's reservoirs stayed more than 80% full during the summer of 2002 (Vickers, 2005). Such a reserve not only helps weather a worsening of the drought (should that occur), it also provides a supply to help ensure that downstream ecosystems receive enough water during the drought as well.

Colorado also experienced one of its worst droughts on record in 2002, but Denver was slow to respond with effective measures (Harmon, 2004). Despite the fact that the water savings achieved during June, July, and August fell considerably short of the drought program's goals, tight water restrictions and a lawn-watering ban were not put in place until September 1 and October 1, respectively. With reservoir levels so low, Denver Water implemented cloud-seeding later that year to try to increase rainfall over its storage reservoirs. Despite this effort, on Feb. 10, 2003, Denver's reservoirs were only 44% full, compared with a normal level of 82% for that point in the year (Gardner, 2004).

Along with mandatory water-use restrictions and bans, a helpful drought-planning measure for some utilities is a dry-year option arrangement. This strategy usually involves a city arranging to purchase water from farmers who are willing to forgo irrigating crops during a drought (which often will mean fallowing cropland) and instead sell water on a temporary basis to the city. If arranged well and in advance, such short-term water transfers may be not only ecologically beneficial but also economically advantageous to both parties: the farmers receive extra income from both the option and the short-term water sale, and the city secures supplies for a drought without having to build extra reservoir capacity. In states where water rights remain with public entities, a regional water authority can help arrange and execute these dry-season transfers.

Another useful tool, and one familiar to most utilities, is to establish a trigger level that in effect sounds the alarm to put the conservation strategy into motion. Utilities typically have a trigger tied to reservoir capacity levels; when capacity drops to a certain point, the drought plan kicks in. By adding a trigger tied to stream-flow levels, utilities can ensure that action is taken when ecological health is at risk. Massachusetts officials have set such a trigger for flows in the Ipswich River, which ran dry during the summers of 1995, 1997, and 1999. When flows in the Ipswich drop to a specified level, towns are required to institute water conservation measures. (Massachusetts officials are now in the process of requiring such triggers in much of the state.) Because it can be politically difficult to restrict water use, even during droughts, the existence of a publicized trigger level can be a useful tool for gaining public acceptance of drought measures.

It bears repeating that water-use restrictions, dry-year options, and other measures are necessary but not sufficient to ensure ecological health during periods of drought. Some portion of the water saved or acquired must be allocated to ecosystems at risk. There is an ethical dimension to this philosophy of sharing with nature: it implies that nonessential human uses of water, such as lawn irrigation, be curtailed before fish and other aquatic creatures begin to suffer and die.

Protect source watersheds. Compared with water used for irrigation or many industrial activities, drinking water must meet high standards of quality. The more polluted the raw source water, the more expensive and difficult it becomes to treat it and make it safe for drinking. As watershed lands are converted from forests and wetlands to urban and agroindustrial uses, there is increased potential for rivers, lakes, and aquifers that serve as drinking water sources to become contaminated with pathogenic organisms, heavy metals, farm and lawn pesticides, and other health-threatening substances.

Because of these threats, the importance of watershed protection is growing. In their foreword to the 2004 report *Protecting the Source,* jointly published by AWWA and TPL, AWWA Executive Director Jack Hoffbuhr and TPL President Will Rogers stated, "For 60 years, the safety of most of America's drinking water has been dependent on technology. Today, water suppliers are revisiting the idea that watershed protection—the first barrier against contamination—needs to, once again, be an integral part of their water quality protection strategy" (Ernst, 2004).

Fortunately, protecting the ecological services provided by healthy watersheds can also protect water quality and, in some cases, water quantity as well. As mentioned previously, natural forests and wetlands are unequivocally effective at cleansing water supplies. They act as nature's water factories, churning out high-quality water that helps keep rivers and streams healthy and also lowers the cost of treating water for domestic use. Healthy watersheds also provide many other valuable ecosystem services at the same time, including recreational opportunities, biodiversity conservation, erosion control, and climate protection. As a result, there are creative synergies to tap when designing and financing watershed protection strategies.

Several major US cities, for example, have avoided the construction of expensive filtration facilities by investing instead in watershed protection to maintain the purity of their drinking water. The US Safe Drinking Water Act requires that cities dependent on rivers, lakes, or other surface waters for their drinking supplies build filtration plants unless they can demonstrate that they are protecting their watersheds sufficiently to satisfy federal water quality standards. Boston, New York City, Portland (Me.), Portland (Ore.), Seattle, and Syracuse (N.Y.) are among the cities that have taken the watershed protection route and are saving their residents millions of dollars (and in the case of New York City, several billion dollars) in avoided capital expenditures (Postel & Thompson, 2005). Because the driving force behind these efforts is water quality protection, however, these cities are not necessarily maintaining the overall ecological health (including environmental flows) of their watersheds. As noted earlier, the management of New York City's reservoir system in the Catskills and Delaware watersheds has diminished and altered river flows in ecologically harmful ways.

An emerging strategy for watershed protection that is showing significant promise in some parts of the world is establishing mechanisms whereby the beneficiaries of a watershed's ecosystem services pay the providers of those services to protect them. In the absence of strong regulations and enforcement of watershed activities, this approach can promote and finance watershed protection by creating a market for buyers and sellers of valuable ecosystem services. There are many variations on this approach. New York City's effort, for example, involves investing $1.5 billion in an array of watershed activities—from forest management to controlling farm pollution to upgrading wastewater infrastructure—in order to avoid the $6 billion cost of a filtration plant (National Research Council, 2000).

Quito, the capital of Ecuador, has taken a different approach to paying for watershed services. The city draws about 80% of its drinking water from two protected ecological reserves, which, though part of the national park system, are also used for cattle, dairy, and timber production. With urging and support from local and international groups, Quito established a watershed trust fund to pool the demand for watershed protection among the various downstream beneficiaries and to finance watershed protection activities. Contributors to the fund include Quito's electricity supplier (which generates about 22% of its hydropower in the surrounding watersheds), a private beer company, and the city's municipal water supplier, which dedicates 1% of its monthly drinking water sales to the fund. The water agency works with The Nature Conservancy (which also supports the fund) to identify projects that offer both high biodiversity values and water supply benefits (Postel, 2005; Echavarria, 2002).

Even where opportunities to combine water quality protection with other ecological services are not apparent or feasible, drinking water utilities usually still have ample reason to invest in watershed protection. Good watershed management offers an extra safeguard against microbial or chemical contamination that may not be removed by conventional treatment plants; keeping such contaminants out of drinking water sources in the first place reduces risks to human health. Therefore, it makes sense to include the cost of watershed protection in consumer water rates (just as routine operation and maintenance costs are included).

CONCLUSION

A new framework for water management is now taking shape. At its core is a reassessment of the value of freshwater ecosystems that considers water's worth not just when it is extracted from the natural environment but also when it is left in place to do the work of nature. These ecosystem services are rarely valued monetarily yet provide enormous benefits to society. Scientists have developed ways to incorporate ecological goals into the management and use of dams, reservoirs, watersheds, and groundwater systems that can help society make wiser choices and maximize the total value derived from freshwater ecosystems, including extractive uses, biodiversity conservation, and other ecosystem services.

As urban populations and water demands increase, drinking water utilities will be on the front line of efforts to rebalance water management to better protect ecological health. Fortunately, numerous tools and approaches are available to meet this challenge.

ABOUT THE AUTHOR

Sandra L. Postel is director of the Global Water Policy Project, Amherst, MA. She is also a visiting senior lecturer in environmental studies at Mount Holyoke College in South Hadley, Mass., and a senior fellow with Worldwatch Institute (Washington, D.C.), which published her 2005 book Liquid Assets: The Critical Need to Safeguard Freshwater Ecosystems. *Her other works include* Last Oasis: Facing Water Scarcity, Pillar of Sand: Can the Irrigation Miracle Last? *and* Rivers for Life: Managing Water for People and Nature *(co-authored with Brian Richter). Through research, writing, consulting, teaching, and public speaking, Postel works to preserve freshwater ecosystems and the life they sustain. She studied geology and political science at Wittenberg University in Springfield, Ohio, and resource economics and policy at Duke University in Durham, N.C.*

REFERENCES

Barten, P.K. & Ernst, C.E., 2004. Land Conservation and Watershed Management for Source Protection. *Jour. AWWA,* 96:4:121.

Beecher, J.A., 1995. Integrated Resource Planning Fundamentals. *Jour. AWWA,* 87:6:34.

Breunig, K., 2003. *Losing Ground: At What Cost? Changes in Land Use and Their Impact on Habitat, Biodiversity, and Ecosystem Services in Massachusetts.* Massachusetts Audubon Soc., Lincoln, Mass.

Council of Great Lakes Governors, 2001. Great Lakes Water Management Initiative: Draft Annex 2001 Implementing Agreements. www.cglg.org/projects/water/annex2001implementing.asp.

Echavarria, M., 2002. Financing Watershed Conservation: The FONAG water fund in Quito, Ecuador. *Selling Forest Environmental Services: Market-based Mechanisms for Conservation and Development* (S. Pagiola, J. Bishop, and N. Landell-Mills, editors). Earthscan, London.

Edwards Aquifer Authority, 2005. www.edwardsaquifer.org (accessed Apr. 12, 2005).

Ernst, C., 2004. *Protecting the Source: Land Conservation and the Future of America's Drinking Water.* Trust for Public Land and AWWA, Washington.

European Parliament & Council of the European Union, 2000. Directive 2000/60/EC Establishing a Framework for Community Action in the Field of Water Policy. *Official Jour. European Communities,* (Dec. 22, 2000) L 327:1.

FitzHugh, T.W. & Richter, B.D., 2004. Quenching Urban Thirst: Growing Cities and Their Impacts on Freshwater Ecosystems. *BioSci.,* 54:741.

Gardner, L., 2004. Surviving the Worst Drought in 300 Years. Proc. AWWA Water Sources Conf., Austin, Texas.

Gilliom, R.J. et al, 2006. *Pesticides in the Nation's Streams and Ground Water, 1992–2001.* Circular 1291, US Geological Survey, Reston, Va.

Harmon, R., 2004. Denver Water's 2002 Drought Response and Water Conservation Campaign. Colorado State University, Greeley, Colo.

Howarth, R. et al, 2000. Nutrient Pollution of Coastal Rivers, Bays, and Seas. *Issues in Ecol.,* 7:1.

Hutson, S.S. et al, 2004. *Estimated Use of Water in the United States in 2000.* Circular 1268, US Geological Survey, Reston Va. February 2005 revisions at http://pubs.usgs.gov/circ/2004/circ1268/index.html.

International Conference on Freshwater, 2001. *Water—A Key to Sustainable Development: Recommendations for Action.* Bonn, Germany.

Millennium Ecosystem Assessment, 2005. *Ecosystems and Human Well-being: Current State and Trends.* Island Press, Washington.

National Research Council, 2002. *Watershed Management for Potable Water Supply: Assessing the New York City Strategy.* National Academy Press, Washington.

Poff, N.L. et al, 1997. The Natural Flow Regime: A Paradigm for River Conservation and Restoration. *BioSci.,* 47:769.

Postel, S., 2005. *Liquid Assets: The Critical Need to Safeguard Freshwater Ecosystems.* Worldwatch Inst., Washington.

Postel, S., 1999. *Pillar of Sand: Can the Irrigation Miracle Last?* W.W. Norton, New York.

Postel, S. & Richter, B., 2003. *Rivers for Life: Managing Water for People and Nature.* Island Press, Washington.

Postel, S.L. & Thompson, B.H., Jr., 2005. Watershed Protection: Capturing the Benefits of Nature's Water Supply Services. *Natural Resources Forum,* 29:98.

Postel, S. & Vickers, A., 2004. Boosting Water Productivity. *State of the World 2004.* W.W. Norton & Co., New York.

Richter, B.D. et al, 2003. Ecologically Sustainable Water Management: Managing River Flows for Ecological Integrity. *Ecolog. Applications,* 13:206.

Richter, B.D. et al, 1997. How Much Water Does a River Need? *Freshwater Biol.,* 37:231.

Silk, N. & Ciruna, K. (editors), 2005. *A Practitioners Guide to Freshwater Biodiversity Conservation.* Island Press, |Washington.

South African Government Gazette, 1998. South African National Water Act No. 36 of 1998. Vol. 398, No. 19182, Aug. 26, 1998.

Tharme, R.E. & King, J.M., 1998. *Development of the Building Block Methodology for Instream Flow Assessments and Supporting Research on the Effects of Different Magnitude Flows on Riverine Ecosystems.* Water Research Commission, Cape Town, South Africa.

United Kingdom Environment Agency, 2003. Watermark Study of Water Use and "Best Practice" Benchmarks at Public Facilities in the U.K. Watermark, Liverpool, England.

US Bureau of the Census, 2004. US Interim Projections by Age, Sex, Race, and Hispanic Origin. www.census.gov/ipc/www/usinterimproj (release date Mar. 18, 2004).

Vickers, A., 2006. New Directions in Lawn and Landscape Water Conservation. *Jour. AWWA,* 98:2:56.

Vickers, A., 2005. Managing Demand: Water Conservation as a Drought Mitigation Tool. *Drought and Water Crises: Science, Technology, and Management Issues* (D.A. Wilhite, editor). CRC Press, Boca Raton, Fla.

Vickers, A., 2001. *Handbook of Water Use and Conservation: Homes, Landscapes, Businesses, Industries, and Farms.* WaterPlow Press, Amherst, Mass.

Wastewater Is Water Too

READERS OF *JOURNAL AWWA* HEAR A LOT ABOUT RISING CUSTOMER DEMANDS, VAST INFRASTRUCTURAL NEEDS, AND THE CONSEQUENTLY HUGE CAPITAL INVESTMENT REQUIREMENTS THAT CHALLENGE AND CLOUD THE FUTURE OF THE US DRINKING WATER INDUSTRY.

BY STEVE MAXWELL
(*JAWWA* September 2003)

We should remember, in the same vein, that our colleagues at the other end of the water management process—the wastewater treatment industry—face almost identical problems and challenges. Indeed, although it is estimated that more than 1 billion people on this planet currently lack adequate access to clean drinking water, more than 2.5 billion people lack access to any kind of sanitation services.

Many people in the water utility business have tended to think about the water and wastewater industries as being two rather different businesses. Sure, in our houses as well as in our factories "clean" water usually comes in through one pipe, and "dirty" water goes out through another. However, the differences and the boundaries between water and wastewater are slowly and surely fading. Today, we hear more and more frequently that "water is water"—regardless of its type or the amount of treatment it may need to undergo before we drink it or use it to manufacture computer chips. Water utilities are increasingly viewing wastewater as another potential source of drinking water. Certainly, the water and "wastewater" industries are more notable for their similarities than for their differences. Lets examine some of those similarities.

First, the overall size of the two different businesses are almost identical. Environmental Business International has estimated the total market, or revenue generated, for municipal wastewater treatment at about $29 billion per year and has put water utility revenues just slightly higher at $32 billion per year. Thus, in terms of their overall impact on the economy and their percentage of the total Gross National Product, the water business and the wastewater business are very similar.

Second, the two industries use many of the same technological approaches and systems. Treatment technologies and management systems are undergoing a similar proliferation and advance in wastewater treatment as in primary water treatment. As regulatory requirements become tougher and more comprehensive, newer technologies are coming into broader use—membrane filtration, advanced oxidation treatment techniques such as ultraviolet radiation and ozonation, and constructed wetlands and other approaches that mimic "natural" treatment systems. More thoughtful strategic and operational planning and better asset management have likewise made possible great efficiency gains in both industries.

Third, the overwhelming need for infrastructure replacement and capital expenditure is probably the most critical challenge facing the wastewater industry, just as it is in the drinking water business. Although studies abound and dollar estimates vary, most observers put the total investment required in the wastewater treatment business over the next 20 years at around the same $300 billion level that AWWA has estimated for drinking water infrastructure.

Upgrading and maintaining the wastewater and sanitation infrastructure are clearly just as critical as upgrading and replacing the primary water infrastructure—to human health, to the environment, and to the functioning and growth of our economy. Users of the service will not be able to pay for all of these costs. The Water Infrastructure Network has estimated that if all of the required investments for upgrading water and wastewater infrastructure were to be funded by local users, the fees that we all pay each month would have to approximately double. That is not likely to happen, so federal and other funding mechanisms must be developed—in water and in wastewater infrastructure.

Fourth, because both industries have the economic characteristics of natural monopolies, they exhibit similar tendencies or imperatives toward consolidation. Bigger regional systems, if operated properly, can often provide service to customers at improved efficiency or reduced costs. Bigger operations can allow greater economies of scale, so there is a trend toward consolidation of systems—similar to the rapid consolidation we have seen among the vendors to the water and wastewater businesses over the past several years.

Because of the confluence of these similar challenges, the wastewater business is also increasingly turning to private operators and private sources of capital in an effort to help meet all of these increased demands. This trend toward greater private participation in the overall water business has attracted many companies in the past several years. Many observers attribute the great "foreign invasion" of the US water industry in the late 1990s to an effort by European players to gain strategic position for the impending privatization of the US municipal business. However, it is now clear that there are limits to privatization—on both the water and the wastewater sides of the business.

The issue of the solid by-products of wastewater treatment—so-called "biosolids"—and what to do with them is perhaps the one aspect of the wastewater treatment business which in fact does make it a bit different from the primary drinking water business. As advanced treatment standards have been regulated into place over recent years, the wastewater treatment business has become more and more efficient and has offset the higher costs of regulatory compliance by developing and marketing an economically valuable by-product: biosolid material.

So, over time, we are coming to realize that *water really is just water*. Wastewater, just like groundwater, surface water, or even pristine mountain lake water, should really be viewed as just another potential source of raw water for water utilities to utilize in the provision of clean water to their customers. In the future, water utilities may well compete fiercely over their rights and access to wastewater streams as various raw water sources become more restricted and hard to find.

So, the boundaries between two businesses that have historically been thought to be quite different continue to blur. The water and wastewater industries should both be fighting for the same ideals. One day we will prob-

ably recognize them as essentially the same industry.

The challenges and needs of this coalescing water resource industry are likely to be one of the most pressing problems facing humankind over the next century. The Johannesburg Earth Summit of 2002 set as one of its goals the halving of the number of people without decent water and sanitation services by the year 2015. Making these kinds of objectives a reality rather than simply "happy talk" will require a vast effort by all the countries of the world. Given that only about 6% of the world's water and wastewater needs are met by private organizations, publicly owned water and wastewater organizations will clearly have the major role to play in achieving these sorts of goals.

It is important to note that this is not just a problem for the rest of the world—it is a problem right here at home too. The vast and pressing nature of the world's water problems about the broader water industry were just highlighted in a special section on water in the July 19 issue of the widely respected international newsmagazine *The Economist*. Concluding that throughout history, water resources have been "ill-governed and colossally underpriced," this review highlighted many contentious arguments and issues plaguing the international water industry. However, the report also stated that there is one aspect of water use about which nearly everyone agrees—and which we would all do well to consider—although we are making gains, the United States is the most wasteful nation on earth in terms of water use.

Supply from the Sea: Exploring Ocean Desalination

BY JEFF SZYTEL
(*JAWWA* February 2005)

ecause local and regional supplies are becoming stretched and overallocated and as the world's population is continuing to expand into areas of limited water supply, planners and politicians will have little choice but to turn to the ocean as a supplemental source of freshwater. Although there are still many barriers to large-scale implementation of ocean desalination, advances in technology, water policy, economic allocation, and public awareness will continue to drive the development of ocean desalination projects well into the foreseeable future.

MULTIPLE TRENDS HAVE ADVANCED THE DESAL OPTION

Ocean desal as a source of drinking water is not a new concept. Egyptian, Persian, Hebrew, and Greek civilizations studied various desalination processes. Aristotle and Hippocrates both advocated the use of distillation in the fourth century B.C. By 2001, there were more than15,000 desal plants worldwide treating various source waters, with a total production capacity of nearly 6.2 bgd (23 GL/d). Most ocean desal facilities are located along the coasts in the energy-rich Middle East. However, ocean desal is a growing market in Spain, England, the United States, and Mexico. Several trends have contributed to the advancement of ocean desal both in the United States and abroad in recent years. These trends include consistent improvements in desal technology, increased governmental subsidy, increased consideration of co-location with power plants, and more frequent use of collaborative approaches that bring private-sector and public entities together for project development.

BETTER TECHNOLOGY HAS INCREASED VIABILITY

Nothing has contributed to ocean desal's increasing viability and growth as much as the continuing improvements in technology, particularly membrane technology. In 1959, the first reverse osmosis (RO) membrane was developed by Loeb and Sourirajan at the University of California, Los Angeles. These original cellulose-acetate membranes allowed researchers to apply high pressure in order to separate ionic species from water molecules, producing freshwater from a salty solution. Since the early days of RO, there have been significant advances in membrane technology to improve salt rejection, reduce transmembrane pressure, and decrease membrane fouling. RO membrane manufacturers continue to refine and optimize the manufacturing, packaging, and fabrication processes. Additionally, researchers and manufacturers continue to improve the efficiency of high-pressure pumping and energy recovery systems and the effectiveness of RO pretreatment. These technological innovations have substantially decreased the capital, operations, and maintenance costs of RO for ocean desal.

REGIONAL SUBSIDIES SUPPORT WIDER USE OF DESAL

Although technological advancements continue to reduce the cost of ocean desal, its economic viability must be justifiable when compared with alternative sources, such as additional freshwater supply or wastewater reuse. However, the inevitability of ocean desal as a viable supplemental freshwater source is driving some regional water agencies, particularly those along the coastlines, to subsidize the development of large-scale desal projects. Examples of agencies promoting large-scale ocean desal projects include Tampa Bay Water in Florida and the Metropolitan Water District of Southern California. Additionally, state and federal agencies such as the Texas Water Development Board and the United States Department of Energy continue to offer grants and financial incentives to encourage ocean desal. These subsidies not only create a market that drives private investment into research and development—they also allow some water suppliers to begin to economically integrate ocean desal into their existing supply portfolio.

CO-LOCATION WITH POWER PLANTS MUTUALLY BENEFICIAL

Throughout North America and Europe, ocean desal plants are almost exclusively co-located with power-generation facilities. This trend can be attributed to the availability of a reliable and inexpensive source of electricity, availability of existing intake and outfall structures, and potential for utilizing a higher-temperature feedwater source. Because energy use is one of the most significant cost factors when ocean desal is considered on a life-cycle basis, reliable and inexpensive energy can dramatically decrease the overall cost of this technology.

Utilizing existing intake and outfall structures limits the environmental impact of a new ocean desal facility and can also substantially reduce construction costs. Overall environmental impact is greatly reduced because new submarine facilities are usually not required during construction, and the return flows from power plant cooling systems offer a significant predilution for the concentrated brine waste discharge. Additional benefits may be gained by drawing the RO feedwater from the power plant's spent cooling water. With a warm water feed, the hydraulic resistance of RO membranes will be lower, allowing the same membranes to be operated at a higher rate (flux) with the same pressure or produce the same flow rate at a lower pressure. The direct benefits include reduced energy consumption and/or footprint requirements. However, with warmer feedwater, salt rejection across the RO membranes is typically reduced, and multiple "passes" of RO membranes may be needed to achieve similar product water quality.

PRIVATE/PUBLIC PROJECT DEVELOPMENT HAS HELPED UTILITIES MANAGE RISKS

Considering the limited large-scale application of ocean desal in the United States, project development typically involves significant risk related to maintaining environmental compliance, implementing advanced technology, ensuring treatment performance, maintaining system reliability, and controlling operations and maintenance costs. Water utilities and public agencies that are interested in developing ocean desal must therefore develop approaches to mitigate and/or allocate these risks. Private developers have attempted to fill this need by promoting their ability to physically deliver new ocean desal plants while applying comprehensive project delivery approaches, such as design–build and design–build–operate, to manage overall project risk. Although third-party private developers have helped promote and implement ocean desal throughout the United States, their role will likely diminish as the market matures and public water agencies become more comfortable with managing these risks directly.

IS OCEAN DESAL COST-PROHIBITIVE?

When considered on a cost-per-volume-treated basis, the cost of ocean desal has decreased dramatically over the past 20 years. For example, the Tampa Bay Water desalination facility is expected to deliver water at a total cost of around $2.00/1,000 gal ($0.53/1,000 L). When

this trend is superimposed onto a curve representing the cost of other new supplies in areas with limited supply options (such as expanded import systems), "crossover" is either eminent or has already occurred. There is no question that ocean desalination is an expensive solution for obtaining additional freshwater supply. However, in some areas of the country where population growth will far outstrip the availability of "conventional" freshwater sources, ocean desal may in fact be the "least-cost" alternative for new supplies.

Although a straight comparison of cost per volume of new freshwater supply is an important first step in considering the economics of new sources of supply, this approach fails to take into account an equally important factor—source reliability. Associated with each new supply is an inherent reliability. Contributing factors may include regional weather patterns, environmental regulations, infrastructure condition, natural disasters, and variability in water quality. Water agencies must consider their ability to provide sufficient freshwater to their customers under a wide range of reliability scenarios.

Financial analysis offers a useful analogy when considering supply economics. As every investor knows, the first rule of financial management is diversification. Each investment has an inherent risk and return. By combining investments of varying risk and return, an optimally "efficient" portfolio can be achieved, which maximizes the return for a given level of aggregate risk. The same approach can be applied to water supply planning. However, instead of considering risk versus return, water planners usually consider reliability versus life-cycle cost. Although ocean desal represents a relatively high-cost supply alternative, it can be highly reliable. Therefore, by combining ocean desal with lower-cost, lower-reliability supplies, water managers can achieve an optimally "efficient" supply portfolio that maximizes reliability for a given aggregate cost.

GETTING THE SALT OUT: COMPARING DESALINATION TECHNOLOGIES

There are two primary mechanisms for ocean desalination: thermal processes and membrane processes. Thermal processes rely on induced evaporation to separate water vapor from a salt solution, followed by a condensation step that returns the water vapor to liquid form. Membrane processes rely on selectively permeable membranes that reject dissolved ions while allowing water molecules to pass through under high pressure.

Thermal processes include multieffect distillation, multistage flash distillation, and vapor compression Membrane processes include reverse osmosis and sequential nanofiltration (the Long Beach Method). Table 1 gives a brief comparison of thermal and membrane processes for ocean desalination.

TABLE 1   Comparison of membrane and thermal processes for ocean desalination

Parameter	Membrane Process (Reverse Osmosis)	Thermal Process (Distillation)
Installation	• Prepackaged modules • High mobility of modular system (ideal for emergency water supply use) • High space/production capacity ratio (400–1,000 gfd [680–1,700 L/m²/h])	• Can be combined with electric power generation
Pretreatment	• Screening • Scaling prevention • Fouling prevention (fine filtration for suspended solids removal plus acid addition for microbial growth prevention) • pH adjustment for compatibility with membranes	• Screening • Scaling prevention
Posttreatment	• Degasification • pH adjustment • Addition of calcium (Ca) and bicarbonate (HCO₃) (to the level of 100 mg/L calcium carbonate [CaCO₃])	• Degasification • pH adjustment • Addition of Ca and HCO₃ (to the level of 100 mg/L CaCO₃)
Waste generation	• Screenings • Brine disposal • Backwash • Used filter cartridge • Cleaning solutions	• Screenings • Hot brine disposal • Cleaning solution • Wash water
Operation and maintenance	• Shut down for cleaning every four months • Membrane replacement every three to five years • Cartridge filter elements replacement about every eight weeks • Preventive maintenance (instrument calibration, pump adjustment, chemical feed inspection and adjustment, leak detection and repair, structural repair) • Main operational concern: fouling	• One annual shutdown of six to eight weeks for general inspection and maintenance (damage repair, scale removal, vacuum system cleaning, pump inspection) • Corrosion and scaling mitigation, monitoring, and removal (increases with temperature) • Preventive maintenance
Water quality	• 300–500 mg/L total dissolved solids (TDS) • Limited removal of volatile organic chemicals • Risk of bacterial contamination of the membranes	• 1–50 mg/L TDS • Cannot remove volatile organic chemicals
Energy requirements	• For pressurization of membrane • 2–10 kW·h/m³ for RO	• For heating • 3–6 kW·h/m³ for multistage flash distillation • 2–4 kW·h/m³ for multi-effect distillation • 8–12 k kW·h/m³ for vapor compression (not including thermal energy)
Costs	• Capital cost: $1,000–$1,600/m³/d ($3.79–$6.05/gal) • Production cost: $0.45–$0.67/m³ ($0.002–$0.003/gal)	• Greater potential for economies of scale • Capital cost: $900–$2,000 m³/d ($3.41–$7.57/gal) • Production cost: $0.46–$1.50/m³ ($0.002–$0.006/gal)

As with any large investment in capital facilities for water supply, implementation of ocean desalination requires the application of thorough and comprehensive management and engineering approaches. Significant considerations may include: supply planning and economics, water quality analysis, site evaluation, technology evaluation (e.g., pilot testing, pretreatment, configuration, fouling), energy considerations, residuals management options, environmental compliance and facility permitting, public outreach, supply integration, project aesthetics, and project procurement options.

ABOUT THE AUTHOR

Jeff Szytel is a project manager with HDR, San Diego, CA.

REFERENCES

Ackerman, L. et al, 2003. Assessment of Seawater Desalination as a Water Supply Strategy for San Diego County. University of California, Santa Barbara, Calif.

Avlonitis, S. et al, 2003. Energy Consumption and Membrane Replacement Cost for Seawater RO Desalination Plants. *Desalination,* 157:151.

Buros, O.K., 2000 (2nd ed.). *The ABCs of Desalting.* Intl. Desalination Assn., Topsfield, Mass.

Einav, R. et al, 2003. Environmental Aspects of a Desalination Plant in Ashkelon. *Desalination,* 156:79.

Ettouney, H. et al, 2002. Evaluating the Economics of Desalination. *Chem. Eng. Prog.,* 98:12:32.

Hellmann, D. et al, 2001. Saving of Energy and Cost in Seawater Desalination With Speed Controlled Pumps. *Desalination,* 139:7.

MacHarg, T., 2004. West Coast Researchers Seek to Demonstrate SWRO Affordability. *Desalination and Water Reuse Quarterly,* 14:3:10.

Pantell, S., 1993. Seawater Desalination in California. California Coastal Commission.

Redondo, J., 2001. Brackish-, Sea- and Wastewater Desalination. *Desalination,* 138:29.

Sallangos, O. et al, 2001. Operating Experience of the Dhekelia Seawater Desalination Plant. *Desalination,* 139:115.

Schölzel, H., 1998. Desalination (A Technical Appraisal for its Application in Pacific Island Countries). South Pacific Applied Geoscience Commission, Suva, Fiji Islands.

Semiat, R., 2000. Desalination: Present and Future. *Water Intl.,* 5:1:54.

Texas Water Development Board, 2002. Large-scale Demonstration Seawater Desalination in Texas: Report of Recommendations for the Office of Governor Rick Perry.

Tsiourtis, N. et al, 2001. Seawater Desalination Projects. The Cyprus Experience. *Desalination,* 139:139.

US Congress, Office of Technology Assessment, 1988. Using Desalination Technologies for Water Treatment. OTA-BP-O-46, March 1994, Washington.

Wangnik, K., 2003. The Development in Seawater Desalination. *European Desalination Society Newsletter,* 18:7.

Wilf, M. et al, 2001a. Improved Performance and Cost Reduction of RO Seawater Systems Using UF Pretreatment. *Desalination,* 135:61.

Wilf, M. et al, 2001b. Optimization of Seawater RO Systems Design. *Desalination,* 138:299.

The Challenges of New Technology Development in the Water Industry

BY STEVE MAXWELL
(*JAWWA* July 2004)

Countless studies have suggested that water challenges may well become the defining political and economic issues in the lives of our children and grandchildren. As water issues reach crisis proportions in more areas of the world, we are likely to see vast changes in the way water is owned, treated, distributed, used, and recycled—in short, we are likely to see a revolution in how we think about water. As we are ultimately forced to pay more and more for water, we will undoubtedly become increasingly discerning and clever in the ways in which we utilize this critical resource.

New technologies—as well as new and innovative applications and understanding of existing technology—will play a critical role in helping the planet to adjust to these growing challenges. In this and Article 27, we will review the challenges facing technology developers in the water industry and outline key strategies for the successful startup, development, and management of technology ventures.

Throughout history, there has always been one camp that warned of impending social or ecological disaster—from the days of Malthus' cautions about overpopulation, on through the influential declarations of the Club of Rome "Limits to Growth" report in the early 1970s, and up to today's fears of impending water shortages. The other "side" of this debate has often suggested that technology can be the answer to all our problems—to overpopulation, epidemics, drought, and world hunger—to name just a few. Although the blind faith of these optimists that technology can solve all of our problems may be a bit idealistic or simplistic, history has definitely shown that technological development and advancements can indeed help.

New technologies inspired by the water pollution control regulations of the past four decades have allowed the United States to make vast advances in many areas of water quality and pollution control. Effluent being discharged from wastewater treatment plants today is frequently cleaner than the water in the natural waterways into which it flows. When the Cuyahoga River in Cleveland caught fire in 1969, it focused the concern of a whole nation on the rapidly deteriorating condition of our natural waterways. The Clean Water Act was passed three years later, and there are no rivers catching on fire today.

More than 10 years ago, the influential report from the National Commission on the Environment, "Choosing a Sustainable Future," called for a new generation of environmental and water treatment technologies and recommended that "the most important and immediate target of US environmental policy is to encourage the development and adoption of technologies compatible with sustainable development." During the intervening decade, hundreds of new technologies have been developed in response to the recognition that a clean environment and clean drinking water constitute one of the true growth industries of the future. New technologies will continue to emerge and evolve—technologies that will enable more efficient purification, treatment, storage, distribution, and use of clean water; recycling and reclamation of wastewaters; and practices that will promote general environmental sustainability.

Despite these obvious gains, however, it is also true that there have been few revolutionary or major breakthroughs (for many decades) in the way water is treated or used, and there does not appear to be any silver bullet on the horizon. Today, new treatment and purification technologies are needed to help municipalities and industry comply with the expanding plethora of control regulations—a regulatory scheme recently described by Pontius in the March 2004 JOURNAL AWWA as ". . . becoming so large and complex that it threatens to collapse under its own weight." Simple sand filters are about as common today as they were 50 years ago. Most observers would agree that there has been a lack of notable success in recent years in terms of innovative water treatment technologies.

One of the reasons for slow progress in the development of truly innovative or revolutionary ideas in the environmental field is the existence of numerous barriers to technological development. Some of these barriers are common to all businesses, but many are unique to the water and wastewater business. The key barriers to technology development in the water industry include the following:

■ *Regulatory barriers.* In the water industry, technological innovation has to occur within a complex—and sometimes contradictory—regulatory and economic framework. It sometimes seems that there is an endless series of regulatory hurdles required for approval of a new technology. Complicated permitting requirements, which often vary by state, can seriously hinder the ability to demonstrate a new technology. Utilizing one technology to comply with one regulatory requirement can sometimes lead to the creation of other new problems or the failure to comply with other regulations—witness the recent concern over the growing use of chloramines, which were originally used to minimize the formation of chlorine disinfection by-products.

The immediate need for a particular new technology is often driven by a specifically defined pollution control regulation, but these regulations are themselves in a constant state of flux and evolution. Regulatory discharge levels, for example, are changing over time, and the level of enforcement of these regulations may vary from region to region. California is often hailed as the most progressive state in terms of environmental protection, but it has its own set of approvals, certifications, and regulatory requirements for water and wastewater treatment and purification technologies. Federal programs geared to promoting new technology development have not been well coordinated with one another in the past, or they have often sent conflicting signals to developers. Many technology developers have steered away from government research and development funding altogether because of its associated complexity, inefficiency, and unreasonably high administrative burden.

■ *Marketing barriers.* Widespread market recognition and acceptance are difficult to achieve in an industry that has traditionally been somewhat conservative, resistant to change, and slow to embrace new technological approaches. Most new technologies have to suffer through years of cautious testing and review by the engineering community before they can begin to flower commercially; many potential users (particularly in the municipal end markets) are hesitant to be "guinea pigs" in the earlier stages of technology testing. Long-standing relationships between major municipal clients and their engineering consultants seem to amplify this problem—few entrenched engineers want to risk a long-standing relationship by recommending an experimental technology, even though it may offer huge potential advantages to the client. One example of this is the 20 years or more it took for membrane filtration to successfully break into widespread application.

In addition, because the water industry is changing rapidly and remains heavily fragmented, it is extremely difficult to estimate the size and growth rate of specific market sectors. The lack of good market information makes it even more difficult for developers of new technology to assess and evaluate markets and to determine where to focus their commercialization and marketing efforts. Finally, as mentioned, the vagaries and unpredictability of the regulatory process add to the uncertainty faced by technology firms—it is very difficult to develop sound long-term marketing strategies when whole markets can be created or destroyed by the stroke of a legislative pen. In sum, from the perspective of the technology developer, the customer is often conservative and skeptical of new approaches, and the future market is a moving target.

■ *Technical barriers.* Interesting ideas or purportedly revolutionary concepts don't count for much in the water treatment technology field. You have to be able to objectively prove that the "gadget" works before people will change their traditional buying habits. Put succinctly, you can't sell it if you can't prove that it works. Customers want to see solid and independently verified operating data, and it can be very difficult to obtain credible data without at least a pilot-scale demonstration unit. As any startup technology company can tell you, this requirement quickly turns into a "chicken and egg" problem—development funds are difficult to acquire because customers are not willing to purchase an unproven product, but the product can never be developed and proven without that initial investment capital. It is difficult to find investors willing to fund commercialization of a technology before it is fully technically proven—most investors are skeptical of investing in ideas.

■ *Financial barriers.* It goes without saying that the funding of new technology is difficult. Although the investment community is showing more and more interest in the water industry, actual equity investments or venture capital–type deals are still fairly rare in this industry. Many equity investors have indicated a strong interest in this market, but few so far have put their money where their mouth is. Furthermore, even when the funds are there and available, it may (from the perspective of the developer) entail giving up too much control of the company.

Some of this reticence or caution on the part of investors, however, is understandable. The water industry has not produced many big "winners" from an investment perspective—either in terms of small private companies or in terms of the larger publicly traded companies. For some of the reasons just discussed, individual company profitability has been comparatively low or extremely slow in coming. Unfortunately, as more attention has begun to be focused on the water industry, so many promoters have flooded into the business that it is often difficult to tell which technologies are real and which aren't. However, despite the relatively unexciting financial performance of water investments to date, venture capitalists and equity investors of all stripes continue to be interested in the business. Funds will increasingly become available to firms with good ideas and good management teams, even if the prototype products or initial profits may take a few years to develop.

It is increasingly clear that there is a need for new technology, for innovative applications of existing technology and know-how, and for government policies and economic incentives that will foster the development of new ideas and approaches. The world will still beat a path to your door if you can build a better mousetrap, but there is a formidable set of barriers and obstacles that must be cleared before commercial and economic success can be achieved.

This page intentionally blank.

Factors for Success in New Technology Development

BY STEVE MAXWELL
(*JAWWA* September 2004)

In this column, what it actually takes to build a successful technology company in the water industry will be described.

The successful development venture in the water industry must, obviously and foremost, have a viable technology, product, or service. As with most other things in life, actions speak louder than words—developers need to have verifiable proof that their technical claims are true. The water industry is, unfortunately, not much different from most others in terms of having its share of promoters and con men who have hyped new technologies that later turned out to be frauds. However, even many seriously intentioned startup companies lack sufficient and independent operating data to support the claims they make. Clearly, acquiring such data early in the development process can be exceedingly difficult. Few potential clients want to be the first to try an unproven product. However, the developer must find ways of demonstrating the product—perhaps even to the extent of providing money-back guarantees to early-stage "guinea pig" clients. Many experts in this field believe that as many as 10 or 15 working sites are required before the ability to achieve full commercialization can be reached. Only working prototypes can provide the sort of full-scale operating data that customers are going to want.

However, contrary to what some scientists and technology developers might want to believe, much more is required for economic success than just a working or innovative new technology. The paragraphs that follow identify the most critical of these other, equally important requirements.

■ *There must be a firm and clearly identified market.* Even the most clever and innovative ideas are useless without a market and a clearly identified customer. As the technology moves to market, the question foremost in the developer's mind should no longer be "Will it work?" but "Will anybody buy it?" To be blunt, if there is no market, you don't have a business. Companies must carefully define their targeted end markets, assess the size and growth rates of those markets, determine what drives customers' buying habits, and understand the technological, economic, and even politi-

cal factors that influence the market. That list can be a pretty tall order. In the water market, many developers have fallen victim to the "tiny market share" syndrome—the vague and often fatal assumption that because the market is so large and growing so rapidly, that a new company needs only a tiny share to be wildly successful. This is flawed reasoning—

even tiny shares in big markets can be difficult to acquire. It is also important that a "need" isn't confused with a "market." In many sectors of the technology market and in many areas around the world, there may be a serious need, but economic and political realities may have thus far prevented the development of a serious market demand. In short, make sure there is a market for your product—then try to figure out how big it is and how fast it is growing.

■ *There must be an understanding of the competitive environment.* Another critical aspect of understanding the market and potential demand for the new technology is to have a clear understanding of your competitors' products and strategies. Many scientists and developers seem to believe that their technology is so unique that no competition exists. This is rarely, if ever, true. Parties who make such claims usually either simply don't recognize the competition or are addressing a nonexistent market.

■ *There must be appropriate and timely sources of funding.* This often seems to be the most challenging and frustrating aspect of developing and bringing a new technology to the commercial marketplace. Good relationships with funding organizations of one type or another are usually a critical prerequisite of success—few startup ventures are capable of funding themselves independently all the way through to commercial success. Once the developer exhausts his own funds—and perhaps those of his uncles and cousins—he must begin dealing with more sophisticated financial organizations or investors that apply rigorous standards in analyzing potential investments. Although many technology developers seem to find it distasteful to have to divulge secrets and share ownership with financial investors, technologies usually don't make it to the market without such relationships in place. A common misconception in this area is that external financing needs gradually diminish as internally generated funds begin to flow; in rapidly growing companies, the situation may be the reverse—the faster you grow, the more financial backing and assistance you need.

■ ***There must be a strong management team.*** This is perhaps the single most important component of a successful venture. At the end of the day, most bankers and venture capitalists bet on the person or the management team associated with a new venture in deciding whether or not to extend financial assistance. Besides the technical expertise, the management team must possess solid business and financial skills and understanding, good common sense, and strong experience in the industry. The management team needs to establish a clear business plan (see Market Outlook, January and March 2004) that contains both short-term goals and long-term objectives as well as the strategies and tactics the company will use to reach them. There is nothing more important than a strong passion for success and the desire for economic reward that comes from having heavily invested, both emotionally and financially, in the effort. Although all of this may sound obvious, a lack of business savvy has caused the downfall of countless otherwise successful companies.

■ ***There must be a strong sales and marketing effort.*** Sales and marketing plans are critical, and they must be compatible with the company's ability to deliver the product. This is especially true for small startup ventures. Such firms must walk a very fine line when promoting their product—companies should not promise what they cannot deliver. Too many good ideas fail because inventors lack the skills or desire necessary to market and actually sell their products. On the other hand, developers who try to sell something that they can't deliver are quickly marked as amateurs.

■ ***There must be flexibility and objectivity.*** This may sound obvious. However, many otherwise successful ventures have quickly failed because of unrealistic expectations or an unwillingness to listen to what the marketplace is saying. Developers must understand both the advantages as well as the limitations of their own technologies, and they must recognize that the marketplace will determine the commercial viability and value of their business. Never love your technology more than your business.

Companies must be flexible enough to pursue alternative growth strategies. They must be open to establishing partnerships or ventures with other entities that can bring financial wherewithal, manufacturing know-how, or market connections. Although such relationships almost always involve the sacrifice of some control or ownership, most ventures don't make it off the ground if they are not willing to consider such mutually advantageous relationships.

Unfortunately, many such ventures fail because of a basic difference in mentality or culture between entrepreneurial developers and larger and perhaps more bureaucratic organizations. When seeking a partner to help finance or market a new technology, smaller companies should look for a larger partner who is committed to the technology, can make relatively rapid decisions, and is innovative and flexible as well.

The preceding list of critical success factors seems long and complicated; however, the water business is a conservative "show-me" type of industry, and a company needs to be well equipped in all of these attributes if good technical ideas are to turn into solid and profitable business ventures. Companies must allow their strategies to be market-driven rather than technology-driven, and they must continually strive to maintain the proper balance between business and technical issues. Economic success requires more than a viable and working technology—it requires a viable and working business.

Article 29 will discuss various sources of financing for technology development or expansion projects and the pros and cons of each.

PATHWAY TO SUCCESSFUL TECHNOLOGY DEVELOPMENT

The general technology development process—in the water business, or in any other business—can be seen as consisting of four major steps.

1. Conceptual development. The early stage of technology development comprises recognition of a need for the product and identification of the problem that the technology is supposed to solve. Ideas evolve and are refined, eventually resulting in the actual invention or patenting of the product or technology. The concept is proven as technologically valid, i.e., the process is actually shown to work. Funding at this stage often comes through private sources or government grants. This stage of development is usually conducted by the inventor or the scientist. This is science, not business.

2. Technical development. In this stage, the concept or the patented idea or process is tested and proven. It is shown to be viable by outside—and presumably more objective—parties. Prototype bench-scale models are constructed and tested. Operational and manufacturing requirements are assessed. Initial cost estimates are developed as commercial and marketing factors begin to be considered.

3. Product development. In this phase, larger pilot-scale prototypes are built and tested, followed by demonstration pilot models, so that critical initial operating data can be gathered. Collection of this data is an absolute necessity before any market credibility can be attained. Following acquisition of this type of information, cost data and competitive economics can be roughly evaluated. Materials, processes, and design improvements are assessed. At this stage, the effort becomes more of a business endeavor, as opposed to a strictly scientific or technological effort. More formal business planning is now needed, and data must be collected to measure performance against predetermined technical or market criteria. A new spectrum of sponsors is needed, including suppliers of equipment, people, technical knowledge, and dollars. In the water industry field, the ability to meet specific regulatory requirements is critical.

4. Commercialization and market development. At this stage, the company is striving to become a viable business entity. The technology or process is proven and perfected or is at least well on its way. Now, other (and more "business-related") issues move to the forefront as the most critical determinants of success. More employees must be hired, additional space must be leased, markets must be carefully analyzed, and priority niches or applications for the technology must be identified. Formal strategic business plans must be put together, and sufficient equity funding and credit must be obtained. Manufacturing capabilities must be established, a professional sales force must be hired, an inventory of product must be built, and an administrative system capable of supporting a larger organization must be designed and implemented. In sum, the ability to consistently deliver the commercial product must be established.

This page intentionally blank.

Technology Trends and Their Implications for Water Utilities

BY EDWARD G. MEANS III, JOSEPH BERNOSKY, AND ROGER PATRICK
(*JAWWA* January 2006)

To characterize and respond to the many societal, business, and utility trends, the Awwa Research Foundation (AwwaRF) commissioned a 2000 study of such trends and what they mean for water utilities. That project tapped the experience of water industry leaders and futurists and identified and documented key trends that have been used by water utilities and their associations in their planning processes. Because of the acceleration of many key trends over four years, the study was updated in 2004 to provide water utility managers with a current strategic planning tool. This latest project used new trend research, futurists, and future scenario development and engaged national water utility leaders to debate these trends and develop strategies for future utility success.

The consultant team of McGuire/Malcolm Pirnie and Competitive Advantage Consulting Ltd., developed a detailed paper documenting trend data in key areas of concern for water utilities, including population and demographics, health and medical advances, regulations, climate change, total water management, employment and workforce trends, customer expectations, information technology, drinking water treatment technology, energy, automation, information technology security, physical security, current economic issues, utility finance/infrastructure, politics, regionalization, and private sector involvement in water.

This article will briefly describe some of the drivers behind the information and treatment technology trends, the technology itself, and some of the issues and considerations associated with use of the technology.

TRENDS ALLOW REMOTE, AUTOMATED, AND ON-THE-GO COMMUNICATION

Fundamental information technology trends. Because of improvements in semiconductors, storage, and other components, price declines in computers (adjusted for quality) have actually accelerated since 1995 (Figure 1). A related trend is the migration of computing into other devices and equipment.

Trends toward improvements in microelectronics are expected to continue. As miniaturization proceeds, it may lead to the emergence of nanoscale devices (devices with structural features in the range of 1–100 nm). Potential applications of nanoscale electronics 10–15 years in the future include sensor systems capable of collecting, processing, and communicating massive amounts of data with minimal size, weight, and power consumption. Disk drives and other forms of information storage reflect similar improvements in cost and performance (Figure 2).

In addition, computers are increasingly connected in networks. The growth in networking is best illustrated by the rapid growth of the Internet, which has become a key factor in information expansion by providing a common protocol for communication among devices. Networking is evolving in several ways: more people and devices are becoming connected to networks, the speed and capacity of connections are increasing, and more people are obtaining wireless connections.

An increasing array of applications is making information technology (IT) more useful. Over the past two decades, innovations in software have enabled applications to expand to include educational software, desktop publishing, computer-aided design and manufacturing, games, modeling and simulation, networking and communications software, electronic mail, the Internet, digital imaging and photography, audio and video applications, electronic commerce applications, groupware, file sharing, and search engines.

US industry has invested heavily in IT, especially over the past 10–20 years. Water utilities have lagged behind the private sector but appear to be catching up, with major investment in customer information systems, computerized maintenance-management systems, geographic information systems (GIS), and desktop computing and emerging investment in interactive voice response systems, web services, automatic meter reading (AMR), asset management, and mobile applications. However, recent research shows that IT investments have little effect on productivity unless they are accompanied by first-rate management practices and that companies can significantly raise their productivity solely by improving the way they operate.

Trends in Internet and e-commerce use. During the past few years, the Internet has continued to revolutionize the way information and services are stored and moved. The Internet and its web-enabled infrastructure will continue to grow, and newly developed applications will make use of this infrastructure.

Dial-up Internet access has also been superseded by "broadband" access—a dramatic change since the last strategic review—when only 3% of individuals had such access. Broadband connections represented 51% of home Internet connections in July 2004.

Businesses are using the Internet to cut the cost of purchasing, streamline logistics and inventory, plan production, and reach new and existing customers more effectively. Business-to-business electronic-commerce (e-commerce) has now become business as usual and has allowed businesses of many types to expand their services and restructure their processes.

The government has taken an interest in the "digital divide" that separates technology haves and have-nots in the United States, but the evidence is that the gap continues to close. E-government is continuing to evolve to meet a multitude of needs, expanding and altering the way citizens deal with public agencies.

A major Internet technology trend under way is the expansion of web services. Web services are accessed by client web browsers. Web services can be accessed from anywhere with a network connection. They can also be offered by application service providers on a pay-as-you go basis.

Development of outsourcing and offshoring. The information technology revolution is now enabling the shift of some types of jobs to other countries. The availability and declining cost of high-speed communications systems, in combination with computing technology, is facilitating this trend. Offshoring is occurring first in countries where many people speak English as a first or second language.

An ongoing debate exists about the overall effect of outsourcing and offshoring on the US economy, with some observers predicting a net benefit and others a net loss. Whatever the result, the trend itself seems to be unstoppable. The list of jobs being sent overseas using IT and telecommunications is truly remarkable and includes not only call centers but a number of other functions, including reading X-rays and MRIs, developing animated movies, processing insurance claims, completing US tax returns, underwriting mortgages, analyzing equity, as well as legal and other types of research and accounting services.

The direct effect of the offshoring trend on the public sector remains to be seen; however, access to such services from the private sector will certainly provide opportunities for public agencies to reduce their input costs and provide improved services.

A variety of tools facilitate collaborative efforts. IT offers more and more opportunities for communities of interest to form, share information, and work toward common goals. Some of these collaborative mechanisms are casual, such as online chat rooms and weblogs.

Other collaborative mechanisms are more formal, such as a team working on a design project. Businesses are increasingly using collaborative software programs to facilitate group discussion and decision-making. Such software allows participants to interact in dedicated online spaces, engage in discussions, and share and track information. It is expected that some of these tools will be important aspects of knowledge management in the future.

Wireless networking and mobile devices are becoming more common. At present, most people connect to the Internet through devices that rely on wires. The wireless trend has accelerated dramatically over the past few years, as more products are available to eliminate wires from the desktop to mobile media players, cell phones, and personal digital assistants. Wireless fidelity will continue to grow in the home and outside with free community/city sponsored hot spots and corporate networks. During the next few years, most mobile phones will be capable of Internet access.

Future is ripe for voice-over-Internet protocol. Instant text-messaging applications were forerunners of voice-over-Internet protocol (VoIP). VoIP voice and data communications services will see explosive growth in the future. Consumers and businesses will benefit because VoIP will enable lower-cost long distance calling and better integrate the services provided by present telephone companies.

Security and malware more important as Internet reliance grows. Network and computer security has become a hot issue. Spending on data system security will increase as companies try to mitigate sharply increasing losses from worms, viruses, espionage, hackers, and data theft.

The growing number of annoying and/or malicious aspects of the networked economy has necessitated a whole new taxonomy under the heading of malware. Malware includes destructive programs, such as viruses, worms, and Trojans; spyware; spam (accounting for more than half of global e-mail); and phishing—which is hacker-speak for fishing for passwords, social security numbers, and other private information through fraudulent e-mail—often by setting up bogus websites made to look like the real thing. Unfortunately, an analysis of this field quickly becomes out of date as the scammers develop new ways to damage confidence in the online world.

Open-source software sees growth. The difficulty of commercial software competing against Microsoft's

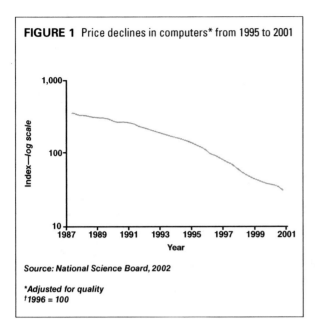

FIGURE 1 Price declines in computers* from 1995 to 2001

Source: National Science Board, 2002

*Adjusted for quality
†1996 = 100

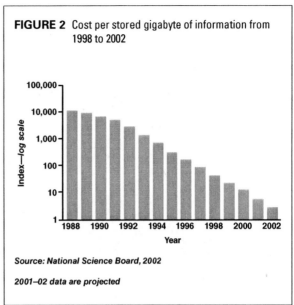

FIGURE 2 Cost per stored gigabyte of information from 1998 to 2002

Source: National Science Board, 2002

2001–02 data are projected

products has spurred rapid growth in noncommercial software under the banner of open source. The majority of these applications run on Linux, an operating system that is itself open source. For example, a cluster of more than 10,000 Linux servers handles Google's more than 200 million searches per day. As it becomes a mainstream business operating system, the market for commercial software running on Linux will expand greatly.

Work spaces are becoming smart. There are a variety of technologies and products that may be combined in the smart work spaces of the future. Smart work spaces will make use of numerous aspects of pervasive computing, that is, computing available everywhere it's needed. Smart spaces offer services provided by embedded devices that are accessed and interconnected with portable devices carried or worn into the spaces. The resulting combination of imported and local devices can support the information needs of users.

Water and other utility IT trends. There is a growing use of information systems to support utility activities and better serve customers. Work-management systems such as mobile dispatch, laptops, and computerized maintenance-management systems are supplementing the more common plant automation and security control and data acquisition systems. Some of these systems are increasingly linked with GIS and other technologies.

There is also an accelerating trend toward using IT to improve customer service. Almost all utilities now have at least a web page with basic information. Increasingly, customers are able to conduct transactions online, including paying bills, tracking service disruptions, viewing GIS data, applying for permits, and applying for service connections.

Utilities that have integrated their GIS with other systems, such as work management systems and AMR

systems, are also seeing many service and efficiency benefits. For example, such integration is improving first call resolution, because the Customer Service Representative can answer many more questions while the customer is on the phone.

Interactive voice response systems are increasingly deployed to improve call routing and provide automated information such as account balances and days to disconnection.

E-commerce is used to a limited extent by water utilities for purchasing and supply. The primary uses are bid notification, communication of purchasing policies, routine purchasing, product searching, information exchange with suppliers, and reference checking. A few water utilities are starting to use the Internet for receiving bids.

Automation is used to a greater extent. As water utilities come under greater cost pressures as a result of the need to replace, upgrade, and build new infrastructure, reduction of operating costs will be increasingly important. Because labor is one of the largest cost categories that utilities must manage, it is reasonable to conclude that automation to reduce labor costs will be a continued focus of water utilities' improvement efforts.

Automation undeniably adds efficiency and reliability to water utility operations. For the most part, this trend leads to a better product, more economical operation, the ability to employ more complex treatment operations, and a safer workplace. However, unintended consequences may exist. Automation changes job descriptions and workplace demands on the water utility staff. An operator who was familiar with every chemical feed pump in the treatment plant may be replaced with an operator who is more comfortable (or at least more familiar) with the electronic displays of the chemical feed operations. Which operator would be better equipped to understand and overcome an unusual emergency situation? It's clear that operator job requirements will grow more complex.

Remote sensing/monitoring/control. The developing requirements to protect water quality at the point of delivery rather than at the central treatment plant has led to the need for more extensive monitoring of water quality at remote locations. In addition, increasing the ability to observe and operate systems remotely will improve staff effectiveness and reduce costs. However, numerous questions remain.

• Are the available remote sensors up to the new challenges?

• What are the key variables upon which an accurate picture of system performance can be constructed?

• Which (and how many) sensors and monitors are needed?

• Are treatment plant and distribution system engineers equipped to design the monitoring systems?

• Can utilities pay the higher up-front costs for installation?

• Are utility staffs going to be trained to make appropriate use of the new technologies, and will they be properly equipped to calibrate and maintain them?

Expert systems useful but come with potential risks. One means of reducing labor costs and increasing the effectiveness of personnel is the implementation of expert systems— automated systems that provide expert-quality advice, diagnoses, and recommendations given real-world problems. Embracing the implementation of expert systems will widen the gap between system designers and system operators. For better or worse, it may not be as necessary for the treatment plant operator to have a well-founded background in treatment plant processes. This may have short-term benefits but create longer-term risks. Reliance on automated systems creates the risk of undetected systematic faults. By what process will the automated expert system be deemed to be a qualified expert, and who will enforce the bounds that limit the applicability of that expertise?

NEW TREATMENT TECHNOLOGY NECESSARY TO MAINTAIN ADEQUATE CONTROL

In many respects, water treatment is the heart of the drinking water industry. It is the most complex and often the most capital-intensive aspect of a water utility. Man has purified water since ancient times, but modern drinking water treatment has been in use for fewer than 200 years. Consisting primarily of conventional filtration, this technique has been the mainstay of the industry since the nineteenth century.

However, that mainstay technology is changing. Diminishing or impaired water sources, increasingly stringent health-based water quality standards, and emerging contaminant issues are driving the use of new or improved treatment technologies. Although they may be effective from a treatment standpoint, these new technologies can produce other major effects including higher capital and operating costs, operations and maintenance requirements, higher energy demands, and residuals/waste management issues.

Technology tends to advance in all fields—but not without drivers. In the drinking water industry, there are several factors behind the effort to develop and implement new treatment technologies.

1996 Safe Drinking Water Act (SDWA) amendments. The 1996 SDWA amendments established a new emphasis on preventing contamination problems through source water protection and enhanced water system management. Congress substantially amended the original 1974 SDWA and the 1986 amendments. The 1996 amendments addressed the maximum contaminant level setting process, monitoring, treatment processes, and actual plant operations. The new regulations provided a process for identifying and regulating emerging harmful contaminants as well as focusing on several known contaminants including disinfection by-products (DBPs), pathogens (specifically *Cryptosporidium* and *Giardia*), radon, and arsenic. The regulations also acknowledged the different risks of and approaches to treating groundwater and surface waters of various qualities. Finally, the 1996 SDWA amendments expanded consumers' access to information about their drinking water.

As regulations were developed, proposed, and promulgated, it became clear that conventional filtration/disinfection could not adequately meet the new water quality standards in all cases. Meeting the Microbial/Disinfection Byproduct Rule cluster would require not only new treatment technologies, but revamped operational strategies as well. Removing arsenic, radon, or radionuclides would drive new approaches to treating drinking water. Thus the suite of regulations arising from the 1996 amendments has been a prime driver for new treatment technology development over the past decade.

Detection of new contaminants in water is ongoing. As water contaminant detection technologies increase in sophistication and detection abilities, contaminants are being detected that were not previously thought to be a concern. The 1996 amendments provided a means for identifying and regulating new drinking water contaminants as they were recognized in the environment or understood to present human health risks. Several chemical and microbiological water contaminants have emerged as potential human health risks in recent years; many of these will not be amenable to control by conventional treatment technologies and thus are also proving to be instrumental in driving the development of new treatment technologies.

Endocrine disrupting compounds (EDCs). EDCs are chemicals that interfere with the normal function of the human endocrine system. Although some are conventional organic and inorganic chemicals, they also include complex pharmaceutical and personal care compounds. Several EDCs are already regulated under the SDWA whereas research into the occurrence, fate, human health effects, and treatment technologies for a host of others is under way. These compounds are not thought to represent a significant human health risk at the concentrations typically found in drinking water, but research continues in this area.

Perchlorate. Perchlorate is both a naturally occurring and synthetic inorganic chemical used as the primary ingredient in solid rocket fuel and to a lesser extent in missiles, fireworks, explosives, and air-bag inflators. Highly soluble in water, perchlorate has been detected as a contaminant in ground or surface water in 20 states. Human exposure to perchlorate can affect the thyroid gland and cause thyroid tumors. Research continues to determine the potential health risks caused from ingesting low levels of perchlorate in drinking water and to improve treatment processes for its removal.

Microbiological contaminants. Many microbiological pathogens have only recently been recognized as being potentially harmful to human health. Under the Contaminant Candidate List (CCL) provisions of the 1996 SDWA amendments, several of these are being investigated for possible inclusion as primary drinking water contaminants including adenoviruses, *Aeromonas hydrophila,* caliciviruses, coxsackieviruses, cyanobacteria (and their toxins), echoviruses, *Helicobacter pylori, Microsporidia, and Mycobacterium avium intracellulare.*

Methyl tertiary butyl ether. An oxygenate additive in gasoline, this compound produces extreme taste and odor (T&O) issues and may pose human health risks. It is persistent in the environment and difficult to treat with conventional methods.

Nitrosodimethylamine (NDMA). A number of studies suggest that NDMA is a probable human carcinogen. It is known to be present in various foods and industrial products, but recent research indicates that it is also a likely by-product of chloramination of drinking water. Control of this contaminant and related compounds may require new treatment technologies or operational process modifications.

Other emerging contaminants. The 2005 CCL 2 contains a host of potentially harmful organic and inorganic chemical water contaminants, certain of which may eventually be regulated. Accordingly, research and development of alternative treatment technologies will need to continue, and the water utility community will undoubtedly play a key role in this important research.

New source waters continually needed. Continued development and population growth impels water utilities to seek additional water resources; finding them has been problematic, especially in arid western states and those with high growth rates. Traditional groundwater and surface water supplies may be overtaxed and some

display signs of environmental stress. Some coastal aquifers have experienced saltwater intrusion. This dilemma has led many utilities to seek entirely new sources. Often these sources are impaired or of lesser quality, such as brackish groundwater. Coastal communities are looking to seawater desalination to address future water needs.

Now municipalities must look to new and alternative treatment technologies to treat millions of gallons of water per day at reasonable costs. As the demand for even greater quantities of water continues to grow, so too will the demand for additional cost-effective treatment options.

Consumers become more demanding. Consumers today are increasingly demanding higher quality water, both from a health and an aesthetic standpoint. Rising bottled water sales can, in part, be attributed to a perceived quality differentiation from tap water. As water professionals continue to understand the nuances of T&O incidents, it is evident that new treatment techniques and operational practices can minimize their occurrence and or effect.

With the aging of health- (and image-) conscious Baby Boomers, there may be an even greater demand for higher quality water. Although conventional treatment/disinfection produces "safe" water, certain areas or source waters may be forced to turn to new treatment technologies to meet their consumers' particular requirements.

New and emerging drinking water treatment technologies. This discussion is intended to provide a brief overview of new and emerging drinking water treatment technologies. Some of these technologies are not really "new," but their application has seen significant growth in recent years (e.g., ultraviolet disinfection and ozone treatment).

Membranes. There are two basic classes of membranes: "loose" or low-pressure membranes operating between 5 and 50 psi (microfiltration and ultrafiltration), and "tight" or high-pressure membranes operating between 125 and 800 psi (nanofiltration [NF] and reverse osmosis [RO]). They provide a barrier to both organic and inorganic contaminants.

Although they are extremely effective treatment technologies, low-pressure membranes are ineffective for removing dissolved organic matter; thus T&O- and color-causing materials will pass through. High-pressure membranes remove these and many other low-molecular-weight contaminants including DBPs.

Advances have reduced capital and operating costs, increased reliability, and produced more efficient membranes resulting in lower power demands. Membrane technology, especially in multi-stage units, is widely accepted in the water industry.

Research is being conducted on what might be termed "smart membranes." Using advanced ion-beam manufacturing technology, nanometer-sized pores are etched into coated polycarbonate membrane surfaces. Employing the same theory as current electrodialysis technology, the membranes can be designed to selectively remove only one (or more) contaminants of interest by specifying the pore size, voltage, and electric field. Researchers believe these "smart membranes" will be extremely energy efficient and may even offer an integral contaminant (e.g., perchlorate) removal mechanism.

Ultraviolet (UV) disinfection. The germicidal effects of ultraviolet irradiation have long been known, and this technology has been used in both the United States and Europe since the 1950s. However, these applications have typically been on a smaller scale and validation issues, source water constraints, and maintenance issues have constrained this technology's use for drinking water until recently. It remains a very promising technology as these issues are addressed; considerable research and development work is continuing.

Ion exchange (IX). IX processes have enjoyed widespread use in the industrial chemical field for many years, but drinking water applications were typically limited to softening applications or nitrate treatment. However, interest in IX processes has surged recently as a treatment technology for inorganic contaminants such as arsenic, radium, hexavalent chromium, lead, and fluoride, among others. Both continuous and fixed-bed column processes are undergoing research and development. Although effective at removing a variety of contaminants, IX processes produce waste streams that, depending on their characteristics (and the location of the process itself), may be difficult and costly to dispose of. Some waste streams (e.g., arsenic) may be hazardous in some jurisdictions.

• *Advanced oxidation.* This term includes ozone, ozone with hydrogen peroxide addition, and UV with peroxide addition. The water industry has known of the disinfectant properties of ozone for many years, and this technology has been used in large-scale water treatment plants.

However, with the regulatory focus shifting to *Cryptosporidium* and *Giardia,* ozone is being considered for use specifically against these pathogens. Combining ozone and UV with hydrogen peroxide can provide an effective treatment for certain organic compounds (e.g., NDMA). The effect of rising energy costs on the expanded use of advanced oxidation treatment technologies remains to be seen. Management of organic oxidation by-products of ozone treatment is also a consideration.

Desalination-specific technologies. Most first-generation desalination units were thermal distillation processes with extreme energy requirements. As such, their use was geographically limited (typically to the Middle East where energy is relatively cheap and water is unavailable) and not cost-effective in other areas.

Thermal technologies include multistage flash distillation, multiple effect distillation, and vapor compression. Since the 1970s, two membrane technologies have been in limited use in this country for desalination, electrodialysis, and RO. Again, cost and performance issues prevented their widespread, large-scale use.

As advances have been made in membrane technologies (especially RO) interest in their use in desalination applications has increased. Approximately 20 seawater RO plants are in various planning stages along the California coast. Cost issues remain as do challenges with pretreatment, permitting of coastal plants, and energy requirements.

However, as the marginal cost of traditional water sources continues to rise and the availability declines, desalination will grow in attractiveness.

ANCILLARY TECHNOLOGIES BROADLY AFFECT THE WATER INDUSTRY

Water treatment technology is not the only field in which great advances will be made in the coming decades. The introduction of new technologies in every area of the water utility—distribution system, source water, front office, back office, and laboratory—will fundamentally transform these functions as well.

Rapid improvements are under way in water quality monitoring and instrumentation. The 1996 SDWA amendments also provided the impetus for improved water quality monitoring instrumentation. Utilities were compelled to gain a greater understanding of their water's chemistry from source water through the treatment plant and out into the distribution system. Enhanced monitoring—requiring new instruments and methods—was required for pathogens, DBPs, and a host of other contaminants.

Water quality monitoring also acquired a new importance following the Sept. 11, 2001, terrorist attacks. Those events not only expanded the number of contaminants of concern to include xenobiotics and chemical/biological/nuclear agents, but there was a need for much quicker, near real-time analysis.

Progress in the water quality monitoring and instrumentation field is ongoing and rapid. Companies are developing multisensor packages for normative water quality parameters (e.g., pH, dissolved oxygen, temperature, conductivity) for use in source waters and throughout the distribution system. A variety of sensors for biological contaminants including rapid immunoassays, rapid enzyme tests, and rapid polymerase chain reaction tests are being developed and field-tested. Field-deployable gas chromatograph/mass spectrometers are also being improved to provide near real-time detection for a host of organic contaminants.

As treatment technologies become more sophisticated, process-monitoring capabilities will have to follow suit. Online instrumentation will monitor performance, equipment utilization, energy demand, and input/output water quality in an effort to efficiently produce the highest quality water. At a cost of only pennies each, miniaturized sensors will be able to monitor virtually every minute aspect of a treatment train, providing extraordinary degrees of process optimization.

More information necessitates better information management. This expanding suite of process and water quality monitoring instruments will produce enormous amounts of data. Humans will simply be unable to analyze and evaluate the information that will constantly be streaming in to the control room. Fortunately, the information technology revolution is keeping pace—if not outstripping—advances in other technologies.

Moore's Law (named for Gordon Moore, founder of Intel Corp.) posits that the number of transistors on a chip (and therefore the absolute processing power of the chip) doubles approximately every 18 months. Since first stated in 1965, Moore's prediction has remained surprisingly valid, and analysts expect it to remain true for another 10–20 years. While processor capabilities and speeds have been exponentially increasing, the costs of information processing technology are also dropping dramatically—sometimes as much as 25% per year. This intersection of rising performance curves and plummeting cost curves has produced widespread information-management capabilities that were previously unknown.

As a result, it is now possible to cost-effectively acquire, store, and process unimaginably vast arrays of data—a process that was virtually impossible before the recent silicon revolution. The power of this technology allows ready acquisition of enormous datasets; artificial intelligence (AI) systems will assist in its conversion to information. But it is still the innate intelligence and abilities of the human operator that will allow that information to be transformed into useful knowledge.

AI systems will analyze source water quality data and optimize water treatment processes to reduce energy costs and chemical use. Real-time distribution system monitors will identify problem areas of low flow or residual disinfectant and then provide correctional input to reservoirs and automated pumping stations. Operators will be able to monitor and control every aspect of the treatment/distribution process from multifunction hand-held communications devices.

TREATMENT TECHNOLOGY INTRODUCES NEW COSTS AND CHALLENGES

All of these sophisticated treatment, monitoring, and information processing technologies are not without concerns—concerns that must be proactively addressed by managers if these devices are to be effectively integrated into tomorrow's water utility. Some of these concerns are generic to the introduction of any new treatment process or equipment; others are unique to the nature of the technology involved. Successfully

meeting these challenges may involve external third parties such as regulatory agencies or new types of enterprise partnerships.

Capital costs are expected to decrease. Many of the new and emerging technologies (UV, ozone, membrane systems) are comparatively expensive when contrasted with conventional filtration. However, much like information-management technology, their costs continue to decline as additional research and development are conducted. This is especially true for membrane systems. Large membrane treatment systems that would have been prohibitively expensive even a decade ago are now well within the realm of reasonable cost comparison with conventional technologies. Costs (both absolute and comparative) will remain a primary consideration, but as new regulations, contaminants, and impaired source waters drive the development of alternative technologies, so too will their costs become far more competitive.

Operational costs may remain high. The operational costs of many new technologies may be considerably higher than current conventional treatment techniques. For example, the relatively high operating pressures required by NF and RO translate into high electrical energy demands for the pumps. Significant advances in membrane technologies have been made over the past 15 years that have reduced required operating pressures, resulting in lower power demand. However, given recent energy price spikes and blackouts, this remains a major operational consideration. Installation of dedicated back-up power generators would mitigate the risk of power outages but add considerably to overall capital costs of the treatment facility.

Similarly many of these treatment technologies engender relatively high maintenance requirements—requirements that may be quite different from and more onerous than conventional treatment trains. Membrane systems require careful performance monitoring to assure that the unit is not fouled, and cleaning must be precisely performed to avoid damaging the membranes and to assure optimal efficiencies. UV units require aggressive cleaning to avoid degrading the lamps and reducing their disinfecting capabilities.

Finally, the technical skills necessary to properly operate, maintain, and repair these advanced monitoring and treatment technologies may not be present in every water utility. Identifying, hiring, and retaining qualified technicians may be difficult and expensive, especially with pending retirements of Baby Boomers and a general reduction in the number of engineering and technical degrees granted in US colleges.

Residuals management creates additional issues. Proactively managing water treatment residuals is a growing concern at all water utilities regardless of the treatment technology employed. Simply disposing waste to the sanitary sewer may no longer be either an appropriate or allowable option.

Such considerations extend to, and may even be more critical for, the technologies described previously. Although the efficiencies of membrane units have increased substantially in recent years, there is still a much larger water loss from their concentrate stream than there is from conventional treatment technologies. The concentrate stream may also contain contaminants at levels that are inappropriate for direct discharge or discharge to a publicly owned treatment works. In some states (notably California) land disposal (via impoundments) is also problematic depending on the wastestream characteristics.

Other technologies, especially IX and related systems, produce brines and solid waste that, depending on the contaminant being removed (e.g., arsenic, vanadium), may constitute a hazardous waste. From an operational and financial standpoint, generating, processing, storing, and disposing of hazardous wastes from a physical facility present enormous challenges. Most water utilities have not experienced the high costs of disposing of large volumes of hazardous waste. Special storage and handling areas may be required as well as specialized training for operators.

Removal of two or more exotic water contaminants such as arsenic and radionuclides may produce mixed waste—waste that is both hazardous and radioactive. Disposal of such waste is difficult and expensive. In fact, the requisite infrastructure to safely handle, transport, and cost-effectively dispose of such waste from a large number of generators may not even exist.

These last issues encompass a whole new host of regulatory requirements for the drinking water industry. Hazardous waste and mixed waste regulations are demanding, and compliance is difficult and expensive.

INNOVATION REQUIRES AWARENESS AND SUPPORT

Improved IT brings new challenges. The rapid, ongoing developments in computing power and applications of information technology imply that water utilities and industry research bodies should boost their awareness of such developments and how they can assist in efficiency and performance improvement. For example, developments in knowledge-management systems indicate that water utilities should examine how they can use IT to capture a hidden opportunity from the upcoming Baby Boomer retirements.

However, IT investment must be accompanied by first-rate management practices to improve performance. Therefore, water utilities need to work out how their prospective IT investments will actually improve operations, even for systems that are replaced because of obsolescence.

Growing maturity in e-commerce and the closing of the "digital divide," particularly in contrast to the situation at the time of the last strategic review in 2000, indicate that utilities and research bodies should

reexamine potential applications and approaches that may have been prematurely written off. Also, they should stay up to-date with how customers and other utility stakeholders are using the Internet in their daily lives. The water industry should maximize the use of the Internet to communicate with stakeholders, coordinate efforts, centralize functions, collaborate, and otherwise make use of network effects. Such potential applications include streamlining procurement, dealing with customers, and collaborating with a wide range of existing and emerging stakeholders.

The rapid growth of IT security and malware issues and the threat of cyber-terrorism compel the water industry to stay up-to-date, develop backup plans, and be flexible in their approach to IT, such as selective outsourcing of security functions to specialists.

Automation of water treatment is likely to expand as new technologies require less hands-on management and water utilities press to reduce labor and operating costs. Automation can increase efficiency, permit better control of complex processes, and help to quickly identify malfunctions that could affect water quality or worker safety. Water utilities will continue to take advantage of these opportunities to improve service but will be faced with changing workforce requirements as a result.

Water services worker training should include a better understanding of system dependencies and interconnections to meet the challenges created by highly automated systems. In addition to worker training, there will be a need to assess the ability of newly automated systems to meet unexpected conditions. The widespread power outage in the eastern United States in August 2003 was caused by a minor malfunction, followed by a multitude of automated systems doing specifically what they were designed to do.

Remote-sensing and monitoring can help alleviate concerns for system security and help maintain high water quality. The water industry needs to be proactive in implementing and improving these systems.

Development of expert systems will reduce the skill level needed to operate treatment and distribution systems under normal conditions, but will increase the skills necessary to identify and address unexpected situations. The development of expert systems should include the identification of boundaries under which they can be employed.

The water industry should actively support the development of IT-rich equipment and devices to improve asset management and optimize operations and maintenance. This is too big a task for any individual utility and therefore should be done in a cooperative manner, at an industry-wide level, and in partnership with equipment suppliers.

Water utilities should seek out IT applications that will reduce cost and improve service, not just to provide a form of basic infrastructure. Obvious cost-saving developments to take advantage of in the near term are VoIP and open-source applications.

Water utilities should focus on management practices that will result in productivity improvements from their IT investment, because research shows that without this, no productivity improvement occurs.

Water utilities should use IT to turn the wave of Baby Boomer retirements from a problem into an opportunity. The emergence of knowledge-management applications and web services are recent examples that could result in better knowledge transfer than is currently the case and cost savings through centralization/outsourcing.

The water industry should consider forming a central clearinghouse to provide advice on IT equipment and application selection because the IT field can be especially confusing and hard to keep up with and is not a core competency within the industry.

Water utilities should reexamine their procurement procedures to maximize benefits from e-commerce, including cooperative purchasing and process streamlining and removing artificial rules and impediments.

Drinking water treatment technology. New treatment technologies driven by regulations and increasing use of marginal quality water supplies will place new demands on operator skills and change the footprint of the treatment system's physical plant. In addition, the investment in these new technologies will place demands on the capability of the utility to acquire capital.

The operational costs of the new technologies may be higher than current conventional technologies (e.g., energy). On the other hand, automation of these new technologies may reduce labor costs. Similarly, maintenance costs may be higher because of the increased complexity of the treatment systems and the need for specialized training of personnel. However, since the new technologies can address new contaminants and regulatory concerns—and can use impaired sources—they are likely to be the preferred option in many cases.

An additional concern with all water treatment technologies is the issue of residuals management. The concentrated brines and resin regeneration brine streams produced from new membrane and ion exchange facilities—as well as the settled solids from conventional settling—may pose difficult disposal challenges.

Water utilities should encourage development of new technologies and assure that they are considered when it is time to add to or modify the utility's facilities. In addition, the utilities should work to assure that regulatory and construction grant programs are structured to encourage innovation.

ACKNOWLEDGMENT

The authors acknowledge the AwwaRF funding and project manager, Linda Reekie, and acknowledge the project advisory committee guidance of Andrew DeGraca, John Huber, and Rosemary Menard. They also thank the utility managers who gave generously of their time and expertise in the Futures Workshop, as well as the experts interviewed at the start of the project. This work reflects their collective wisdom and active engagement. Important logistical support by Lorena Ospina, Gloria Rivera, Ryan Reeves, and Nicole West is appreciated.

ABOUT THE AUTHORS

Ed Means III is senior vice-president of McGuire/Malcolm Pirnie, Irvine, CA. Means has 26 years of experience in water utility management and water quality. Roger Patrick is the president of Competitive Advantage Consulting Ltd., and has 15 years of experience as a management consultant helping public and private organizations improve their performance. Joseph Bernosky is a project manager with Leonard Rice Engineers Inc.

REFERENCES

Najim, I & Trussell, R.R., 2002. *New and Emerging Drinking Water Treatment Technologies*. The National Academy Press, Washington.

National Science Board, 2002. Science and Engineering Indicators. National Science Foundation, Washington.

National Science Foundation, Division of Science Resource Statistics. www.nsf.gov/statistics/showpub.cfm?TopID=9&SubID=22 (accessed Oct. 25, 2005).

NIST (National Institute of Science and Technology). www.nist.gov (accessed Sept. 23, 2005).

R.W. Beck Inc., 2002. Technical Memorandum B.7—Demineralization Treatment Technologies.

U.S. Economic and Statistics Administration. Digital Economy 2003. https://www.esa.doc.gov/results.cfm?criteria=Digital+economy+2003&Go=Go (accessed Oct. 25, 2005).

This page intentionally blank.

Sources of Financing for Your Water Technology Business

ONE OF THE MOST IMPORTANT ATTRIBUTES—AND CHALLENGES—OF ANY NEW COMPANY IS ITS FINANCIAL BACKING AND STRENGTH. ATTRACTING AND SECURING THE NECESSARY FUNDS TO DEVELOP A NEW TECHNOLOGY, TO EXPAND OR PERFECT AN EXISTING PRODUCT, OR JUST TO SUSTAIN A SUCCESSFUL ONGOING BUSINESS CAN BE ONE OF THE BIGGEST CHALLENGES IN RUNNING A BUSINESS.

BY STEVE MAXWELL
(*JAWWA* **November 2004**)

Although there is a broad spectrum of financing sources—and hundreds of equity investors who are excited and interested in the water industry—attracting the right type of capital and obtaining it at an acceptable "cost" can often be difficult and time-consuming. The previous three articles have dealt with the challenges of developing and building water-related technology or service ventures. This article will provide a quick overview of the different sources of financing that are available to entrepreneurs and owners in this industry as well as some of the pros and cons of each.

PERSONAL INVESTMENTS MAY TAKE YEARS TO PAY OFF

Almost all entrepreneurs and developers of new technologies get their start by committing their own personal monies and assets to the business concept and startup expenses as well as the long, hard, and unpaid months or years of labor that it typically takes to get a new venture off the ground. It may also include the entrepreneur and his or her colleagues contributing their own personal savings, life insurance cash values, second home mortgages, and the like as they struggle to provide the necessary financing to get a new business up and going.

It is also common to see such startup ventures draw on similar financial contributions from friends, family, and business associates. This can turn into an emotional tightrope for the technology developer because this type of financing can have obvious positive or negative (and long-lasting) implications for friendships and personal relationships. In addition, many startup firms use the unpaid contributions of specialized labor from friends and associates (sometimes termed "sweat equity") often in return for options or shares of stock in the company. These types of investments may not be worth much for many years to come.

Another initial source of funds for the startup technology business may be from personal credit cards or lines of credit from commercial banks. It is easy to borrow against personal credit cards, but they typically limit the total borrowing, and in addition they usually carry high interest rates. Much more attractive from the overall perspective are personal lines of credit, under which more reasonable interest rates and terms can often be negotiated.

WEALTHY INDIVIDUALS, ANGELS, AND VENTURE CAPITALISTS OFFER POTENTIAL SOURCES OF FUNDING

As a small firm begins to perfect and implement its business model, to have pilot-type projects in operation, and to demonstrate the commercial viability of its technology or service idea, a new range of potential financing sources begins to open up. High net–worth individuals, or "angels" as they are often called, can sometimes be attracted to invest equity capital into small firms at a fairly early date. In many communities there are networks or organizations of such individuals who are seeking interesting and potentially profitable long-term investments, and the entrepreneur can seek to connect with such organizations.

Obviously, the entrepreneur has to convince the angel of the merit of his or her ideas and approach. On the other hand, there are countless wealthy individuals looking for investment ideas that are more attractive—if also more risky—than "standard investments," such as those available in the public stock markets. A typical angel investment might range from a small amount up to several hundred thousand dollars. Such financing can be made in the form of a loan, but it's more typical for an angel to receive equity or stock in the company, which may—if and when the company reaches commercial success—carry a very attractive return. However, all such early-stage investments inherently carry a high degree of risk.

As the business grows and as more and more commercial success is achieved, larger and perhaps more sophisticated investors may become interested in investing in the venture. There is a gradational spectrum among wealthy individuals, angels, and true venture capitalist organizations. Venture capitalists, or early-stage private equity investors, typically enter the mix as the technology development venture becomes more proven and as its products or services begin to become commercially viable. Venture capital firms are typically organizations that raise a pool of capital from wealthy individuals or institutions and then invest it over some given period of time. Venture capitalists have been instrumental in the successful commercialization of many businesses in this country, particularly in the high-tech, medical, and environmental technology marketplaces. There are hundreds, if not thousands, of true venture capital firms in the United States, managing funds that range in size from a few hundred thousand or million dollars at the low end to billions and billions of dollars at the high end.

Most technology developers, if they stay on a successful development track, will bump into the venture capitalist community at some point. Venture capital can offer an array of benefits to the entrepreneur, but like anything else, it also brings some costs and risks. For the right deal, venture capitalists are willing to invest millions of dollars in order to get a venture going, and in return they typically receive stock and key decision-making participation on the company's board of directors. Venture capitalists can often bring valuable business acumen and expertise to the table because they have been through many similar startup efforts in the past. In addition, venture capitalists often bring good networking contacts and credibility in the financial community or at least among key potential customers. Investments by certain types of highly successful and respected venture capitalists can represent a real "feather in the cap" for the startup firm, conveying instant credibility in the marketplace. On the other hand, some entrepreneurs feel that venture capitalists undervalue their businesses and attempt to exert too much strategic and business control.

There are numerous directories and listings of venture capital organizations that an entrepreneur can contact at the appropriate time. Different venture capitalists have different investment minimums, revenue expectations, and deal parameters that they will accept. Some venture capitalists will invest in any type of promising business, whereas others develop an industrial specialty or a specific end-market focus. At the moment, there are a number of venture capital groups that are specifically interested in the burgeoning water and wastewater treatment industry, and almost all investors seem to be positively inclined toward the water business. Thus, for a good service or product idea in the water industry today, it is not difficult to find venture investors who will seriously listen to you.

AS REVENUES AND EARNINGS BUILD, PRIVATE EQUITY INVESTMENT GROUPS BEGIN TO APPEAR

After venture capitalists—as the business continues to build revenues and earnings—a much broader array of private equity investment groups come into play. Although these are generally similar groups looking for attractive returns on their investments, they tend to shy away from the earlier-stage and riskier investments that true venture capitalists traditionally seek. There is a vast number of private equity investment groups in the country, providing a dizzying array of different types of investment vehicles and options and focusing on different geographies, end markets, or technologies. However, most such groups are looking for established businesses with substantial revenue and earnings. Equity investors often step in when the business founder is retiring, to assist in a partial or total ownership transition, or when the entrepreneur decides it is time to roll together with a consolidation effort in the industry. The exit strategy for such investors is typically an initial public offering (IPO) or a sale to a large intra-industry or "strategic" buyer.

Suppliers or customers can also represent an implicit source of financing for startup ventures. Careful management of short-term assets and liabilities—an art at which most entrepreneurs quickly become pretty accomplished—can help implicitly utilize the capital of both suppliers and customers to help finance the business. Delaying the payment of bills as long as practical can help turn the supplier into a short-term financier of the business. Careful attention to rapid collection of customer receivables and the maintenance of low inventory levels can allow the entrepreneur to keep as little cash tied up as possible. More significant may be the opportunity for more explicit financing from outside parties—if customers really want to see a new product developed, they may be willing to help provide funds in a more direct way as well. Likewise, if the startup venture could be a substantial new customer, vendor firms may also be willing to chip in.

Traditional commercial banks and other types of credit institutions are typically used by almost all small companies as well as larger and established ones. At an early stage, some firms may only be able to obtain a short-term, line-of-credit type of financing arrangement with the bank—an arrangement in which the bank lends money on a short-term and variable basis to help the firm smooth out fluctuations between its receivable and payable balances, meet payroll requirements, and help with inventory and purchasing when the company's own revenues and financial well-being may be somewhat variable week to week or even day to day.

Short-term debt is typically used to describe loans that are repaid within a year, whereas long-term debt may involve a larger sum of money to be repaid over a several-year period. Banks and other credit institutions may also lend money in amounts that are tied directly to one or another of the assets of the company—for a small firm the primary asset is often the accounts receivable of the business. Lenders may "factor" the receivables (i.e., assume that a certain percentage of the receivables could be collected in the event of a business failure) and be willing to lend that amount of cash to the business.

INCREASED SUCCESS = LARGER ABILITY TO BORROW

As the company grows and builds its equity value and earnings, its ability to borrow also expands. As firms grow and as their credit rating and financial reputation begin to solidify, they may undertake larger and longer-term debt from commercial banks and lending agencies. Examples of larger or one-time financing needs might include the construction of a new manufacturing facility or the acquisition of another firm to complement the existing business. Interest rates on long-term loans vary as a function of the credit-worthiness of the borrower, but they also tend to fluctuate up and down as a function of other macroeconomic factors and trends. The use of long-term debt varies widely among companies. Some more conservative firms prefer to operate on equity capital and keep their loans to a minimum, whereas others believe they can grow more quickly and create more shareholder value by borrowing more to finance faster growth.

The federal government, as well as state and local governmental bodies and agencies, also has money available for funding new technologies and to help new businesses get started. A full review of the various types of grants, technology development programs, and technology certification efforts provided by various federal agencies is beyond the scope of this column. There are, however, numerous grants and sources of financing available from the government, particularly in an area like the water purification business, which has pressing needs.

Although relatively rare, some smaller and startup type ventures are able to obtain financing through IPOs. Typically, in today's market, a firm needs to have at least $100 to $200 million in revenue before it can even consider conducting a sale of stock to the public and becoming traded on the stock exchanges. However, occasionally—if a new firm has great technological promise or if it addresses a booming new market—early-stage IPOs may be possible. However, IPOs are much more typical years later, after the business has matured and grown, when the initial founders and investors are seeking a way to liquidate their hard-earned investments.

CONCLUSION

This has been a very brief and cursory review of different types of financing for startup or growing businesses. Finding the right kind of funding is never easy. The right kind of financing depends on individual preferences and the specific situation of the company, and almost no money comes without some sort of strings attached. As has been discussed in previous columns, the widely anticipated general economic significance of the water business and its projected future growth have thrust this industry into the limelight. Fortunately for technology developers in this industry, strong investment interest is evident from all corners. Entrepreneurs with a good idea and strong business plan for the water industry should not go hungry.

This page intentionally blank.

Assessment and Renewal of Water Distribution Systems

BY NEIL S. GRIGG
(*JAWWA* February 2005)

Should utilities make a large capital investment to renew aging and deteriorating water distribution systems, or should they be reactive and wait for failure before investing? This article reports on the results of a project funded by the Awwa Research Foundation (AwwaRF) about three aspects of system renewal—setting priorities, assessing condition, and performing renewal tasks—and how these aspects pertain to asset management (AWWARF, 2004). The project included a knowledge synthesis, three workshops, utility surveys, and interviews. This article presents a framework for condition assessment and prioritizing capital improvement programs, and reviews options for renewal technologies. It also includes examples of how leading utilities are handling assessment and renewal activities today.

DEFINITIONS OF THE ASSET MANAGEMENT PROCESS ARE NEEDED

The water supply industry needs a defined vocabulary for the asset management process, including infrastructure renewal. Table 1 gives proposed definitions drawn from a composite of sources.

For the purposes of the AwwaRF project and this article, asset management is defined as "an information-based process used for life-cycle facility management across organizations." Every utility should have an asset management program, whether formal or informal. As shown in Figure 1, an asset management system comprises a collection of work activities that a utility also does as part of other work. These activities may be defined in different ways, but the common tasks of inventory and condition assessment apply to all processes.

"ONE SIZE FITS ALL" DOES NOT APPLY TO CURRENT RENEWAL METHODS

Views about how to set priorities, assess condition, and perform renewal tasks in water distribution systems are in flux. Some utilities believe that the replacement problem is urgent, requiring more national investment. Others believe that the problem is not serious and that pipes will last longer than projections show. Some

utilities believe that they should rehabilitate or replace pipes on the basis of observed failures, rather than try to assess the condition of their pipe inventory. Some use advanced methods for planning and prioritization; others are only beginning to think about implementing formal methods. "One size fits all" does not apply to current distribution system renewal methods.

UTILITIES RECOGNIZE THE NEED FOR SYSTEM RENEWAL

Infrastructure is a mixture of old and new. The water distribution system infrastructure in the United States comprises old and new materials. Most pipes installed from the late 1800s through the late 1960s were made of cast iron and later reinforced with cement–mortar lining (CML). Other improvements include ductile-iron pipe (the material of choice to replace cast iron), asbestos–cement (AC) pipe, and polyvinyl chloride (PVC) and high-density polyethylene (HDPE) pipe technologies, which are being used more now (AWWSC, 2002).

Pipe inventory has increased 11%. Results extrapolated by population from AWWA's 2002 distribution survey (2003a) indicate the current pipe inventory in US water distribution systems is 980,000 mi (1.6×10^6 km). This is an 11% increase from the 1992 inventory of 880,000 mi (1.4×10^6 km) reported by Kirmeyer et al (1994).

Needs studies highlight lack of consensus. Driven by the Safe Drinking Water Act and media attention to infrastructure, several needs studies have been completed during the past 16 years (ASCE, 2001; USEPA, 2001; Stratus Consulting, 1998; Kirmeyer et al, 1994; National Council on Public Works Improvement, 1988). However, there has been a lack of consensus on the needs because of the absence of measurement standards and general difficulty in making estimates.

In addition, the institutional structure of distribution system management does not favor large investments for renewal. Utilities recognize the need for system renewal, and some replacement occurs, but at a slow rate. In spite of the needs, AWWA's 2002 distribution survey results (AWWA, 2003a) suggest that pipe-replacement rates have slowed from the replacement

Table 1   Definitions of renewal activities

Term	Proposed Definition
Asset management	An information-based process used for life-cycle facility management across organizations; similar to "capital management" or "infrastructure management"
Condition assessment	Evaluation of the readiness of a component to perform its function
Infrastructure integrity	Integrated measure of condition, capacity, and soundness of infrastructure component
Maintenance	The work of keeping components in proper condition, including preventive maintenance and corrective maintenance
Needs assessment	Estimate of needs for renewal and new capacity in infrastructure systems
Prioritization	Ranking of projects according to fixed criteria
Rehabilitation	To restore or upgrade existing components through extensive work
Renewal	To renovate or restore functionality of a component, including major repair, rehabilitation, or replacement
Repair	To restore functionality of existing components after damage or breakage
Replacement	To provide a substitute for a distribution pipe or other component

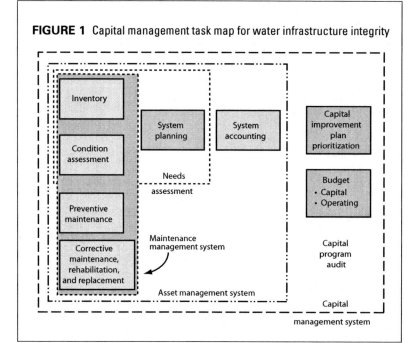

FIGURE 1 Capital management task map for water infrastructure integrity

rate of once every 200 years as reported by Kirmeyer et al (1994). Repair and rehabilitation also occur but are not reported with systemwide statistics. Given the high cost of renewal and the need to use scarce capital for other purposes, boards are often reluctant to allocate much funding to distribution system renewal. Add to this the absence of direct regulatory controls on pipe integrity, and the once-every-200- years replacement rate (or slower) is easy to understand.

UTILITIES MUST PRIORITIZE RENEWAL PROJECTS

Prioritization is important in the renewal of water distribution systems because renewal projects compete with other projects for expansion, reliability, service, and capacity and to meet regulatory requirements. How prioritization works within the context of capital improvement planning is outlined in *Water Utility Capital Financing* (AWWA Manual M29, 1998) and *Capital Planning Strategy Manual* (AwwaRF, 2001a).

A voting method can be used for relatively simple prioritization decisions. For more complex decisions, a spreadsheet can select and weight evaluation criteria, develop performance measures, and rank projects by scores. More complex problems that require a documented process with multiple stakeholders and competing interests should use multicriteria decision analysis (AWWARF, 2001a).

Criteria include return on investment and risk allocation, as well as public sector and social issues (AWWSC, 2002; AwwaRF, 2001a). Utilities should select their criteria from broad categories such as the following:

• Technological and managerial—Does the plan meet goals effectively? Does it improve service and reduce interruptions? Does it reduce risk and vulnerability?

• Legal and institutional—Can the owner implement the plan in the long term? Does it pass a test of legal

TABLE 2	Examples of prioritization methods
Method	**Reference**
Priority action numbers for mains	Hope, 2003
Private water company priorities based on performance, customer service, and water quality	Connor & Mason, 2004
KANEW (predictive distribution system condition assessment) model offers method to determine a water main replacement rate	Deb et al, 1998
"Nessie curve" predicts when replacement waves will occur; provides "decay curves"	AWWARF, 2001b
Genetic algorithm	Dandy & Engelhardt, 2001
Individual pipe models	Andreou et al, 1987
Comprehensive decision support models	Burn et al, 2004; Hafskjold, 2003; Deb et al, 2002; Kleiner & Balvant, 1999

feasibility and achieve compliance with regulations?

• Political and social—Is the plan politically feasible? Does it support other plans? Are people impacted negatively? Does it reduce or eliminate threats to public health?

• Financial and economic—Does the plan have an acceptable return rate? Does it produce revenue? Does it provide service to new customers? Does it reduce operating costs? Does the plan pass a cost-versus-benefit test? Does it support economic development and benefit the local economy?

• Environmental—Does the plan introduce negative environmental consequences?

In the end, decisions require trade-offs. AwwaRF's *Capital Planning Strategy Manual* (2001a) states:

"Realistically, the existence of political 'deal-making' is probably a given in most utility environments. Capital planning decision processes and models are not intended to replace or eliminate the traditional role of politics. Rather, these models can complement the political process; they provide a starting point for negotiations and greater insight into the potential tradeoffs between alternatives."

Many prioritization methods that consider factors such as pipe age, failure mechanisms, installation history, leak history, water quality, criticality, hydraulics, corrosion, material, pressure, location, soil type, groundwater, and loads are available. Examples of these methods are listed in Table 2.

Several utilities have developed methods for distribution system capital prioritization. Although their approaches vary, they collectively demonstrate the

required features of a planning system—data, analysis, and plans. Examples of utility planning models are listed in Table 3.

After reviewing these methods and examples, it is apparent that any valid prioritization program comprises a sequence of inventory and condition assessment, ranking by criteria, decision analysis, and programming of improvements (Figure 2).

CONDITION ASSESSMENT NOT PERFORMED CONSISTENTLY

As defined in the AwwaRF project and for this article, condition assessment means to evaluate the readiness of a component to perform its function. Although this evaluation is required for asset management, it is not applied consistently in the water utility industry and the term does not appear in texts about distribution systems (AWWA, 2003b). However, condition assessment does occur in the water utility industry, but it is not as well developed for water systems as it is for road pavement, for example, which is a visible infrastructure component and has a politically sensitive performance indicator (poor roads).

As systems deteriorate, condition information is required to plan and schedule repair, rehabilitation, and replacement (Horn, 2003; Margevicius & Haddad, 2003; Mavrakis, 2003). Water quality complaints also create the need for condition information to know which procedure to use (e.g., flushing or cleaning). In distribution system operations, condition information is required to schedule tasks such as testing and valve operation. The budget office requires information on priorities that begins with condition assessment (Becker, 2003). Government Accounting Standards Board (GASB) Statement Number 34 requires condition information in utility capital accounts (APWA, 2002).

A framework for condition assessment proposed. By drawing from a study of performance indicators (Deb et al, 1995) and other studies of condition assessment, a framework for condition assessment can be organized. Thus a framework is proposed with the following four levels: system goals and criteria, condition indicators, integrated system parameters, and decisions relying on system parameters (Figure 3). In addition, location and registration

TABLE 3	Examples of utility-planning models	
Model		**Reference**
East Bay Municipal Utility District (Calif.) model based on AWWARF research utilizes district mapping system.		Clinton, 2003
New Orleans (La.) water "Master Plan" uses GIS*, hydraulic software, and KANEW model.		Becker, 2003
Los Angeles Department of Water and Power (Calif.) uses age-related deterioration scores to program replacement.		Mavrakis, 2003
Washington Suburban Sanitary Commission (D.C.) uses MMIS, data, and GIS; includes weighting factor according to type of maintenance.		Tucker & Corriveau, 2003
Columbus (Ga.) did GIS-based replacement study, which proved that GIS and age-dating are useful, access to data bases is required, and getting data represents progress itself.		Horn, 2003
Sydney (Australia) has risk-consequence matrix to rank pipes by risk consequences and likelihood of failure; uses KANEW model to forecast financial needs and develop risk procedure.		Radovanovic, 2003
Portsmouth Water (Portsmouth, England) replaces 1% per year with point-system prioritization program.		Neve & Hedges, 1998

*GIS—geographic information system, MMIS—maintenance management information system

of components are required to locate underground assets and record data about them. As shown in Table 4, the framework uses nine basic condition indicators: three for the capital side and six for operations.

The system parameters require local standards or national or regional standards. They yield information about six integrated system parameters that can be reported to management and the public: integrity, security and reliability, water quality, losses, flow, and pressure. The final level of the framework involves the use of the information for renewal work, operations and maintenance, and budget and finance. Although advanced methods are available for pipe location and for the parameters of condition assessment, they may not be satisfactory or cost-effective in all cases.

Effective pipe location methods are necessary. Utilities need high-quality records and effective equipment to determine depth of assets, distinguish gas pipes from water pipes, and not be affected by other utilities (Deb et al, 2001). Methods include probing and digging, and technologies such as electromagnetic, ground-penetrating radar and sonic, acoustic, and infrared thermography. To assess the status of pipe location methods, AwwaRF and the UK Water Industry Research Ltd. (UKWIR) held a research workshop on the location of multi-utility buried pipes and appurtenances (AwwaRF & UKWIR, 2002). This workshop produced a good snapshot of the current level of development in pipe location methods.

Structural assessments of pipes can use destructive evaluation or nondestructive evaluation (NDE) methods. Direct inspection, sounding, coupon sampling, controlled destructive evaluation (CDE), the remote field

eddy current (RFEC) technique, and acoustics are the main methods in use. They apply in different ways to metal or concrete pipe, and little information is available for their application to plastic pipes (Hughes, 2002; Smith et al, 2000; Ellison, 2001).

Destructive evaluation methods. The main destructive methods for pipe testing use coupon sampling and CDE. The Los Angeles Department of Water and Power (LADWP) has experience using such destructive pipe location methods with a large-scale program (Ellison, 2001), but few reports about CDE other than from LADWP have been noted. Coupon sampling is a tried-and-true technique but is not always reliable.

NDE methods. NDE methods are more popular than destructive technologies when they work effectively. Older testing techniques, such as hydraulic or water quality testing, are NDE methods, but the research focus has moved to newer technologies.

Jackson et al (1992) found that the RFEC technique is the most common and the most effective NDE method for iron-pipe testing. Dingus et al (2002) reported that NDE for concrete is also improving. RFEC can be used to detect broken wires in a prestressed concrete cylinder pipe (PCCP) (Mergelas & Kong, 2001). One example of a utility using the RFEC technique in a PCCP is the San Diego County Water Authority (SDCWA), which uses internal inspections and RFEC to assess the integrity of empty pipe. When the pipe is in service the utility uses acoustical monitoring (Galleher, 2003). In the acoustic method for PCCP, sensors listen for broken wires and signals, which can be sent over the Internet (Woodcock, 2003). The use of internal inspection or evaluation tools, or "smart" pigs, for RFEC studies may increase after methods are found to enable them to pass constrictions. For example, in oil and gas applications, such tools monitor longer runs without constrictions.

It is still to be determined how extensively utilities will use NDE methods. Dingus et al (2002) are optimistic: "a shift from using empirical data to NDE methodologies will allow system operators to make optimal decisions based on pipe condition, leading to rehabilitation and replacement of damaged pipes only.

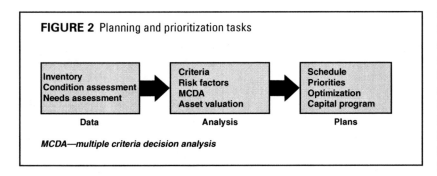

FIGURE 2 Planning and prioritization tasks

Inventory Condition assessment Needs assessment	→	Criteria Risk factors MCDA Asset valuation	→	Schedule Priorities Optimization Capital program
Data		**Analysis**		**Plans**

MCDA—multiple criteria decision analysis

.... Future NDE [methods] will allow utilities to examine pipes and create a life expectancy timetable from the data.... No method for predicting failures will ever be 100% accurate, but this will give a closer approximation than . . . today's information."

Condition assessment measures should be utilized more effectively. Measures are available currently for the condition assessment framework presented earlier. Utilities regularly track repair frequency of pipes, operability of valves, and leaks and breaks. Water loss accounting is an accepted method to test a network for physical integrity (Committee Report, 2003), and hydraulic and pressure testing yield useful information about pipe condition (Cesario, 1995; Walski, 1984). Reliability can be assessed by customer data and interruptions. A supervisory control and data acquisition (SCADA) system enables utilities to collect energy data and make it available to a condition information system.

However, utilities report that they should utilize available information much better than they do. There is no standard procedure to record data on leaks, breaks, and condition indicators. Utilities should manage existing data, analyze pipes without digging them up, get good drawings and inventory information, and use software to organize data effectively. Some utilities advocate a national comprehensive database of US pipes, but our inquiries suggest little support for one. Ideally, a composite condition index could be compiled, but single indexes are not feasible for water distribution infrastructure. Condition assessment can be expensive, and some utilities believe that it is most needed for critical mains or those that might cause large-scale failures. However, some utility concerns, such as fire flow and water quality, may be equally important for smaller mains.

In the future, advanced condition assessment applications might include real-time assessment such as a smart tool that is sent down pipes and retrieved with data to download into computer software; small chip sets in pipes that give the type of pipe, date of installation, and so on; magnetic-encrypted identification tools; sensors to record sounds of breaks; fiber optics to record breaks in light; and improved metering capabilities to locate sources of leaks. Many research needs remain for condition assessment, mainly

to continue improvements in NDE, leak detection, and loss control. Research needs also remain for determining causes for infrastructure deterioration and corrosion and determining better in-situ methods. References listing research needs include Ellison, 2001; Smith et al, 2000; and Kirmeyer et al, 1994. AwwaRF (2001b) has funded a number of studies relating to distribution systems and continues to maintain an inventory of research needs.

RENEWAL TECHNOLOGIES OFFER OPTIONS FOR DISTRIBUTION SYSTEM CAPITAL MANAGEMENT

Water distribution system renewal activities include corrective maintenance, major repair, rehabilitation, and replacement. Making technical decisions about options for them requires knowledge of installation conditions, materials, and methods. Because of the complexity of water main systems, most large-scale rehabilitation or replacement projects use traditional renewal technologies such as CML and cut-and-cover replacement. However, utilities also use newer methods such as epoxy lining, cured-in-place pipe (CIPP) lining, slip lining, and pipe-bursting.

Maintaining water service and preventing contamination during renewal are important. Usually service connections must be reconnected manually. If a robotic method could be found to reconnect water services, trenchless technologies might be more attractive.

Comparisons are made. Several reports compare renewal technologies (Selvakumar et al, 2002; Deb et al, 2001; Ellison, 2001; Heavens & Gumbel, 2001; Smith et al, 2000). Usually the comparisons cover advantages and disadvantages of technologies such as conventional replacement, CML, epoxy lining, spray-on lining, cathodic retrofit, testing plus spot repair/ replacement, HDPE slip lining, pipe-bursting, polyester-reinforced polyethylene, CIPP lining, reinforced CIPP rehabilitation, and joint seals. The comparisons may also cover different size ranges, materials, and cost factors.

The debate over whether to be reactive or proactive in distribution system maintenance is part of a larger question of whether to renew aggressively or to wait for failure before renewing pipes. Goodson (2002) proposed a "reliability-centered maintenance" approach as an alternative to proactive maintenance. This "run-to-failure" approach may lower cost and place a focus on critical components, which are defined as water quality, customer service, regulations, and safety.

Deb et al (2000) presented overall guidance for the maintenance of distribution systems. The authors concluded that maintenance emphasizes flow and pressure

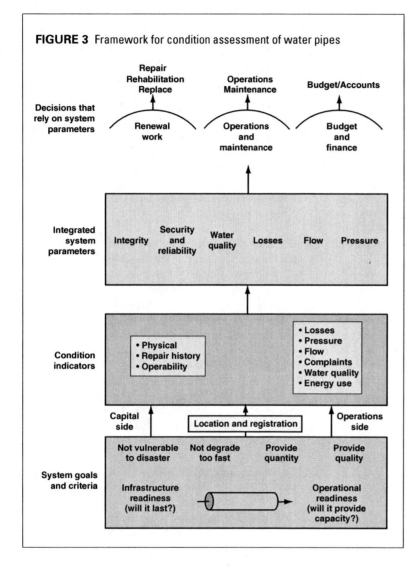

FIGURE 3 Framework for condition assessment of water pipes

tasks. New technologies and methods are emerging for locating problems, excavating, and applying repair materials. For example, technologies for locating leaks are improving through water loss control programs (AWWA Water Loss Control Committee, 2003).

Repair sleeves. The procedure for using a repair sleeve involves locating the break or leak, notifying customers, calling a "safe digging" utility notification system if available, shutting off valves, excavating, and installing the repair sleeve, as explained by Von Huben (1999). Special precautions are taken to disinfect the sleeve and to protect the pipe from contamination. Spot repairs may only require clamps or patches (Ellison, 2000).

Joint sealing. Joint sealing technologies are displayed regularly by vendors at water industry trade shows. Internal joint seals can be used to eliminate leaks in joints for pipes 16 in. (410 mm) and larger. They can also be used to bridge over cracks and similar problems in pipe segments or old connections. A new but expensive method for large-diameter pipe is to use carbon fibers (Ellison, 2000). Linings for PCCP repairs might include adding prestressed tendons, shotcreting, and slip lining with welded-steel pipe.

Rehabilitation methods include cleaning and lining. The principal methods of rehabilitation are pipe cleaning and lining. Replacement methods include open-cut replacement and trenchless replacement using microtunneling, pipe-bursting, or slip lining, which is sometimes thought of as rehabilitation and sometimes as replacement. Repair and replacement work is also explained in *Rehabilitation of Water Mains,* which covers performance criteria, cleaning, lining, joint seals, and pipe bursting (AWWA Manual M28, 2001).

Pipe-cleaning technologies. Technologies for pipe cleaning have been summarized by Smith et al (2000) and Ellison (2000) and include flushing, foam swabbing, air scouring, high-pressure water, pressure scraping, chemical cleaning, and abrasive pigging. Flushing, including unidirectional flushing (UDF), is the most common method. The AwwaRF report, *Investigation of Pipe Cleaning Techniques,* explains existing pipe-cleaning and relining techniques and the effectiveness of these techniques on water quality improvement, infrastructure reliability, risk management strategies, customer service,

but neglects reliability and water quality. Water quality is maintained by a chlorine residual, the structural integrity of pipes, and system cleanliness, including prevention of tuberculation, corrosion, and sediment accumulation. Reliability is addressed by preventive maintenance, a trained workforce, better design and construction, and effective capital improvement programs.

In addition to comparing renewal technologies, two reports by Deb et al (1999, 1990) provide information about how the technologies perform. In a 1990 report, Deb et al explained assessment methodologies, and in a 1999 report, the authors explained how the technologies work and presented a decision tree for selection of methods.

New repair technologies are emerging. The concept of a "repair" actually covers a range of work, from minor, maintenance-related repairs to major repair work (Grigg, 2002; Jordan, 2000). A repair requires locating the problem, organizing resources, and completing the work

TABLE 4   Condition indicators and decision applications

Condition Indicator	Decision Application
Infrastructure readiness	
Physical integrity of pipe	Maintenance and renewal
Repair history	Maintenance and renewal
Operability	Maintenance and renewal
Operational readiness	
Water loss	Monitoring and improvement
Pressure	Maintenance, renewal, operations
Flow	Monitoring and improvement
Complaint frequency	Maintenance, renewal, operations
Quality of water	Maintenance, renewal, operations
Energy	Monitoring and improvement

and maintenance of reasonable friction factors (AwwaRF, 2002).

Lining technologies. Liners range from nonstructural (cement mortar or epoxy) and semistructural (close-fit pipe, woven hose, CIPP, or spirally wound pipe) to structural (continuous pipe or segmented pipe). Comparisons are explained in Ellison (2000), Smith et al (2000), and Gummersheimer and Heavens (1998). The experience base with CML is extensive—more than 50 years. It is not without its critics, but many recommend CML as a rehabilitation technique. After the pipe is cleaned, winches pull the lining machine through the pipe and a drag trowel to provide a finish. Service laterals less than 2 in. (50 mm) are blown free of mortar before the lining sets. CML provides an alkaline environment next to the metal to resist internal corrosion. It extends pipe life and reduces leak rates, but how long it lasts remains to be established. Experience with CML is reported by Russell (2003) and Van Dyke (2003).

Epoxy lining has emerged more recently. It can be used where soft water will deteriorate cement lining. Conroy et al (1995) reviewed literature on epoxy lining, including results for UK utilities, and tested a main in the United States using closed-circuit TV and laboratory testing. Testing showed that leaching of the epoxy in contact with water to be well within water quality regulations. The report recommends regulations to allow epoxy lining to compete with CML. Currently AWWARF is sponsoring new projects on the evaluation of epoxy lining.

Polyurethane lining is also becoming more popular. One reviewer of this article commented on its merit as a competitor to epoxy lining, but it was not reviewed in the project.

Trenchless technologies. Trenchless technologies to open spaces for renewal include microtunneling and pipe-bursting. Deb et al (1999) explained three pipe-

bursting methods—hydraulic, pneumatic, and static pull—and documented the applications with video. After microtunneling or pipe-bursting, new pipe can be pulled through to complete an installation. This includes ductile-iron pipe (Carnes et al, 2003; Sterling, 2003).

Cathodic protection. Cathodic protection is a different kind of renewal, but retrofits may extend the life of metallic pipe. On new pipe, the cost is less than 1% (Ellison, 2000). Alternating and direct currents affect corrosion, and possibly water quality, and may create safety hazards.

Utilities reluctant to use new technologies. The project's working group on renewal technologies observed that in contrast with utilities in the United Kingdom, North American utilities are reluctant to use newer technologies because of lack of awareness, perceived cost-effectiveness, liability, and uncertainty about sustainability. They suggested the development of case histories, especially of European practices. Another utility suggested a project to evaluate the barriers to the use of new technologies and develop potential solutions.

Deb et al (2002) and the American Water Works Service Company (AWWSC, 2002) presented decision trees for the choices. The decision trees seemed almost identical, with the exception that AWWSC added design factors such as the number of connections and the type of soil conditions. In reviewing these decision trees, it was concluded that the picture is more complex than can be depicted on a simple "either-or" display, and that a more detailed chart would be required to illustrate the decision paths.

Renewal in water distribution systems is more complex than in wastewater because of the valves, service connections, fittings, and other hardware, but new methods are evolving all the time. Practices among utilities vary greatly. All utilities repair broken pipes and leaks, but some utilities are more aggressive than others. Some utilities rehabilitate pipes regularly, and others only replace pipes without practicing much rehabilitation activity.

CURRENT INVESTMENT NEEDS PRESENT FORMIDABLE BARRIERS TO UTILITIES

A number of questions about the current status of water distribution systems renewal were raised in this article. Evidence reveals that the once-every-200-years rate of replacement may be slowing even further. Why are utilities replacing pipe at such a slow rate? Some utilities' conservatism may be based on financial constraints and a lack of proof that the benefits of pipe renewal outweigh the costs. Condition assessment technologies must be improved if utilities are to reach other conclusions. Techniques such as coupon sampling only give fragments of evidence about condition, and utilities need more convincing evidence before major investments are approved. More economical renewal methods are also needed because current investment needs present formidable barriers to utilities.

How will utilities respond to these future challenges in managing their water distribution infrastructure? After all, infrastructure renewal is only one of many challenges facing water utilities in the twenty-first century.

Regardless of a utility's approach, their capital budgets are "driven by the need for new capacity, regulatory pressures, and the need for infrastructure repair and replacement. . . . What is needed . . . is an organized, documented process, driven by stakeholder priorities and coordinated with other financial and master planning activities. . . . This process must include state-of-the-art analytical tools that allow prioritization of projects based on return on investment and aligned with the utility's vision" (AwwaRF, 2001a).

The renewal process must begin with inventory and condition assessment. Because pressurized water distribution systems are buried and may be contaminated by probes, it is difficult to obtain actual information about condition. However, indirect information such as a record of leaks and breaks, nearby pipes, and age and installation conditions can be used to compile valuable condition indicators.

Prioritization is the heart of the planning process and must consider critical versus noncritical links, the right level of renewal, meeting performance standards, how much to spend, and affordability. It must blend with management strategies such as "opportunity-based" work, e.g., when breaks and leaks are discovered or when road work is scheduled by others.

Although there isn't a magic bullet for renewal technologies, utilities have a rich menu of options that includes technologies ranging from new methods for excavation to trenchless technologies. Research is needed about the failure mechanisms with different pipes, the causes for deterioration, and how to integrate results to provide guidance for utilities. The goals of such research should be to develop more accurate, user-friendly test methods to determine condition of the pipe; expand understanding of causes for deterioration, leaks, and breaks; and prevent problems and predict length of life under various conditions.

DECISIONS ABOUT INFRASTRUCTURE RENEWAL ARE RISK-BASED

Regardless of the renewal method used, decisions about infrastructure renewal are risk-based. Leading utilities are considering probability of breaks, expected time-to-failure, maintenance requirements, and water quality to make investments that avoid the consequences of failure. Even utilities without renewal programs are making implicit risk-based decisions to accept the possibility of failures. If the utility is regulated, its risk-based decisions must be defended in rate cases.

For the future of the utility industry, most writers predict that utilities will be more businesslike and thus more interested in distribution system management. Regardless of future pressures on water utilities, alternative futures suggest that utilities will continue to manage distribution systems much as they have in the past, but they will include many more monitors, safeguards, and protective systems.

In the long run, this may lead to the buildup of an echo-boom of investment needs, as predicted in economic models. On the other hand, future developments such as new technologies, new understanding of health effects, new regulations, and new business practices may offer options that are not now available. Taking a "watchful waiting" attitude about renewal may offer a strategy that preserves utility assets and options for the future. A current replacement rate of once every 200 years may just be a short-term view of a rate variable that can change drastically when new information becomes available.

Should a utility be proactive or reactive in its distribution system renewal program? It turns out that this is not a decision at all, but a question of how well the utility understands its risk calculation and the costs and benefits of pipeline renewal. This understanding will depend on the effectiveness of the utility's management systems for inventory, condition assessment, and planning. If the utility is regulated, the results of these management systems will show up in rate decisions. However, infrastructure decisions affect more than just the bottom line. Utilities must consider issues that go beyond the financial impact and extend to customer satisfaction and public health.

ACKNOWLEDGMENT

This project was funded by the AWWA Research Foundation (AwwaRF) as project 2772 and conducted under the direction of AwwaRF project managers, Jian Zhang and Kate Martin. The author would like to acknowledge the participants of the project for reporting on their practices and guiding the project's research. Utility participants include Charleston (S.C.), Cleveland (Ohio), Columbus (Ga.), Denver and Fort Collins

(Colo.), Kobe (Japan), Los Angeles Department of Water and Power (Calif.), Louisville Water Company (Ky.), Marina Coast Water District and Metropolitan Water District of Southern California (Calif.), New Orleans (La.), Northern Colorado Water Conservancy District (Colo.), Olathe (Kan.), Ottawa (Canada), Salt Lake City Corporation (Utah), San Diego County Water Authority (Calif.), Sioux Falls (S.D.), South Florida Water District (Fla.), Sydney Water (Australia), Tampa Bay Water (Fla.), Washington D.C., and Washington Suburban Sanitary Commission (D.C.). The author extends special thanks to Michael Woodcock of the Washington Suburban Sanitary Commission who contributed substantially to the project from its inception to its finish and to Peter Fitzwilliams of the Dallas Water Utilities (Texas) and Diane Sacher of the Winnipeg Water and Waste Department (Canada) for serving on the AwwaRF Project Advisory Committee.

ABOUT THE AUTHOR

Neil S. Grigg is a professor in the Department of Civil Engineering at Colorado State University, Fort Collins, CO. He received his BS degree from the US Military Academy in West Point, N.Y.; his MS degree from Auburn University in Alabama; and his PhD from Colorado State University. A member of AWWA, ASCE, and APWA, Grigg has 20 years of experience in infrastructure management and more than 30 years of experience in public works management.

REFERENCES

Andreou, S.A.; Marks, D.H.; & Clark, R.M., 1987. A New Methodology for Modeling Break Failure Patterns in Deteriorating Water Distribution Systems: Theory. *Adv. Water Res.*, 10:2.

APWA (American Public Works Association). Policy on GASB 34. www.apwa.net (accessed Feb. 9, 2002).

ASCE (American Society of Civil Engineers), 2001. Report Card for America's Infrastructure. www.asce.org/reportcard (accessed Feb. 13, 2002).

AWWA, 2003a. *Water:\Stats 2002 Distribution Survey.* AWWA, Denver.

AWWA, 2003b (3rd ed.). *Principles and Practices of Water Supply Operations: Water Transmission and Distribution.* AWWA, Denver.

AWWA Manual M28, 2001 (2nd ed.). *Rehabilitation of Water Mains.* AWWA, Denver.

AWWA Manual M29, 1998 (2nd ed.). *Water Utility Capital Financing.* AWWA, Denver.

AwwaRF (AWWA Research Foundation), 2004. Assessment and Renewal of Water Distribution Systems. AwwaRF, Denver.

AwwaRF, 2002. Investigation of Pipe Cleaning Techniques. AwwaRF, Denver.

AwwaRF, 2001a. *Capital Planning Strategy Manual.* AWWARF, Denver.

AwwaRF, 2001b. Financial and Economic Optimization of Water Main Replacement Programs. AwwaRF, Denver.

AwwaRF & UKWIR (UK Water Industry Research Ltd.), 2002. Proc. Multiutility Buried Pipes and Appurtenances Location Workshop. London.

AwwSC (American Water Works Service Co.), 2002. Deteriorating Buried Infrastructure Management Challenges and Strategies. White paper. www.epa.gov/safewater/tcr/pdf/infrastructure.pdf (accessed Feb. 13, 2004).

Becker, J., 2003. Water Distribution System Assessment and Hydraulic Model: Sewerage and Water Board of New Orleans. Presented at Distribution System Infrastructure Integrity Workshop, Fort Collins, Colo.

Burn, S. et al, 2004. PARMS: An Approach to Strategic Management for Urban Water Infrastructure. IWA Meeting on Strategic Asset Management, San Francisco.

Carnes, K.; Singh, R.; & Carpenter, R., 2003. Boring a Hot Trend With Ductile-Iron. *Opflow*, 29:7:1.

Cesario, L., 1995. *Modeling, Analysis, and Design of Water Distribution Systems.* AWWA, Denver.

Clinton, P., 2003. EBMUD Pipeline Replacement Program. Presented at Distribution System Infrastructure Integrity Workshop, Fort Collins, Colo.

Committee Report, 2003. Applying Worldwide BMPs in Water Loss Control. *Jour. AWWA*, 95:8:65.

Conroy, P.J.; Hughes, D.M.; & Wilson, I., 1995. Demonstration of an Innovative Water Main Rehabilitation Technique: In Situ Epoxy Lining. AWWARF, Denver.

Dandy, G.C. & Engelhardt, M., 2001. Optimal Scheduling of Water Pipe Replacement Using Genetic Algorithms. *Jour. Water Res. Planning & Mngmnt.*, 127:214.

Deb, A.K. et al, 2002. Decision Support System for Distribution System Piping Renewal. AwwaRF, Denver.

Deb, A.K. et al, 2001. New Techniques for Precisely Locating Buried Infrastructure. AwwaRF, Denver.

Deb, A.K. et al, 2000. Guidance for Management of Distribution System Operation and Maintenance. AwwaRF, Denver.

Deb, A.K.; Hasit, Y.J.; & Norris, C., 1999. Demonstration of Innovative Water Main Renewal Techniques. AwwaRF, Denver.

Deb, A.K.; Hasit, Y.J.; & Grablutz, F.M., 1998. Quantifying Future Rehabilitation and Replacement Needs of Water Mains. AwwaRF, Denver.

Deb, A.K.; Hasit, Y.J.; & Grablutz, F.M., 1995. Distribution System Perfor mance Evaluation. AwwaRF, Denver.

Deb, A.K. et al, 1990. Assessment of Existing and Developing Water Main Rehabilitation Practices. AwwaRF, Denver.

Dingus, M.; Haven, J.; & Austin, R., 2002. Noninvasive, Nondestructive Assessment of Underground Pipelines. AwwaRF, Denver.

Ellison, D., 2001. Distribution Infrastructure Management: Answers to Common Questions. AwwaRF, Denver.

Galleher, J., 2003. PCCP—Lessons Learned. Presented at Distribution System Infrastructure Integrity Workshop, Ft. Collins, Colo.

Goodson, A.F., 2002. Utilization of Reliability Centered Maintenance. Proc. 2002 AWWA Infrastructure Conf., Chicago.

Grigg, N.S., 2002. *Water, Wastewater, and Stormwater Infrastructure Management.* CRC Press, Boca Raton, Fla.

Gummersheimer, D.A. & Heavens, J.W., 1998. Evaluation of Two Novel Techniques for the Structural Renovation of Water Mains. 1998 AWWA Missouri Section Ann. Conf., Lake of the Ozarks, Mo.

Hafskjold, L.F., 2003. CARE-W: Computer-aided Renewal of Water Networks. Presented at Distribution System Infrastructure Integrity Workshop, Fort Collins, Colo.

Heavens, J.W. & Gumbel, J.E., 2001. To Dig or Not to Dig: Design, Specification, and Selection Issues in the Trenchless Renovation of Water Mains. *Assessing the Future: Water Utility Infrastructure Management* (D.M. Hughes, editor). AWWA, Denver.

Hope, R., 2003. Establishing the Estimated Annual Replacement Rate for Water Mains. Workshop for Infrastructure: Above and Below Ground. Proc. 2003 AWWA Ann. Conf., Anaheim, Calif.

Horn, T., 2003. Small Line Replacement and Prioritization: Columbus Water Works. Presented at Distribution System Infrastructure Integrity Workshop, Fort Collins, Colo.

Hughes, D.M., 2002. *Assessing the Future: Water Utility Infrastructure Management.* AWWA, Denver.

Jackson, R.Z.; Pitt, C.; & Skabo, R., 1992. Nondestructive Testing of Water Mains for Physical Integrity. AwwaRF, Denver.

Jordan, J.K., 2000. *Maintenance Management for Water Utilities.* AWWA, Denver.

Kirmeyer, G.J.; Richards, W.; & Smith, C.D., 1994. An Assessment of Water Distribution Systems and Associated Research Needs. AwwaRF, Denver.

Kleiner, Y. & Balvant, R., 1999. Using Llimited Data to Assess Future Needs. *Jour. AWWA,* 91:7:47.

Margevicius, A. & Haddad, P., 2003. Catastrophic Failures of Cleveland's Large-diameter Water Mains. Presented at Distribution System Infrastructure Integrity Workshop, Fort Collins, Colo.

Mavrakis, G., 2003. Evaluating Condition of Water Facilities Using Predictive Life Cycles. Presented at Distribution System Infrastructure Integrity Workshop, Fort Collins, Colo.

Mergelas, B., & Kong, X., 2001. Electromagnetic Inspection of Prestressed Concrete Pressure Pipe. AwwaRF, Mississauga, Ont.

National Council on Public Works Improvement, 1988. Fragile Foundations: A Report on America's Public Works. Final report to the President and Congress.

Neve, A.R. & Hedges, M.R., 1998. The Portsmouth Water Approach to Network Renewals and the Influence Upon Leakage. *Pipes & Pipelines Intl.,* 46:5.

Oram, P.; Connor, S.; & Mason, S., 2004. Mains Rehabilitation Strategy: Developing a Defined Program for Prioritization and Expenditure. Proc. 2004 AWWA Ann. Conf., Orlando, Fla.

Radovanovic, R., 2003. Critical Water Mains: The Need for Risk Analysis. Presented at Distribution System Infrastructure Integrity Workshop, Fort Collins, Colo.

Russell, B., 2003. CML Planning/Contracting at the Louisville Water Company. Presented at Distribution System Infrastructure Integrity Workshop, Fort Collins, Colo.

Selvakumar, A.; Clark, R.M.; & Sivaganesan, M. 2002. Costs for Water Supply Distribution System Rehabilitation. *Jour. Water Res. Planning & Mgmt.,* 128:4:303.

Smith, L.A. et al, 2000. *Options for Leak and Break Detection and Repair of Drinking Water Systems.* Battelle Press, Columbus, Ohio.

Sterling, R., 2003. Rehabilitation of Drains and Sewers: Book Reveiw. *Trenchless Technology,* 12:6:30.

Stratus Consulting, 1998. Infrastructure Needs for the Public Water Supply Sector. Report prepared for AWWA, Denver.

Tucker, M. & Corriveau, A.R., 2003. Leveraging Legacy Maintenance Data and GIS to Optimize Infrastructure Replacement and Rehabilitation. Washington Suburban Sanitary Commission, Laurel, Md.

USEPA (US Environmental Protection Agency), 2001. 1999 Drinking Water Infrastructure Needs Survey: Modeling the Cost of Infrastructure. EPA 816-R-005. Washington.

Van Dyke, T., 2003. Cleaning and Lining, Sioux Falls. Presented at Distribution System Infrastructure Integrity Workshop, Fort Collins, Colo.

Von Huben, H. 1999 (2nd ed.). *Water Distribution Operator Training Handbook.* AWWA, Denver.

Walski, T.M., 1984. *Analysis of Water Distribution Systems.* Van Nostrand Reinhold, N.Y.

Woodcock, M., 2003. Research on Pipe Breaks and Renewal. Presented at Distribution System Infrastructure Integrity Workshop, Fort Collins, Colo.

The Paradox of Water Dollars: Connecting the Supply of Funds With Investment Demand

THERE IS SIGNIFICANT CURRENT INVESTMENT INTEREST IN THE WATER INDUSTRY—INVESTORS OF ALL STRIPES ARE LOOKING TO ESTABLISH A POSITION IN THE BUSINESS. MAJOR INDUSTRIAL COMPANIES, PRIVATE EQUITY FUNDS, AND INDIVIDUAL INVESTORS ARE FLOODING THE MARKET WITH DOLLARS. AT THE SAME TIME, WE HEAR MORE AND MORE ABOUT THE VAST EXPENDITURES THAT WILL BE REQUIRED TO ADEQUATELY MAINTAIN AND EXPAND OUR WATER INFRASTRUCTURE AND THE SHORTFALL IN CURRENT SPENDING. SOMEHOW, THIS "SUPPLY" OF INVESTMENT DOLLARS IS NOT SUFFICIENTLY CONNECTING WITH THE "DEMAND" FOR INVESTMENT DOLLARS. WHAT CAN BE DONE TO IMPROVE THIS SITUATION?

BY STEVE MAXWELL
(*JAWWA* April 2007)

There are two key themes that are very consistent across the water industry—themes that have been prominently emphasized and re-emphasized in this article over the past several years.

First, from an investment perspective, the water industry is definitely hot. Investment interest in all facets of the water industry has been booming. Investors of all stripes have been rushing to investigate and participate in the water industry—corporate and strategic investors like General Electric and Siemens, hundreds of private equity groups like the Carlyle Group and Clayton, Dubilier & Rice, and thousands of private individual investors. The stock prices of water companies have been driven to high levels, merger and acquisition transactions are occurring at stratospheric valuations, new companies are rushing to the public markets, and various types of new water-specific investment vehicles and mutual funds are being set up. As much as $100 billion was raised in 2006 for new infrastructure investment funds—dollars that are often directed toward water investment opportunities. However, it often seems that this stampede of investor interest is chasing after relatively few real and attractive investment vehicles. Investors lament the paucity of pure-play water stocks in the United States, and they are increasingly turning to foreign stock markets in order to invest in attractive water companies. In short, as global water concerns are

better understood and as commercial opportunities are better defined, there is a huge amount of investment money seeking a home in the water industry.

Second, at the same time that this frenzy of investment interest has intensified, we hear incessantly about a drastic need for new investment—capital expenditures that will be required to maintain and expand the drinking water and wastewater infrastructure in North America and around the world. As mentioned in previous columns, every month seems to bring forth a new study of future water needs outlining the hundreds of billions of dollars that will be required to continue to provide clean water. Article 7 cited the most recent US Environmental Protection Agency (USEPA) estimate of capital requirements, just for the drinking water side of just the US infrastructure system, at some $277 billion over the next two decades. The controversial "spending gap"—the difference between what experts estimate should be spent to correct these problems and what we are actually spending—continues to increase. This seems to portend a potential disaster somewhere in the future. Similar estimates of investment needs, but in even larger amounts, apply for most of the rest of the world as well. The infrastructure is crumbling in most of the economically developed world and doesn't even exist in many parts of the less developed world. In short, the needs in this industry are truly vast—and the situation cries out for new dollars to be invested.

So, what's wrong with this picture? On one hand, we seem to have lots of dollars looking for a place to be invested, and on the other, there are huge infrastructure and capacity needs crying out for new dollars. On one side, hungry investors complain about the lack of good investment opportunities, and on the other, the public clamors for the rebuilding and expansion of dilapidated water systems. Is there some kind of major structural disconnect or dislocation? If not, what is going on? It seems as if there should be a ready solution here.

As outlined in the following discussion, this apparent paradox can be explained by several factors, and there may be some steps that could be taken to help bridge the disconnect.

MOTIVES OF WATER INDUSTRY INVESTORS

First, let's examine the motives or objectives of the investors looking at the water industry. Typically, these parties are interested in putting their money or corporate resources into assets that can earn a handsome return, regardless of whether they are individuals investing in public companies or giant corporate investors making multibillion-dollar strategic investments. Individual and private equity investors usually also want to put their money into relatively liquid (sellable) assets, although their recent forays into toll roads and urban real estate reflect a growing pressure to simply put these funds to work. Generally, however, investors want superior returns on their investments, they typically want the returns as fast as possible, and they want to be able to exit the investments when they choose. This often means buying the stock of a publicly traded company or acquiring a company that presumably can be built up and strengthened (later to be sold to someone else for a profit).

One of the hallmarks of the water industry has been its fragmentation—the huge number of relatively small companies competing against one another in nearly all segments of the market. As a result, there simply aren't very many large, attractive, and available (publicly tradable or acquirable) companies in the US water industry. Prices have thus been bid to high levels for these few companies, and many investors are turning their attention to foreign countries, where water investments are often more readily available.

WHERE ARE CAPITAL EXPENDITURES FOR WATER NEEDED?

Now let's examine the other side of the water investment issue—where is it that the real need for capital expenditures resides? Most of the drinking water and wastewater infrastructure around the world is owned and operated by government or municipal agencies such as cities, towns, regional authorities, or central government agencies. Because of the universal need for water, because water is typically viewed as part of our natural heritage, and because water has the economic characteristics of a natural monopoly, it has typically been managed by governments for the benefit of the people—not necessarily to turn a profit and certainly not for market efficiency. It is within these public agencies where the lion's share of these huge capital expenditures is required and where the dollars are going to be spent.

When private companies need money, they can go to the public markets and raise new funds by selling additional shares of ownership to new investors. They then use that capital for the expenditures they require for growth. They have diluted their ownership by bringing in new shareholders, but they have presumably improved their growth and future prospects and thus have become a more attractive place for investors to put their money. Municipal or government agencies cannot do this—and this is the short answer to the paradox mentioned previously. Water infrastructure around the world is largely owned and controlled by government agencies, and there aren't many avenues for profit-motivated private parties to truly invest in municipal or government operations. Investors can "loan" money to municipal agencies (see the following discussion on municipal bonds), but there simply aren't many financial tools or mechanisms through which private equity capital can be invested in governmentally owned entities. In short, you can't buy shares in the government.

In practice, of course, the situation is not quite that simple. First, not all the water and wastewater infrastructure that serves the public is controlled by government agencies. Many countries have, with varying degrees of success, experimented with privatization of the water infrastructure—outsourcing operations and sometimes turning outright ownership over to private entities. *Global Water Intelligence* recently estimated that approximately 10% of the world's population currently gets its water from private operators and that this figure is likely to grow to around 16% by 2015. According to the publication, more than 45% of Western Europe is served by private operators, but only about 10% of the US population gets its water from privately operated systems of one sort or another (and an even smaller portion is served by privately operated wastewater systems).

One of the primary drivers behind privatization (a word that is generally synonymous with the terms outsourcing or public–private partnership) relates to the issue being discussed here—a means of attracting private capital to solving public water problems. Investors *can* invest in private operating or infrastructure services companies, thus helping to provide the capital for infrastructure expansion. These companies have been some of the most rapidly growing over the past decade: SUEZ and Veolia Environnement in Europe, and in the United States, Aqua America, and the soon to be publicly traded American Water. However, many observers believe that public resistance to privatization is increasing in the United States. If private operation of water systems is politically or socially constrained, it won't represent much of a solution to the paradox of the water dollars.

ROUTING PRIVATE DOLLARS TO PUBLIC PROJECTS

There are several methods that water agencies and financial investors have used to bridge this gap—mechanisms by which private investment dollars can help fund public projects. Many water infrastructure investments (as well as such other public investments as highways and schools) are funded by means of municipal bond issues or various forms of public debt instruments in which private parties can invest. These are typically long-term fixed-income vehicles in which an investor is essentially lending money to a public agency over a long-term period (15 or 20 years) and receiving a coupon or interest on that principal during the intervening years. Municipal bonds are viewed as being very safe investments, a dependable way to loan money at a fixed interest rate with little chance of losing the principal. Because municipal bonds tend to be low-risk and low-return investment vehicles, they typically don't attract the more aggressive investor, who is looking to take greater chances but who in turn wants a much higher return on the dollar. So this doesn't represent a significant solution to the paradox either.

There is another type of approach that is sometimes used to attract private dollars to fund infrastructure capital requirements on an individual project finance basis. In this situation, a particular project, even though it may be part of a municipal system, is carved out as a separate entity that can attract and employ private money. Even though such a project may be integrated with and ultimately owned by the public utility agency (such as Tampa Bay Water's desalination facility in Florida), financial structures are created and managed in order to make use of private funds and to allow private investors to gain a competitive rate of return on their investments. The financing of such projects, however, does not usually present many liquidity options to investors, nor is it always as transparent as investors might wish it to be.

Finally, there is the entire spectrum of design–build–operate, build–own–operate, build–own–transfer, and build–own–operate–transfer types of water projects. Under these mechanisms, private entities contribute the expertise and capital to build and start operation of what will ultimately be a municipal facility—allowing the private parties to invest money and make an acceptable return before transferring ownership and/or opera-

tion of the facility back to the public agency. Needless to say, these types of projects are gaining in popularity as public investment needs increase.

In the overall scheme of water treatment and infrastructure investment needs, there are a couple of other key factors to keep in mind. First, it is not just municipal agencies that need to invest in water delivery and wastewater treatment technologies. Manufacturing and industrial companies—particularly those that have specialized water needs (e.g., pharmaceutical, oil and gas, and electronics manufacturers)—must make similar infrastructure investments. Financing issues are more straightforward in this arena, and investment and purchasing decisions tend to be made more quickly. This is why many equipment and service vendors to this industry have opted to focus on industrial rather than municipal markets. However, the municipal market will ultimately be so large that few vendors will be able to ignore it.

The other factor to generally keep in mind is that private investors are, of course, not the only sources of investment and funding for the water industry. There are various means of both local and US government financing. In the 1950s and 1960s, a US construction grants program made funds available to local municipalities—a program under which much of the currently aging water and wastewater infrastructure was originally built. Today, the state revolving fund loan programs have made cheap debt capital available to utilities, and there are numerous other programs available—most of them under the aegis of the USEPA. However, even if taken together, all of these programs represent only a few drops in the bucket in terms of overall need. Many observers believe that at some point in time, the US government will have to step back into the role of providing major financing for water and wastewater infrastructure. Unfortunately, it may take some sort of public health crisis to force the issue. In general, US policy today dictates that infrastructure needs should be met at the local level by municipal bond financings and increased user fees.

To summarize, in today's marketplace, most of the water dollars can't really connect very efficiently with the agencies that have major capital requirements. Instead, we have a situation in which most investment interest is directed into companies that are vendors to the actual water providers, companies offering treatment equipment, chemicals, monitoring equipment, and the like. Stated another way, most private investment is going into *vendors to the real spenders*—private companies that sell services and equipment to the utilities and municipal agencies that actually provide water to the public. Instead of being able to flow easily toward the real needs, too many of these dollars are going toward pumping up the values of the vendor companies. In somewhat of another paradox, if the actual water providers don't have enough capital to invest, then ultimately the vendors are not going to do as well either.

Given the urgency of the world's water problems, this is a situation that cries out for new and revolutionary approaches and more creative financial mechanisms that will allow private investors to put their money to work for the public good and concurrently be able to earn a competitive rate of return on those monies in the process. However, if this does not occur on any widespread scale through outsourcing or public–private partnerships, then we need to find an alternative for the short term. There is a significant interest in water investment, and water providers clearly have a significant need for funds. We simply must figure out better ways to connect this supply of water dollars with the obviously increasing demand for these dollars.

How Healthy Are Your Utility's Financial Fundamentals?

BY HOWARD W. HOOVER AND GEORGE A. RAFTELIS
(*JAWWA* April 2005)

During the past decade, more and more communities have changed their perspective regarding their water utility as they recognize that it is truly a business that is owned by the citizens of the community with the advantage of a strong customer service motivation. This evolution in perspective has required water utility managers to extend their focus from strictly operations and engineering/capital planning to one that encompasses organizational, financial, and customer service aspects. Today customers and elected officials expect their water utility managers to exhibit to all stakeholders that the utility can operate as a self-sustaining enterprise that responds to the needs of its customers and creates value to the community. The utility's financial strength is a principal benchmark factor in demonstrating its ability to meet its short- and long-term obligations. Utilities are finding that the historic goal of revenues meeting expenses no longer provides proof of a sufficient level of financial strength and that there are many other factors that come into play. To improve a utility's financial strength, it is important to consider these factors and consider the benefits that financial strength can bring to the community.

OVERVIEW

As utilities look toward the coming decades, there are numerous financial issues that can affect their financial strength. This article will focus on three of the more important issues that utilities must pay attention to in order to increase the health of their financial fundamentals:

• financial planning and management,
• effective pricing, and
• affordability.

By increasing the emphasis on these issues, utilities will increase the likelihood of experiencing improved financial performance and assure their ability to provide value to their owners. Though these issues have always been important, the past decade has seen them become higher priorities for utilities. Operations and maintenance (O&M), water quality, capital planning, and other engineering issues are not the only challenges utility managers will face. By addressing financial planning and management, effective pricing, and affordability a utility will be focusing on its financial fundamentals, which will not only have a direct positive effect on its self-sufficiency as an enterprise fund, but will increase public and stakeholder confidence as well. Through this commitment, a utility is laying a solid foundation for a prosperous future.

FINANCIAL PLANNING AND MANAGEMENT

Effective financial planning and management should be emphasized by a utility and serve as a springboard in making its transition into the next decade.

Benefits are many. Through financial planning and management, a utility can attain many benefits that will help it reach both its short- and long-term goals and objectives.

Generating adequate revenues. Revenues through user rates, charges, and fees that are sufficient to meet revenue requirements, especially O&M, are the first step in a utility meeting its basic needs.

Maintaining adequate reserves. Surpluses generated through rates are a significant source of funding reserves and allow a utility flexibility in its financial planning.

Supporting O&M and capital budgeting. A financial forecast that is developed through the financial planning process allows for estimating O&M and capital budgeting needs into the future and for rates that may be adjusted to address those needs.

Providing effective financial reporting. A full disclosure of financial performance is necessary and provides more information for auditors and rating agencies, allowing them to make better-informed assessments of the utility and its ability to finance debt.

Providing comparative benchmarks. Planning offers a means for utilities to better gauge their financial performance by comparison with other similar utilities.

Ensuring accountability to stakeholders. By having a viable financial planning and management system in place and a platform for communicating this system to the public, stakeholders gain more confidence in utility management and are better informed of the utility's

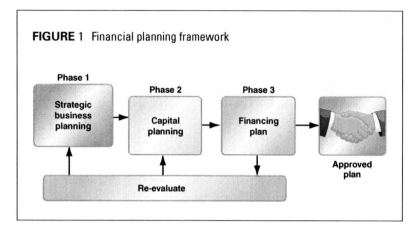

FIGURE 1 Financial planning framework

performance.

Support of the investment process. As mentioned earlier, financial planning allows investors to become better informed of a utility's potential and provides rating agencies with more accurate information, supporting more timely and accurate ratings.

Monitoring expenditures. With such a system in place, the accounting of revenues and expenditures and the improved financial reporting that goes along with it provides the utility with more data to review expenditures, which can then be carried forward through the budgeting process.

These benefits allow a utility to plan more effectively and become better prepared in meeting future challenges that face the water and wastewater industry. Specifically, financial planning is key to the success of a utility's financial performance both in the short and long term. In order to implement financial planning into the utility's strategic and tactical mind-set, a financial planning network is useful in laying out the steps that should be taken. As the flowchart in Figure 1 indicates, setting up

a financial plan may be accomplished in three phases. A discussion of these phases follows.

Phase 1—Strategic business planning. In understanding Phase 1, the utility must first recognize its mission and core values. As shown in Figure 2, strategic business planning is a dynamic process that requires a utility to examine itself internally and determine how those qualities may be leveraged in benefiting its external environment.

The mission, as shown in Figure 2, is the foundation of the strategic business planning process, and as such, it is important to effectively communicate the mission to stakeholders. This, in turn, will allow the utility to establish goals or high-level qualitative statements that contribute to accomplishing the mission. After establishing goals, the utility should set objectives that are measurable and can be used to gauge success. The intended results should be that the utility has in place a set of strategic action plans that will allow it to realize its goals and objectives within a reasonable time frame. Strategic business planning is a dynamic process that requires periodic updating because the utility industry and its customer base are evolving and in a constant state of flux. For example, new regulations and population growth have a direct effect on the utility and should be addressed in its strategic plan. Strategic plans that are not revisited tend to lose their effectiveness, and so adjustments are necessary, especially in cases in which the original plan has not addressed some issues that have surfaced. Every strategic plan will have initial shortcomings, but it is never too late to address them; the important thing is that they be addressed.

For some utilities, verbally expressing what has been their innate strategy for years can be a challenge. Figure 3 shows a partial strategic planning framework. The framework is essential in getting not only management on the same page, but staff as well, so the utility is working with a common purpose that everyone can rally around.

Phase 2—Capital planning. Phase 2 in a comprehensive strategic business planning process is capital planning. Capital facilities represent a major investment by water and wastewater utilities, and planning for this infrastructure must be driven by the utility's strategic business plan. Supply, treatment, transmission, and distribution facilities are needed to provide potable water to homeowners, businesses, institutions, and industrial customers. Investments in collection, transmission, treatment, and disposal facilities are required for wastewater. Capital investments are necessary to maintain high-

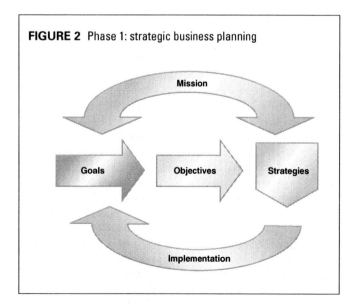

FIGURE 2 Phase 1: strategic business planning

quality service to existing customers and to provide facilities for growth and economic development. To assure that these major investments are prioritized correctly and address the most important needs of the community, the capital plan should be consistent with and supportive of the strategic business plan.

The capital plan identifies the type of facilities that are required over a long-range planning horizon for expanding service, upgrading water and wastewater treatment quality, ensuring system reliability, replacing dilapidated and deteriorated water and wastewater infrastructure, and providing for smaller recurring capital needs. In addition, financial requirements related to the capital plan are identified by year, and appropriate sources to finance these capital items are developed.

The capital planning process should identify the most appropriate capital items in which to invest, phase the purchase or construction of the items appropriately, and ensure that the utility maximizes available financial resources. The capital items that a utility should consider include

- major "backbone" facilities,
- water and wastewater extensions,
- water and wastewater service installations,
- capital equipment and other minor capital items, and
- capitalized operating costs.

Once the utility has addressed which items should be included in the capital plan, it can turn to the capital planning process, which may be divided into three steps:

Step 1—Review requirements in the strategic business plan. The strategic business plan will describe the utility's business priorities and will address issues such as

- the need for enhanced product or service quality;
- expansion opportunities;
- the relative importance of cost, quality, environmental issues, rate sensitivity, and other issues; and
- specific infrastructure programs that are strategic imperatives.

Step 2—Develop a comprehensive facility master plan. The comprehensive water and wastewater facility master plan identifies the capital facilities required for expansion, upgrade, reliability and rehabilitation of the water and sewer systems. Since construction time can be lengthy, and economies of scale can result from using larger facilities to meet long-term demand; the planning horizon of a master plan is usually 20–30 years. In some cases, the planning time can extend to 50 years or

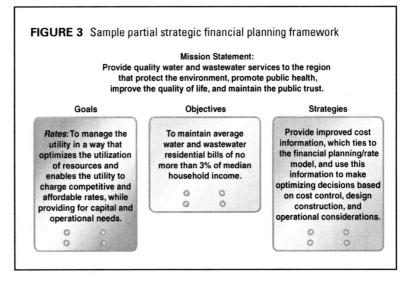

FIGURE 3 Sample partial strategic financial planning framework

Mission Statement:
Provide quality water and wastewater services to the region that protect the environment, promote public health, improve the quality of life, and maintain the public trust.

Goals	Objectives	Strategies
Rates: To manage the utility in a way that optimizes the utilization of resources and enables the utility to charge competitive and affordable rates, while providing for capital and operational needs.	To maintain average water and wastewater residential bills of no more than 3% of median household income.	Provide improved cost information, which ties to the financial planning/rate model, and use this information to make optimizing decisions based on cost control, design construction, and operational considerations.

longer. This time frame is longer than a typical strategic business plan, which has a horizon of 3–10 years. Therefore, the master plan must start with the requirements of the strategic business plan but must be a longer-term view of the future. Additionally, the master plan will be essential for securing appropriate financing.

During the master planning process, an engineer (either a staff or a consulting engineer) evaluates alternative technological solutions and selects an appropriate configuration to address a community's requirements. In some master plans, detailed support for the proposed capital program can be extensive. Aerial photographs and conceptual facility drawings are often required. The plan might also include comprehensive demand studies as well as a thorough evaluation of the environmental, financial, legal, political, and operational effects of adopting the proposed master plan. The facility master plan will normally provide broad estimates of cost for proposed capital facilities, with cost estimates for immediate facilities being more precise than those for long-range facilities. In many jurisdictions, the master plan is divided into five-year increments over the long-term planning horizon.

Step 3—Determine and schedule capital requirements. After a facility master plan is developed for a long-range planning horizon, the next step is to identify the first increment of capital needs to be addressed. These needs are included in a comprehensive capital budget, usually developed for a five-year period. In addition, the five-year capital budget should include the costs of water and wastewater extensions and the costs of minor capital items not identified in the master plan. The year and time when capital facilities become operational, as well as construction time frames associated with capital facilities, must be carefully considered. Recommended financing methods are included as part of the capital improvement plan.

The full capital improvement plan would include additional detail, such as locations, maps, aerial photographs, project descriptions, and demand projection for supporting the facilities included. In addition, it is typical to have separate plans for water and wastewater, with joint facilities, such as administration buildings, equipment, and vehicles, integrated into the plan.

Before a capital plan can be finalized, the utility must understand the economic implications of the capital plan as reflected in phase 3, the financing plan.

Phase 3—Financing plan. Phase 3 is important in that the strategic business plan and the capital plan can only be finalized after an economic analysis of their effect on utility financial performance and rates and charges. This analysis is formalized into the utility's financial plan, which presents projected financial statements for the utility and the economic effect on customers of implementing strategies and achieving the goals of the planning process. The financial plan involves the following steps:

Step 1—Develop capital financing plan. The first step in the process is to evaluate alternatives for financing water and wastewater facilities in the capital plan. Several financing sources should be considered:

- bonds,
- state revolving funds,
- state bond banks,
- grants,
- short-term financing (loans and anticipation notes),
- developer contributions,
- assessments,
- privatization,
- lease/purchase,
- dedicated capital and bond coverage funds,
- operating revenues, and
- investment income.

Short-term financing and bonds, system development charges, and privatization are other options available to a utility. However, in most utility financing plans, rate revenues are a primary source of recovering capital costs when other financing sources have been exhausted or have been deemed inappropriate.

Step 2—Determine annual operating and capital revenue requirements. The next step in the financial planning process is to translate the strategic business plan and the proposed capital plan into annual revenue requirements for the five-year period. Wastewater and water revenue requirements should include expenditures required to implement the business strategies—such as O&M costs associated with existing and proposed facilities—and costs associated with relevant financing techniques, such as existing financing requirements (existing bonds), proposed bonds, and rate-funded capital.

The increases or decreases in revenue requirements to be recovered through rate revenues will be useful in determining whether capital programs should be

scheduled differently or whether rate increases should be phased in over time. These increases and decreases, however, are not projected rate adjustments. To determine rate adjustments, the effects of increased demand must be considered.

Step 3—Calculate fees and charges. After annual operating and capital revenue requirements are determined, customer fees and charges can be determined. At this point, fees and charges are typically estimated to determine the preliminary feasibility of the capital plan. If the economic impact is too severe, then modifications to the plan may be required.

Step 4—Evaluate effect on customers. The next step in the capital and financial planning process is to evaluate the effect that a proposed capital plan will have on customers and others required to pay under various financing scenarios. When evaluating the effects of financing plans on customers, it is important to examine various customer classes at various consumption levels. Doing so will provide valuable insights into how the majority of customers will be affected by various capital programs. In addition, it may be beneficial to evaluate rate consequences during different times of the year to provide a realistic picture of how different customers will be affected. It is also important to show rate consequences for several years to evaluate the long-term effects of various capital plans.

In some cases, the preliminary capital plan produces such a significant effect on customers that it will have to be reevaluated. Certain capital projects may be abandoned altogether, certain projects might have to be restaged, and others might need to be downsized or reconfigured. In making these trade-offs, however, it is important to consider economic, legal, operational, regulatory, and political factors as well as the utility's strategic direction. As a result of going through this process several times, the utility staff and governing body can develop a capital and financial plan that optimizes compliance with community objectives, given the constraints imposed on the utility.

It is important to recognize that the entire planning process is iterative and that business strategies may need to be reviewed and revised as a result of infrastructure analysis or financial evaluation. Additionally, infrastructure plans may need to be modified based on financial performance and/or customer or governing body requirements.

EFFECTIVE PRICING

As indicated in the financial planning framework, the final pricing scenarios that result from Phase 3 can have significant ramifications on the planning process. Pricing is an important tool for utilities to use not only to meet their objectives (as the sample strategy in Figure 3 suggests), but also to gain a better understanding of future needs. The utility's rate structure and its ability to recover relevant costs should be a focus of the utility. Specifically, pricing will be increasingly important in the next decade because

• It is the principal means a utility has to generate sufficient revenue for it to maintain its self-sustaining status.

• It is the direct link to a customer's pocketbook, and therefore it cannot be overstated that a utility should be knowledgeable of how its pricing affects the typical customer's bill. Otherwise, rate increases will be regarded with suspicion by the community leadership and the public.

• It serves as a measure of affordability and provides the necessary information to establish a benchmark. This enables the utility to make adjustments in the financial planning process and maintain pricing at an affordable level and have less effect on the typical customer's bill.

• It can affect economic development through its recovery of costs in serving customer classes and the demand each places on the system. Through pricing, the utility is better able to communicate to the customer classes how their demand affects the system. The utility is therefore better able to adjust its pricing to correspond with the level of service, thereby promoting further economic development for its retail classes. This can be important for commercial and industrial customers whose peaks are less defined than those of residential customers.

• It can aid in demand management by using a rate structure that puts more of a burden on high-volume users. The usual approach is to apply an increasing block rate structure that penalizes customers who use water less judiciously, thereby decreasing demand. This is important during periods of drought or in arid areas, especially in the southwestern United States where water is at a premium.

The two overall goals of an effective pricing structure are to design a rate structure that achieves financial sufficiency while also being responsive to utility and stakeholder objectives. With these two goals achieved, the utility should be able to operate as a self-sustaining enterprise. The first goal lends more to the utility being considered a business unit rather than a governmental department. The second goal adds to overall confidence in the utility when the objectives set in the planning process are addressed and expectations are met.

Rate structure objectives. Following are some typical objectives that the utility and its stakeholders may find important and may want to see addressed by the rate structure:

Cost of service recovery (equity). As mentioned previously, the rate structure's ability to represent the recovery of the utility's costs is central in its being accepted by the public and the municipal leadership. Equitable recovery should be central to a utility's presentation of a rate structure that represents the demand each customer

class places on the system.

Easy to update. Utilities that prefer to update rates annually should decide whether the rate structure they implement can be readily updated with existing financial and user data.

Easy to understand. When presenting a new rate structure to the approving body/authority and the public, the easier it is to understand the more likely it is to be accepted. Often, rate structures that are complex can frustrate decision-makers and make them less inclined to approve it. This is also the case with the public, which may not trust a rate structure that is not easily explainable.

Ease of implementation. The community billing system may be the limiting factor in how complex a rate structure can be. No matter how equitable the recovery, without the appropriate billing support the rate structure cannot be adopted unless appropriate adjustments are made.

Competitiveness/economic development. The degree to which the structure is competitive with those of similar and adjacent communities and the potential effect on existing and prospective commercial and industrial customers must also be considered.

Minimizing customer consequences. With every change there are trade-offs between those who benefit from the change and those who are put at a disadvantage. A rate structure is not a panacea for all users, and some customers will come out worse off after all is said and done. The utility must minimize these instances whenever possible and be knowledgeable of their existence so they can be explained and reasons can be given for the situation.

Legality (litigation potential). Case law has a significant number of instances in which the issue of customer equity as it pertains to rate setting was decided by the courts. Reasonableness tests now have been established through the outcome of these cases. In order to prevent litigation from occurring, utilities should be well-versed in these tests and ensure that their rate structures have met these standards.

Conservation/demand management. For those utilities that have a relatively finite water resource or are in drought-prone areas, conservation and demand management are important issues. Utilities must strike a balance between maintaining their self-sustaining status and implementing pricing that ensures the ability to meet demand. Conservation/demand management has been effective in setting a pricing structure for high-volume users so as to bring usage under control.

Affordability. One of the more significant issues today is the rapid escalation of water rates and the affordability they present to low-income users. The common mission for utilities is to provide a safe, reliable source of water, however this has become more and more complex over the years. Increased regulations and

TABLE 1   Scoring alternative pricing structures

Pricing Criteria	Existing Structure	Alternatives		
		1	2	3
Essential				
Cost of service	C	A–	A–	A
Effect on customers	B	A	A–	B–
Legality	C	B	A–	B–
Very important				
Demand management	A–	A	A	A+
Affordability	A–	A	A–	A+
Revenue stability	C	C	B–	C–
Important				
Equitable contributions	B	B	B	B
Easy to understand	C	B–	B+	C
Easy to update	C	B–	B+	C
Less important				
Ease of implementation	A	A	A–	C+
Rate stability	C–	B–	B	C+
Economic development	B–	B–	B	B

A—excellent, B—good, C—acceptable, D—poor

deteriorating infrastructure have made this mission more challenging and caused rates to increase at an alarming pace. Though in the past water rates have been negligible when compared with other types of utilities such as electric and gas, affordability has now become more of an issue.

Revenue stability. Once rates are implemented, it is important to determine how stable will be the generated revenue. Weather patterns are a major concern with regard to this objective because wet years can decrease demand.

Rate stability. The implementation of rates should be met with confidence that the rate structure will generate sufficient revenue for the utility to meet revenue requirements. Avoidance of multiple increases and decreases is a must for any utility. The confidence the public has in rates can be severely weakened when the rates are not stable. In addition, the ability to provide any sort of reliable financial forecast can also be hampered by instability.

Though rate-setting is more art than science and no rate structure can optimize all objectives, there are some objectives that a utility and the community should find more important and that better serve their needs. These objectives should be prioritized, and efforts should be made toward public education, which is essential to make users understand the trade-offs and the components of rate-setting. Table 1 shows how these pricing objectives can be prioritized and allows for a much clearer thought process when determining what is and isn't important to the community.

RATE AFFORDABILITY

As the third issue that can directly affect a utility's financial performance, rate affordability has become an increasing concern to the water industry. This has been readily apparent during the past few years as customer bills have increased faster than inflation. As shown in Figure 4, since 1996 the increase in water rates has ranged from 2% to 7.5% greater than the consumer price index (CPI), in which the values for 1998 represent the percent change between 1996 and 1998. In addition, wastewater rates have increased from 3.5 to 10.1% from 1996 to 2000 whereas the percent rate change in the CPI decreased by 0.2%. In 2002 and 2004 the percent increase in wastewater rates was significantly greater than the CPI.

When the annual average rate increases from 1996 to 2004 for water (4.25%) and wastewater (4.09%) are calculated and compared with the CPI (2.3%) during the same time period, the water and wastewater rates are approximately double that of the CPI. At this rate, at some point the bills for the typical user will become unaffordable. For this reason it is important for utilities to consider their financial strength and ways they can improve their financial planning.

Factors affecting affordability. What makes water and wastewater ratemaking more of an art than a science is the number of factors that influence the rates' affordability. For many, safe, dependable water and the resulting collection and treatment of wastewater at an affordable price are taken for granted. However, with the recent trend of rising costs, many factors that have had a significant effect on the affordability of higher-priced commodities are gaining emphasis in the water and wastewater industry:

Unemployment rates. High unemployment in some communities puts a tremendous focus on the affordability of rates and can affect the rate-setting practices of the respective utilities and their ability to meet revenue requirements.

Median household income. In poor and impoverished communities lower median household incomes prevail, making it difficult for customers to absorb an increase in rates.

Poverty income levels by household size. Poverty levels should be viewed in terms of both absolute numbers and in relation to poverty areas regionally and nationwide. By comparing poverty levels with median and average household income data, policymakers can estimate the degree to which their customer base is economically stratified. This data will be useful in evaluating possible assistance programs and rate design options.

Community bond rating. If a community is planning to issue bonds to finance capital projects, its bond rating will definitely affect its ability to do so. This will then put the burden on obtaining grant money or, barring that, turning to rates and reserves to determine the community's funding ability.

Community debt. A community with a high level of debt may be forced to review other ways of servicing its debt, which will affect its overall bond rating and make additional financing more difficult to secure.

Property tax revenues as a percentage of total property market value. This ratio provides another indication of the community's ability to generate increased future property tax revenues. For many communities, property tax revenue data and total property market value may be extracted from the state property tax statistical report published by the state revenue department.

In a recent project performed for the Birmingham Water Works Board (BWWB)[1], residential bills at different levels of usage were compared with different income levels necessary to pay the bill. How these factors affect affordability for those customers within the BWWB service area can be seen in Figure 5. This type of multidimensional analysis is valuable because of its ability to reveal potential affordability challenges that might escape notice under a more simplistic "single measurement" approach.

Figure 5 demonstrates how an affordability assessment can include a range of customer dynamics rather than focusing on a single "typical" median household income customer. This expanded analysis reflects the fact that rates that are affordable for a systemwide average customer may not be affordable for different household

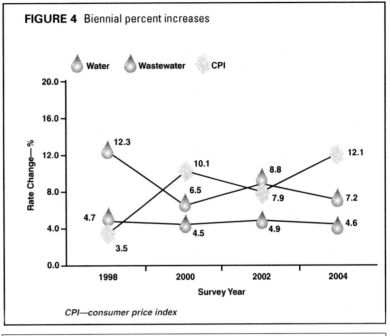

FIGURE 4 Biennial percent increases

CPI—consumer price index

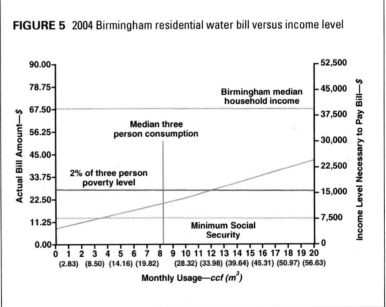

FIGURE 5 2004 Birmingham residential water bill versus income level

sizes at lower income levels. The analysis graph incorporates the following variables:

• monthly bills for different demand levels (sloped line),

• monthly demand and bills for a household of three people (vertical line),

• a 2% affordability benchmark for poverty-level households of three persons (horizontal line measured against the right axis),

• 2% of household income consisting entirely of Social Security payments (lowest horizontal line), and

• median household income over the entirety of the

service area (upper horizontal line).

The vertical demand line intersects the sloped monthly bill line at a point indicating the average monthly bill for a household of three people, and the minimal Social Security level. If this intersection lies below that household's affordability benchmark (horizontal line), then rates are deemed affordable for that household size. Otherwise, rates are judged to be unaffordable for that household size.

By examining affordability through a range of demand and income levels, management is able to evaluate how rate recommendations will affect different segments of its customer base. By combining this knowledge with local census data and customer billing records, utility managers can estimate the number of customers that might have difficulty in paying their bills under current or new rates.

For instance, the bill for the median consumption in a three-person household at an approximate monthly usage of 8 ccf (22.7 m³) is shown to be about $20. At the minimum Social Security income level, a customer can only afford just more than $11.25 for a usage of between 2 and 3 ccf (5.7 and 8.5 m³). When compared with 2% of the income level for a three-person household at the poverty level, such a household can afford a bill with a usage of approximately 11.5 ccf (32.6 m³).

With growing environmental regulations that have increased the need for more vigilant operations and maintenance and major capital needs, water will become less affordable to more and more users without proper financial planning.

As affordability becomes more of an issue, a key indicator of where a utility stands among comparable utilities is through benchmarking. This can be very helpful to give the utility the information it needs to show where it stands in providing affordable service. As indicated in the 2004 Water and Wastewater Rate Survey (AWWA/RFC, 2004), a survey of utilities across the United States, the median monthly charge based on 10 ccf (28.3 m³) of billable flow for water and wastewater is $19.08 and $22.23, respectively. Because all utilities face different issues when performing a benchmark analysis, it is important to include only those utilities of either comparable size and operating characteristics or those located in the same geographic area.

CONCLUSION

A comprehensive financial plan is essential to ensuring a utility operates on a self-sustaining basis. As a first step, strategic business planning is important in developing a financial plan. As the financial plan is being developed, pricing plays a powerful role in addressing economic and noneconomic objectives, and what the community views as important. However, rates and rate structures are not a panacea for all issues, and great care should be taken to develop pricing objectives and

selecting optional rate structures. Financial planning is more important now than it has ever been, considering that during the past eight years water and wastewater rates have increased at a rate of about twice that of the CPI. With charges increasing faster than inflation, affordability to disadvantaged customers is becoming an increasingly important and challenging issue. The past decade has placed affordability and the utility's ability to address it with sound financial planning at the forefront of issues facing the water and wastewater industry. These will be some of the more important considerations utilities will face in the next decade.

ACKNOWLEDGMENT

The authors wish to recognize Peiffer Brandt, vice president of Raftelis Financial Consultants Inc. (RFC), the staff of RFC and AWWA, and those utilities that participated in the 2004 AWWA/RFC Water and Wastewater Rate Survey. The data from the survey were invaluable in the analysis provided in this article.

FOOTNOTE

[1]Raftelis Financial Consultants Inc., Charlotte, N.C.

REFERENCE

AWWA, RFC (Raftelis Financial Consultants Inc.), 2004. *2004 Water and Wastewater Rate Survey*. AWWA, Denver.

ABOUT THE AUTHOR

Howard W. Hoover is a senior consultant with Raftelis Financial Consultants Inc. (RFC), Charlotte, NC. He has a BS from the US Merchant Marine Academy, a BSCE from the University of South Florida, and an MBA from Mercer University, Atlanta, Ga. He is a licensed professional engineer in North Carolina and Georgia. George Raftelis, is president of RFC. He has more than 25 years of experience in the water and wastewater industry. The majority of his projects have dealt with cost of service and rate analysis. Raftelis is chair of AWWA's Management Division and is a former chair of AWWA's Rates and Charges Subcommittee. Prior to establishing RFC, he was the partner-in-charge of Ernst and Young's national environmental consulting practice. Raftelis holds a BS in mathematics and economics from Eckerd College, St. Petersburg, Fla., and an MBA from Duke University, Fuqua School of Business.

Implementing an Integrated Water Supply Program in a Water-Scarce Environment

BY PETER D. BINNEY
(*JAWWA* August 2006)

Aurora, Colo., is a primarily residential community of 300,000 people in the eastern Denver metropolitan area. The service population is expected to exceed 500,000 people within the next 30 years, and the city's water utility, Aurora Water, has been charged with increasing the reliability of the current system as well as providing for the water needs of the future community.

The city's water needs are served by a renewable surface water system developing supplies from three major river basins that include the Colorado, Arkansas, and South Platte rivers. Limited development of alluvial groundwater and nonrenewable deep aquifers complement the surface water supply portfolio. More than 90% of the city's water resources are developed from snowmelt in the May–July period with year-round deliveries from stored waters in 12 reservoirs. Yields from the available water rights are highly variable because of weather cycles in semiarid Colorado and the relatively lower priority of water rights that have been available for development or acquisition by the city in fewer than 55 years that the utility has supplied water.

Colorado administers its water allocations on a "first in time, first in right" Appropriation Doctrine and much of the state's surface water resources were developed by agricultural water users in the late 1800s or core cities in the early twentieth century. Little unappropriated water was available for the newly developing communities or suburbs that represent 85% of the metropolitan area's planned population growth in the next 30 years. Aurora, which represents the largest consolidated demand center that must meet these challenges, has developed a portfolio of demand-management and supply-side programs to reliably meet the community's water needs.

The highly variable productivity of the city's water supply has been demonstrated by performance in the drought conditions experienced in Colorado and the western United States since 2002. Although average annual precipitation in the city is about 12 in., conditions in 2002 caused extremely low water supply conditions that were estimated to occur only once in 300 years. Reservoir storage levels dropped to 26% of capacity— less than a year of water supply in the system.

The wet conditions of the late twentieth century were not typical of the long-term water conditions in Colorado. The stress on the water system yields in the past four years must be considered by the utility as more representative of what could be observed in the future as it acquires new water sources and constructs the delivery systems and treatment plants that will be necessary.

Aurora Water has implemented highly productive and comprehensive water conservation programs to reduce customer demands, especially during the outdoor irrigation season when typically half the annual water demand is to sustain lawns and trees. The utility also acquired permanent and interruptible water supplies from agricultural uses to augment the low-yield productivity of the core water rights portfolio. A combination of demand management and new source development approaches have provided near-term relief and allowed for recovery of carryover storage levels in the city's reservoirs.

Because continued growth will exacerbate the projected water supply shortages, the utility has initiated a long-term integrated water supply program to ensure that adequate service can be provided to customers even under sustained drought conditions. A comprehensive assessment of the reliability of the water supply system, including core infrastructure, led Aurora Water to develop a 10-year capital improvements program exceeding $1.4 billion in system improvements and expansions to serve the community's needs. It has also embarked on a major expansion of the raw water supply system by recovering reusable return flows downstream from the city and expanding new agricultural water leasing programs to provide supplemental yield in drought years. The city is also acquiring new water rights, largely through transfers of agricultural water sources, to provide firm yield.

As new sources of water are developed on river systems downstream from a major metropolitan area, the utility is implementing multiple-barrier water treatment approaches to protect public health and meet acceptance levels from a demanding customer base that has traditionally been served from first-use mountain water sources. This investment is causing significant and sustained increases in water rates and connection fees because the utility provides services on a "cost-of-service" basis and, in accordance with Colorado's constitution, cannot access alternative tax sources to reduce the required increases in utility bills. This confluence of natural events, demand for increasing levels of service because of growth, a fundamental shift from development to reallocation of previously developed beneficial water uses, and local funding of required capital improvements creates many challenges. This article describes how Aurora Water is progressively addressing these challenges in providing adequate levels of service to current customers while positioning itself to meet future needs.

AVAILABLE WATER RESOURCES ASSESSED FOR FUTURE DEVELOPMENT

Traditionally, municipal and agricultural water supply development projects occurred along the mountainous Continental Divide that collected snowfall and then contributed high snowmelt flows in spring months. This period of high peak flows could only be successfully managed by the construction of large reservoirs that delivered year-round flows and provided seasonal carryover when dry periods were experienced.

These reliable sources of high-quality mountain water were fully developed at least 50 years ago, and little firm yield remains for new water projects along the divide. Developable waters do remain on more remote river basins to the west of the divide but would require multibillion-dollar projects that would have to meet stringent environmental protection standards and overcome political opposition.

Water resources administration in Colorado. The state of Colorado administers the allocation and beneficial use of water based on an Appropriation Doctrine that assigns a usufructuary property right that respects a "first in time, first in right" principle. As previously described, the allocation of water followed economic drivers in the state, and more than 90% of the beneficially used water was allocated to agricultural uses that were the cornerstone of the state's economy in the late 1800s and early 1900s.

Major commitments of the state's surface waters were made in a series of interstate compacts that required specified deliveries of water to downstream states for their respective uses. These include the Colorado River Compact (1922), South Platte River Compact (1923), Upper Colorado River Compact (1948), and the Arkansas River Compact (1948). About half of the state's available surface water resources are allocated to meeting downstream compact commitments. Municipal water uses represent about 4% of the state's water uses with an emerging commitment to recreational and in-stream uses creating greater need for the development of new water sources.

Successive reuse of transmountain or foreign waters. Foreign water is classified as waters brought by artificial means into an area from a different watershed. Under the appropriation system in Colorado, the doctrine of "developed waters" applies to the adding of a new supply of water to a stream. Return flows from transmountain water are not a part of a stream and therefore not subject to appropriation (Denver, 1973). The agency that imports the water can freely and without legal constraint reuse the water, move the point of diversion along the stream, and sell or lease it to others. A Colorado statute (Colo. Rev. Stat. Ann., 1969) states: "Whenever an appropriator has, heretofore, or shall hereafter lawfully introduce foreign water into a stream system from an unconnected stream system, such appropriator may make a succession of uses of such water by exchange or otherwise to the extent that its volume can be distinguished from the volume of the streams into which it is introduced. Nothing herein shall be construed to impair or diminish any water right which has become vested."

Various courts have sustained the right to recapture water brought from another watershed for successive uses to the extent that possession has not been abandoned. This characteristic of transmountain waters that are used and then returned to the South Platte and Arkansas rivers by municipalities provides an important source of water that can be incorporated into the plans of public water providers as they develop long-term reliable water supplies. This ability to recapture and successively reuse these developed water sources increases the efficiency of Front Range (an area ranging from Fort Collins in the north to Pueblo in the south, just east of the foothills of the Front Range of the Rocky Mountains) municipal water systems and reduces the amount of new water that must be imported from the Western Slope (the region of Colorado west of the Rocky Mountains) or other drainage basins or from transferred agricultural uses. Each introduced acre-foot of water develops approximately 0.5 acre-feet of reusable return flows, so multiple successive reuse can almost double the value of the transbasin diversion if the infrastructure is available and water quality issues can be successfully addressed.

A detailed integrated resource plan has been developed by Aurora Water as it has considered the level of service capacity of existing systems plus the recent experiences with severe drought conditions and the projected needs of the growing city.

SUSTAINED AND SEVERE DROUGHT AFFECTED COLORADO'S WATER SUPPLY

Beginning in 2002, Colorado experienced a sustained and severe drought, with a critically dry year in 2002 that has been estimated to have a recurrence interval of once in 300 years. The cumulative moisture deficits over the five years are not as severe as those seen in the 1930s or 1950s droughts, but they had a detrimental effect on the operation of the water supply systems of newer cities that have not developed robust water rights portfolios. Figure 1 illustrates the snowpack levels in the primary river basin that provides about 50% of Aurora's water supplies.

The successive years of below-average runoff from snowmelt have to be compensated for through withdrawals from reservoir accounts sustained by carryover storage from the wet years experienced in the 1990s. Figure 2 shows the fluctuations in total storage available for city use from 2000 through the present. Although a desired level of storage would serve three to five years of annual water demand, the city's storage levels dropped to 26% (or less than a one-year supply) of capacity after the first year of critically dry conditions.

This unacceptable circumstance required a rapid response and was addressed through a combination of demand-management strategies as well as short- and long-term supply strategies to provide acceptable levels of service to 80,000 customer accounts. Aurora has set a minimum operating storage level of 60% of total contents prior to runoff before nonrestricted water use is considered. The year 2006 will be the fifth consecutive year during which comprehensive restrictions on all water use in the city will have been in effect.

The area experienced very dry conditions during the snowmelt period this year, and Figure 2 shows the likely reservoir contents anticipated in the coming 12 months, with a strong bias developing toward the minimum forecast. About half of the recovery of reservoir levels after 2003 was a result of demand management whereas the other half was the result of several large short-term leases of agricultural water from fallowed cropping. These drought-related fallowing programs

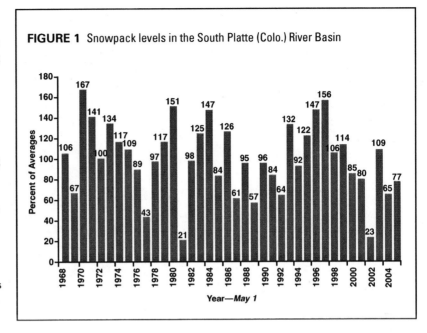

FIGURE 1 Snowpack levels in the South Platte (Colo.) River Basin

required new legislative authorities that the city supported in cooperation with statewide lawmakers.

NEAR-TERM DEMAND-MANAGEMENT ALTERNATIVES ARE REQUIRED

When the severity of the drought was recognized in the spring months of 2002, Aurora Water immediately updated its water management plan, which was based on literature reviews of the experiences of California water utilities during drought conditions in the 1970s. Experience with this comprehensive program of best management practices, aggressive tiered rate structures, rebate programs, and enforcement has demonstrated a different response pattern and level of water-use reduction than are suggested for programs in more temperate climates. The city has seen successive years of more than 30% reductions in annual water use, with the majority of those savings realized from modified outdoor watering patterns.

Indoor water savings were relatively steady, with reductions of 5–10%, but, as shown in Figure 3, major changes were seen in outdoor watering. Although there were short-term reductions as a result of rainfall, a major change in customer behavior was seen between 2002 and 2003–04 when customer response was at its highest. Aurora Water is carefully reviewing those response patterns in 2006 as memories of the severe drought conditions are fading and customer watering patterns seem to be driven more by the "greenness" of their lawns than by the magnitude of their water bills.

Short-term agricultural leasing program. In addition to demand management programs, the city also initiated the largest short-term agricultural leases in the state's history. Effectively these programs paid farmers not to

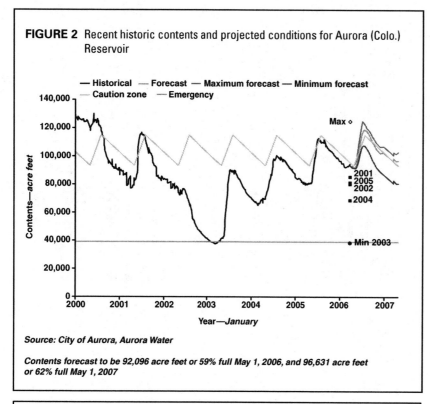

FIGURE 2 Recent historic contents and projected conditions for Aurora (Colo.) Reservoir

Source: City of Aurora, Aurora Water

Contents forecast to be 92,096 acre feet or 59% full May 1, 2006, and 96,631 acre feet or 62% full May 1, 2007

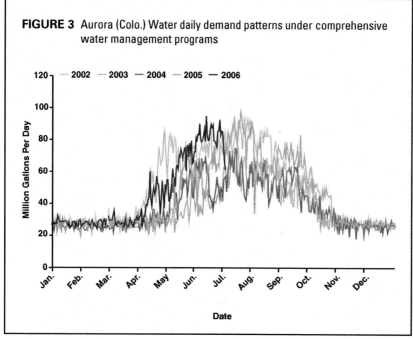

FIGURE 3 Aurora (Colo.) Water daily demand patterns under comprehensive water management programs

irrigate crops and to sell their reductions in consumptive uses to the city. The adopted state laws allowed for this short-term displacement of agricultural economies for no more than 3 years in 10 to minimize long-term and potentially irreversible changes in rural communities. Because these irrigated areas were traditionally served by common ditch systems, it was necessary for the city to enter into a 34-point stipulated agreement to ensure that nonparticipating farmers were not adversely affected by the reduced water deliveries at the ditch headgate.

The city assumed the risk of variable water production as defined acreages were withdrawn from crop production and paid the farmer for loss of profit and ditch assessments as well as soil stabilization and weed control. Initial skepticism about fallowing lands was quickly overcome, and the city was oversubscribed in its ability to move water by more than 100%, which required a reduced allocation of offered acreages. The program was considered to be highly attractive to the farmers and to the city, which developed firm yield under severe drought conditions without investing large capital programs to serve a similar but infrequent need through new, large reservoirs. Aurora Water now has plans to institutionalize short-term agricultural leases for use under severe drought conditions by developing water banks in certain locations under cooperative programs with willing farmers and where the "wet" water can be delivered for use in the city.

Water for return flows swap. Aurora Water also considered traditional methods for enhancing the yields of the water supply system under the extreme drought conditions experienced in 2002 and 2003. In an innovative program, the city acquired the first use of transferred agricultural water rights from a river basin tributary to the city's water supply system that had been developed by a sister city but used under lease by Aurora for a number of years. In a cash and water deal, Aurora Water acquired the first-use rights to those water sources, assigned the return flows for indirect use by the neighboring city, provided makeup

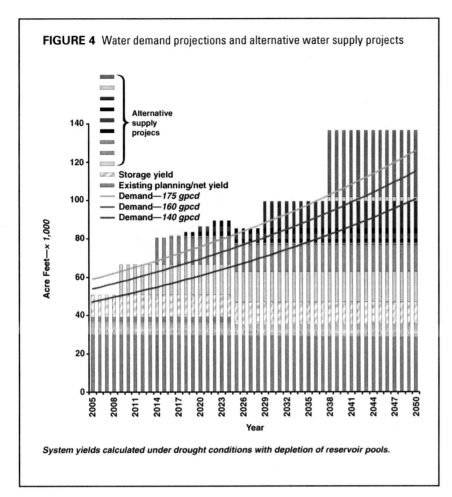

FIGURE 4 Water demand projections and alternative water supply projects

Acre Feet—x 1,000

Alternative supply projecs

Storage yield
Existing planning/net yield
Demand—*175 gpcd*
Demand—*160 gpcd*
Demand—*140 gpcd*

Year

System yields calculated under drought conditions with depletion of reservoir pools.

Demand projections based on forecast population increases and representative water-use factors under various drought-stage scenarios were used to forecast future water needs for the community. Although per-capita consumption rates in excess of 170 gpcd were frequently observed in the water-rich 1990s, water consumption did drop to a low of 123 gpcd under a severe water restriction program.

Current supply capacity under sustained drought conditions (modeled after the 1954–56 drought period) were identified using a comprehensive model of the physical and legal availability of water sources. These water rights produce highly variable yields, with a low of less than 10,000 acre-ft in 2002 to more than 120,000 acre-ft in 1999. This exemplifies the value and the necessity of having large carryover storage accounts for a city whose current normal water consumption is approximately 55,000 acre-ft/year. Figure 4 shows the yields modeled for direct-flow rights and drawdown of reservoir contents over a three-year drought cycle. The figure also shows that without new sources of water, the city would have to institute more severe water restrictions than two-day-per-week irrigation cycles by 2010 if a 1950s-type drought were to occur.

This assessment of demand and available supply illustrated the need for developing a reliable and significant new source of water within a five-year period if acceptable levels of service were to be maintained. A representative development cycle that was generated from similar major water projects initiated by other water utilities were typically found to take at least several decades before delivery. The large number of projects that were shelved or denied under federal regulations (primarily the National Environmental Policy Act) in recent decades was therefore a major criterion for the utility as it considered viable options for addressing the short-term as well as long-term water needs of the community.

water from other downstream sources to keep the neighboring city's water balance whole as well as cash to build a water treatment plant capable of effectively treating this new source water.

Short-term industrial water leasing. Denver Water supported Aurora's reservoir recovery program by instituting a creative water trade program to deliver additional water to the city. A remote industrial site that was not fully utilizing its water rights delivered those waters into the Denver Water system. Because of the physical interrelations between these two major municipal water systems, Aurora water was able to take delivery of a similar amount of water, less conveyance and other losses, as a supplemental supply.

Integrated water resource planning. Although these short-term demand-management and supply strategies were appropriate for maintaining an adequate level of service during the most severe drought conditions, it was necessary for Aurora Water to identify its long-term strategies for additional firm-yield development in order to support the future community. All short-term strategies have been institutionalized in the city's water management plan as commitments to wise water stewardship in this semiarid climate.

MULTIPLE PROJECTS CONSIDERED FOR ALLEVIATING SHORTAGE

More than 50 projects were considered for developing in order to meet the identified shortage. These projects were prioritized based on individual characteristics that included delivery schedule; cost; ability to receive local, state, and federal permits; community support; and implementability. There have been many proposals for addressing the major water supply deficits that are forecast to occur along Colorado's populous Front Range in the coming decades, and these proposals vary from the heroic to the wildly speculative. Estimated project costs of many billions of dollars or decades for the development cycle eliminated most of the larger transbasin projects from further consideration. Development of local nonrenewable and declining deep groundwater resources used by neighboring communities was rejected because of observed significant reductions in well pumping capacity. The area around Aurora was not amenable to historic irrigation, so there were no locally available agricultural water rights for transfer. Conditional water rights held by the city will eventually be developed, but because the permitting cycle for such a project could take more than a decade, it was considered to be a second-generation potential project. Expansion of the short-term agricultural leasing programs was constrained by intergovernmental agreements and restricted availability in severe drought periods.

Consideration of alternative water sources pinpointed the most viable water development program to be the physical recapture of the city's reusable return flows downstream of the city. These water rights have a market value of more than $300 million and would require significantly more investment if the city had to acquire a similar volume of water from other sources.

Prairie Waters Project. This program is a major departure from the city's current water delivery system, which brings highly treatable mountain water sources by gravity to two direct-filtration plants. The water source is physically remote from interconnection points to the water distribution system in the south end of town and will require rigorous and effectively managed treatment to maintain the high level of customer acceptance of water delivered to the tap.

Figure 5 depicts the physical delivery system that will be constructed as part of an overall program that

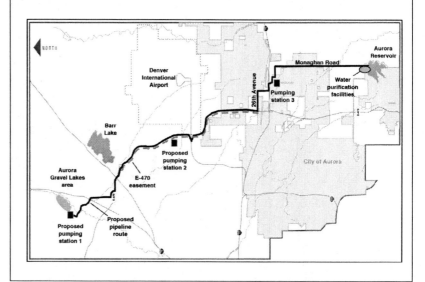

FIGURE 5 Prairie Waters Project, Aurora, Colo.

Aurora Water has named the Prairie Waters Project.

Major elements of this project include the North Campus (Figure 6) facilities where water will be diverted from the river and pretreated before being pumped through a 34-mi pipeline by three pump stations. A new purification plant will soften and purify the water to exceed all drinking water regulations before it is introduced to the water distribution system.

North Campus. This reach of the South Platte River is downstream from a major metropolitan area and is the receiving stream for the regional Metro Water Reclamation Plant's discharge of more than 125 mgd of advanced secondary treated effluent. Water in the South Platte River has historically been diverted for irrigation of row crops and sustains an important riparian corridor that is increasingly viewed as an urban amenity in the form of a park and trail system. Rapid urbanization is also occurring in this area of the metropolitan complex, so retention of greenways that can be integrated with the developing water supply function is important to the utility.

Because of variable and vulnerable water quality parameters in the surface waters, Aurora Water worked with its primary consultant, CH2M HILL, to develop a multiple-barrier diversion and pretreatment program to minimize the purification requirements. A series of alluvial wells located at least three days' travel time from the riverbank is being used to separate aquifer recharge areas from the river. This approach minimizes the potential disruption of the diversion steps from spills in the river. The travel time of alluvial groundwater also provides a significant pretreatment step that substantially lowers the concentrations of nutrients (primarily nitrates) and biodegradable organics. An ongoing pilot-

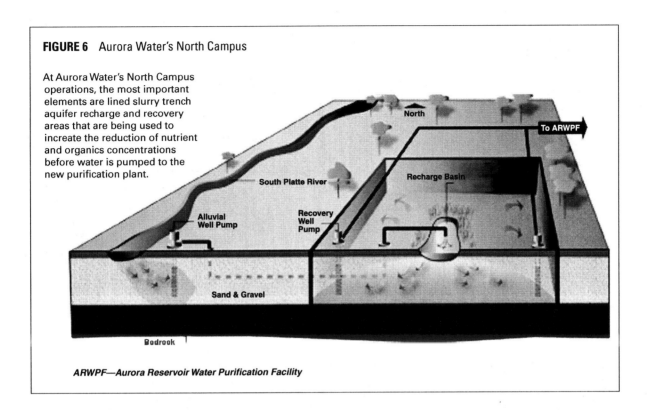

FIGURE 6 Aurora Water's North Campus

At Aurora Water's North Campus operations, the most important elements are lined slurry trench aquifer recharge and recovery areas that are being used to increate the reduction of nutrient and organics concentrations before water is pumped to the new purification plant.

North

To ARWPF

South Platte River

Recharge Basin

Alluvial Well Pump

Recovery Well Pump

Sand & Gravel

Bedrook

ARWPF—Aurora Reservoir Water Purification Facility

test program is being used to demonstrate the treatment value of this step in the program for consideration by regulatory agencies.

Traditionally, water diverted from the South Platte River has been stored in slurry trench–confined open storage areas and released back to the river to offset the hydrologic effects of "out-of-priority" alluvial well diversions. Although this function is incorporated into the North Campus operations, the most important elements are lined slurry trench aquifer recharge and recovery areas that are being used to further reduce nutrient and organics concentrations before water is pumped to the new purification plant. This sustainable biodegradation process is commonly used as a river-bank filtration technique. This particular application has a uniqueness warranted by the source water char-acteristics and the cost-effectiveness of this approach versus the use of membrane technologies with their inherent disadvantages of brine disposal. Monitoring of water quality at various locations through the North Campus facilities will allow the utility to move water to the new purification plant only after it is stable and meets designated influent water quality parameters. Au-rora Water will conduct a series of pilot-test programs over the next three years to confirm surface water infiltration rates, aquifer flux rates, and the consistency of water quality parameters. Depending on the results of these tests, the city may have to acquire additional land holdings to develop the required water quality improvements before delivery to the purification plant.

Aurora Reservoir Water Purification Plant. Because of the source water quality developed at the North Campus facilities, the city has carefully assessed the capabilities of the two existing direct filtration plants to adequately treat water before it is delivered to the distribution system. An early recommendation by the consulting team was the need for a more robust and sequential treatment train that could produce a finished water that replicates the high quality that customers currently receive.

Aurora Water staff worked closely with CH2M HILL treatment specialists to complete the conceptual design of the process—including a softening stage, a combined ultraviolet–advanced oxidation stage, and a granular ac-tivated carbon stage before disinfection. The remaining constituent, total dissolved solids (TDS), will be man-aged by dilution with existing mountain water sources stored in an adjacent reservoir. The goal will be to not exceed a TDS concentration level of 400 mg/L, under the limiting condition of water deliveries during severe drought—under that condition, the proportion of high-TDS water from the North Campus to low-TDS water from the mountain sources is at its highest.

Implementation. All major easements, rights-of-way, and real property have been acquired. Design of the major facilities was completed in the middle of 2006, and construction of the treatment plant will commence later this year. Preordering of long-lead-time equip-ment will allow wet testing of the plant using mountain source water in late 2009.

The North Campus facilities will be pilot tested in 2007–09 to ensure the stability and reliability of the water quality enhancement features of the aquifer recharge areas. On the basis of the availability of funds, the pipe and pump stations could deliver water to the purification plant in 2010 and increase the city's available water supplies by more than 20%. Later phases of the program will be adequate to serve the city's future water needs for at least the next 50 years.

UTILITY REVENUES AND BONDS COVER CONSTRUCTION COSTS, ASSET ACQUISITION

As previously mentioned, Aurora Water is a "cost-of-service" enterprise with rate and fee structures being set to cover operating, debt service, and capital projects needs. There are no external major funding sources available to the utility, so the projected $850 million asset acquisition and construction costs must be covered by revenues from the sale of services and a series of major revenue bond sales. Debt service coverages require the utility to raise its rate and fee structures significantly.

From 2002 to 2006, the unit cost of water has been raised from $2.02/1,000 gal to $4.04/1,000 gal, and a connection fee has increased from $6,711 per residential tap to $16,641 per residential tap. Additional increases will be required in coming years to maintain the necessary cash flow to support anticipated capital improvements. Aurora's City Council has demonstrated a high level of resolve to make this major investment in the community in spite of customer concerns about rising utility rates.

A continuing and comprehensive community outreach program has been implemented to inform and educate the customer base on the benefits and necessity of this investment at a time when neighboring communities are not making similar investments— more costly and complex solutions will have to be implemented in those communities once their customers and elected boards address the need for sustainable and reliable water sources. Without the high level of support from the community, Aurora Water would not be able to implement the Prairie Waters Project, and the community would be subjected to increasingly severe and sustained water restrictions and a potential limitation on the future issuance of taps.

CONCLUSION

Major increases in population and the resulting increase in urban water demands in water-limited areas such as Colorado and other areas, especially in the southwest, are requiring new and costly approaches to municipal water development. These steps require reallocation of existing water uses (primarily agricultural to municipal water uses) and the construction and operation of new water delivery systems. Source water quality is often significantly degraded for traditional sources and multiple-barrier, highly effective water treatment processes are required to deliver acceptable water quality at the tap. Rate and fee structures will increase significantly to cover operating and construction costs of these water systems.

Today's water utility manager must operate effectively in a highly politicized setting and be acutely aware of the public policy, social, environmental, political, and economic aspects of meeting customer needs. Technical problem-solving is often the easiest step in planning and implementing major water supply projects in this setting.

REFERENCES

Denver v. Fulton Irrigating Ditch Co., 179 Colo 47, 506 P.2d 144 (1973).

Colo. Rev. Stat. Ann. para. 148-2-6 (1963) as reenacted and amended by Laws, 1969, ch. 373, para. 21, Rev. Stat,. Ann. Para. 148-2-6 (supp. 1969).

ABOUT THE AUTHOR

Peter D. Binney is the director of Aurora Water, Aurora, CO, and is responsible for the water and related utility needs of the community. Since joining the city in 2002, he has guided the city through a series of water conservation and short-term supply development programs. He is currently developing long-term acquisition and capital development programs to provide safe and reliable water supplies to the community for the next 40 years. He has a BE and ME from the University of Canterbury, New Zealand, and an MS degree from the University of Colorado.

State of the Industry Report 2006: An Eye to the Future

BY JON RUNGE AND JOHN MANN
(*JAWWA* October 2006)

In 2004, AWWA's first State of the Industry (SOTI) survey provided an unprecedented snapshot of the critical issues facing the water industry. Fast-forward two years and two surveys later, and the snapshot is developing into a motion picture tracing the trends and priorities characterizing the North American water industry.

In March 2006, AWWA conducted the third annual SOTI survey. This comprehensive and diagnostic report provides the association and its members with essential information about the state of the water industry as reflected by more than 2,000 water professionals across the United States and Canada. "After three years, we're beginning to see interesting shifts in perceptions of what's important today and in the future," said AWWA Executive Director Jack Hoffbuhr. "The survey information becomes more valuable with each new edition."

SURVEY DESIGNED TO CAPTURE VARIED PERSPECTIVES

Survey objectives. The primary objectives of the 2006 SOTI survey included (1) expanding on the association's historic role as a provider of valuable information that shapes a common agenda for the industry, (2) developing insights regarding the key issues facing the water industry, (3) identifying issues not being addressed adequately and spurring future action, and (4) tracking industry trends such as technology needs and utility capital spending on a continuing basis. For both AWWA and the community it serves, SOTI information fosters a proactive approach to the future. According to AWWA President Terry Rolan, "For people who work in the water community, the State of the Industry survey can be an important tool in determining how to apply their energies and their resources in the coming year."

Study methodology. Mann Consulting Inc., a marketing strategy and research consulting firm based in Littleton, Colo., developed the self-administered survey, which was mailed to a random sample of more than 12,000 AWWA members across the industry. This year's survey sample was augmented to provide a reportable base of responses from the Canadian water industry. In all, the 2006 SOTI survey generated 2,020 responses—1,845 from the United States and 175 from Canada.

The 2006 respondent base represents an experienced, highly educated sample of professionals, with more than half of respondents reporting more than 20 years of experience in the water industry. Two thirds of the respondents indicated they had earned a four-year college degree or higher, and almost one third hold a master's, doctorate, or other advanced degree. Of the respondents, 60% were from utilities, 27% were service providers, and the remainder were individuals associated with the water industry through academia and the government.

One of the most revealing aspects of the three SOTI surveys has been the information garnered from open-ended questions that provide respondents with an opportunity to expand on their thoughts. "The survey does a good job of letting water professionals explain the issues in their own words," Rolan noted. "Given the depth of experience and wisdom among the 2,000 respondents, I think we can feel confident in the insights being provided. It's a great service to the membership."

SURVEY RESULTS REFLECT INDUSTRY CHALLENGES AND CONCERNS

Industry soundness. To capture the water industry's big picture, every year the SOTI survey asks respondents to rate industry soundness today, five years ago, and five years into the future on a scale of 1–7, in which 1 signifies "not at all sound" and 7 indicates "very sound." Since the survey's inception in 2004, US respondents have rated current industry soundness higher than the past five years but lower than they expect the industry to be five years in the future. This suggests an ongoing optimism about the future of the water industry. However, with three years of data now in hand, a slight but clear downward trend in the overall level of soundness has emerged, indicating that the base of optimism is eroding somewhat from year to year.

This year's survey saw an across-the-board drop in relative soundness ratings by US respondents from 2004 to 2006 in each of the relevant time frames

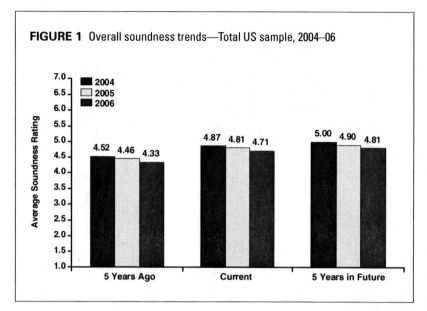

FIGURE 1 Overall soundness trends—Total US sample, 2004–06

Legend: 2004, 2005, 2006

Y-axis: Average Soundness Rating (1.0 to 7.0)

5 Years Ago: 4.52, 4.46, 4.33
Current: 4.87, 4.81, 4.71
5 Years in Future: 5.00, 4.90, 4.81

surveyed (Figure 1). For example, current soundness has slipped from 4.87 (on the 7-point scale) in 2004 to 4.81 in 2005 and 4.71 in 2006. This drop, although not precipitous, nevertheless is statistically significant given the survey's large sample size. Similar declines in past and future ratings are evident as well. Although there is no question that the industry has benefited from its hard work in the past, respondents clearly perceive challenges on the horizon that will command the collective attention of the water community.

Top five issues. Following the soundness question, respondents provided written answers to three open-ended questions asking them to identify critical issues that the industry would face in the next one to two years (near term) and the next three to five years (long term) as well as those issues that were inadequately addressed. Utility respondents were asked a battery of additional questions regarding capital spending trends, technologies, and areas of focus at their specific utilities. The survey concluded with several demographic questions that were used to help categorize responses.

US respondents. US respondents identified the five most critical issues facing the water industry today as

- infrastructure,
- regulatory factors,
- business factors,
- source water supply and protection, and
- workforce.

The compiled survey not only identifies and ranks critical issues but also overlays that information with the additional dimension of whether those issues are adequately addressed (Figure 2). Analysis of these two dimensions zeroed in on several issues that respondents identified as both highly critical and relatively inadequately addressed—infrastructure, business factors, and

source water supply. Though less frequently cited as critical, the issue areas of consumers, workforce, and industry leadership were also deemed relatively inadequately addressed for the level at which they were mentioned as concerns. The critical issue most often cited in the near term—regulatory factors—was judged less of a concern in the future and was seen by respondents as relatively well addressed by the industry.

This year saw a new emphasis on workforce issues, i.e., retaining and replenishing water industry employees. When all responses (near term, long term, and inadequately addressed) were combined, workforce issues edged into the top five areas of critical concern for the first time. Since 2004, the number of respondents citing the industry workforce as a critical concern in both the near and long term has doubled. Furthermore, respondents now consider workforce issues as the fourth most inadequately addressed area. Source water supply and protection and business factors slipped slightly compared with 2005 levels of concern, and security declined slightly as well. Overall, fewer issues were perceived by respondents as inadequately addressed. Most industry issue areas were judged to be more adequately addressed in 2006, with the exceptions of workforce, industry leadership, and energy issues.

Critical issues by demographic group. US utilities (Figure 3) and service providers (Figure 4) both stressed business factors, source water, and infrastructure as critical issues that were not well addressed. Not surprisingly, utilities also reported increasing concern about workforce issues. Service providers also rated workforce as a relatively inadequately addressed issue that will become even more critical in three to five years. About one third of service providers rated business factors as a critical issue in the near term and long term, and more than one quarter indicated the issue is inadequately addressed. Service providers ranked infrastructure second among inadequately addressed areas. For this demographic group, the issue of security has continued to decline, while most other major issues have stayed about even since 2005.

Canadian perspective. Canadian perceptions of the top critical issues paralleled those of US respondents, although some important differences emerged (Figure 5). With regard to the top five critical issues, source water supply was ranked notably higher over the long term and as an issue currently not adequately addressed. Business factors were ranked highest in the near term, with regulatory factors a close second among Canadian respondents. By contrast, in the United States, where the regulatory environment is more complex, regulatory factors grabbed the top position as a near-term issue. Workforce issues were seen as even more critical in Canada than in the United States, and consumer issues were considered inadequately addressed, perhaps a residual effect of the tragic Walkerton, Ont., incident in 2000.

RESPONSES POINT TO DRIVERS BEHIND TOP FIVE CRITICAL ISSUES

A closer analysis of the five issues ranked as most critical illuminates some of the factors driving these concerns.

Infrastructure. Among critical issues, infrastructure already rivals regulatory factors, and furthermore, infrastructure is perceived as a growing concern and one that is relatively inadequately addressed. Combining long-term concerns regarding aging infrastructure (cited by 24% of respondents) with business factor worries about financing its replacement (12%) boosted infrastructure and its associated costs to the top long-term issue at 36%. Taken together, the issues of infrastructure and business factors would top the list of inadequately addressed issues at 28%. Even in the near term, combined concern about infrastructure replacement and financing was the issue area mentioned second most often.

Infrastructure issues and their implications translated into concerns about distribution water quality as well. Clearly, the "aging" infrastructure has become the "failing" infrastructure in the minds of some respon-

dents. A manufacturer from the northeastern United States summed up the issue this way: "Cities and towns are still not able to pay much attention to the crumbling networks of old pipes.

Regulatory factors. Since 2004, respondents have ranked regulatory factors as the most critical issue in the near term. In 2006, regulatory factors remained the most critical near-term concern, but this trend may have run its course. This year, regulatory factors were seen as relatively well addressed at 13%, down from 14% in 2005 and 15% in 2004.

Nonetheless, regulations continue to worry utilities, especially smaller operations. A manager from a drinking water utility in the southeastern United States claimed, "Useless regulations. . . . expend excessive time and money with minimal improvement in water quality. . . . It is critical [that] the water industry not allow regulatory agencies to require utilities to perform duties beyond our control, like the Lead and Copper Rule."

One third of respondents cited complying with new regulations as a concern, and 2% went further and questioned the basis for certain regulations. A US consultant wrote, "The maximum contaminant level goal of 0 ppb for some of those chemicals is totally absurd in the real world. Our analytical ability has far exceeded any common sense." Small systems in particular expressed concern about their ability to meet new and more complex regulations with even fewer discretionary resources compared with those of larger utilities.

Business factors. Business factors encompass a wide array of concerns associated with finance and business, including funding, rates, and privatization. In 2006, responses reflected two dominant themes: (1) financing of repairs, replacements, and/or upgrades to infrastructure and (2) the imbalance between the costs of producing water and the rates charged to customers. A water utility executive from the northeastern United States stated, "The [infrastructure replacement] bill is coming due, and I don't see the resolve from both the utilities and regulators to pay for it." An engineering consultant from the same region expanded on the concern: "Aging infrastructure coupled with chronic underfunding of maintenance and system improvements to accommodate ongoing development will lead to system failures, inability to meet fire demand, and unreliable service."

Almost one third of respondents cited business factors as critical

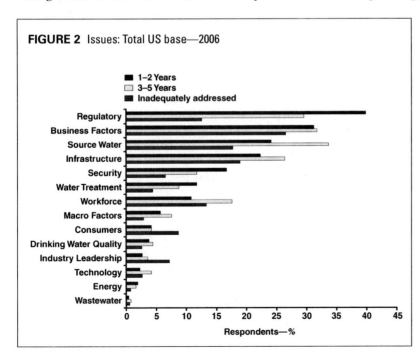

FIGURE 2 Issues: Total US base—2006

Legend:
- 1–2 Years
- 3–5 Years
- Inadequately addressed

(Categories, top to bottom: Regulatory, Business Factors, Source Water, Infrastructure, Security, Water Treatment, Workforce, Macro Factors, Consumers, Drinking Water Quality, Industry Leadership, Technology, Energy, Wastewater)

Respondents—%

issues in the near and long term, with concerns fairly evenly divided between financing and cost imbalances. In addition, 27% of the group believed the issue is not adequately addressed. A manager from a not-for-profit organization in the western United States cited a "need to address issues of securing funding through rates [and the] full cost of service reflected in rates." Another water utility executive in the western United States explained, "The cost of service is low, resulting in a perceived low value of our product—drinking water—which deserves

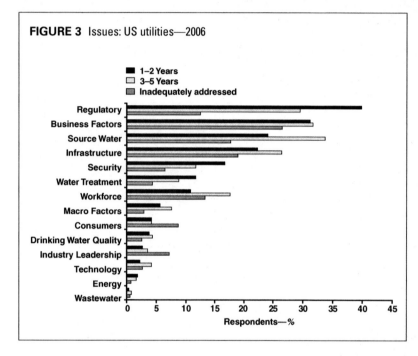

FIGURE 3 Issues: US utilities—2006

an incredibly high value." Several individual respondents agreed, citing similar frustrations over their inability to charge "fair market value" for their product.

Source water supply. In 2006, source water supply and protection were seen as areas of increasing concern, especially among respondents in more arid climates with explosive population growth. Source water supply is one issue area in which concern about the future (34% of respondents) significantly outweighed near-term concern (24%). This suggests that unless significant progress is made, the issue of source water supply will only become more critical over time. A water utility executive in the midwestern United States commented that the next three to five years will see "overgrowth of arid regions. . . . There is not enough water for these regions, and therefore rapid rate of growth is reckless! Sorry, we can't pipe our Midwest water to them. No way." A utility executive in a neighboring state agreed: "[The water industry] cannot allow regions without adequate water supply to continue development. [It is] not a good idea. [There is] too much farmland being developed."

With regard to source water protection, a utility consultant from the northeastern United States expressed concern that "modern lifestyle contamination (degradation) of water supplies [and] reduced water quality from pesticides, pharmaceuticals, [and] runoff" will have a significant effect on the water industry in the next one to two years.

Workforce. Concerns about the water industry workforce have increased significantly over the three years the SOTI survey has been conducted. This growth has been fueled by the overlapping issues of workforce retirement ("brain drain"), ever-increasing requirements for existing workers, and a lack of new qualified and motivated workers in the water industry, a problem frequently attributed to low wages and low perceived esteem in industry jobs. A US water utility manager from the Southwest stated, "Several of us are within sight of retirement, and we are having trouble recruiting high-quality talent to our industry. The technology is becoming more sophisticated even as we are less and less able to attract highly educated and skilled workers to deal with it."

Since 2004, the number of respondents concerned about the workforce in the near term has more than doubled, and one in five utility respondents cited workforce issues as a critical concern in the next three to five years. Workforce factors have become the fourth most inadequately addressed issue facing the water industry today.

A drinking water utility manager from the US Midwest explained the effect of the aging workforce, noting that "not many young people [are] entering the water/wastewater field. College graduates have too many other options that pay better. . . . The workforce is shrinking, and small water systems cannot afford to pay qualified operators the wages they deserve in order to keep them." Among utility respondents, 38% indicated attracting and hiring skilled workers is an area requiring significantly more attention at their utilities.

OTHER KEY ISSUES SPARKED COMMENTS AND COMPLAINTS

Security. Since the first SOTI survey in 2004, security's rank as a critical issue has dropped every year. In 2006, five years after September 11, security fell out of the top five issues for the first time. Only 17% of respondents mentioned security as a near-term issue, and the number dropped to 12% when respondents were queried about the next three to five years. Just 7% of respondents cited security as being inadequately addressed. Still, an industry consultant from the northeastern United States predicted, "Source protection from terrorist attack (either organized or more likely from the 'lone wolf' disgruntled person/former employee)" will significantly affect the water industry in the next one to two years.

Water treatment. Although water treatment continued to be an issue in 2006, near-term concern declined from 15% last year to 12% this year. Similarly, long-term concern fell from 14% in 2005 to 8% in 2006. In 2006, water quality at the point of use popped up on the radar screen as an issue that posed difficulties for water suppliers because of their limited control once water leaves a utility system and enters the end user's system. A few vocal respondents questioned the rationale behind treating all water to drinking-water-quality standards when only a fraction of that water is ingested. One drinking water utility executive from the southeastern United States asked, "How to provide less expensive water for irrigation, car washing [and the like] So much of our natural resources are wasted to provide 'drinking water' quality for these uses." Many of these water professionals suggested the industry take a serious look at developing more point-of-use treatment options.

Macro factors. Macro factors, a category introduced with the 2005 survey, ranked eighth on the 2006 list of critical issues. Macro factors are defined as significant influences from the broader sphere that affect the water industry yet are not controlled by it; factors mentioned included global warming, natural disasters, environmental activism, and population growth. For both US and Canadian respondents, macro factors were seen as more of a long-term concern and one that was fairly adequately addressed.

Consumers. As in previous years, consumer issues were dominated by a few recurring themes centering on consumer education and awareness. The industry is concerned that consumers do not understand the true value

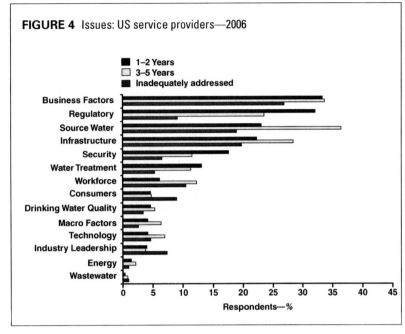

FIGURE 4 Issues: US service providers—2006

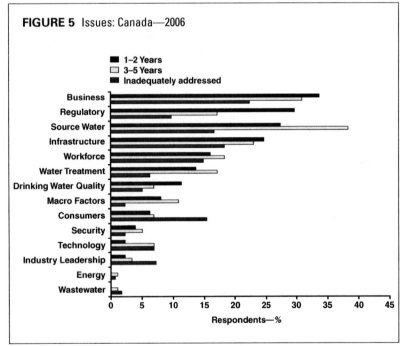

FIGURE 5 Issues: Canada—2006

of water, are reluctant to pay more than current rates despite rising quality and related treatment costs, and lack confidence in the safety and quality of the water supply. Survey responses suggested that bottled water is a lightning rod issue for the water industry. As a drinking water utility manager in the southeastern United States lamented, despite "untrue, unfair, and unchallenged claims by [the] water purification and bottled water industry," the popularity of bottled water has continued to soar, perpetuating myths that bottled water is somehow safer or more regulated than tap water.

Industry leadership. The 2006 SOTI survey uncovered three aspects of industry leadership that concerned respondents: (1) the politics of rates and rate increases, (2) the planning ability to address major industry issues such as infrastructure, and (3) the desire to have regulations more obviously rooted in sound science. An operator at a drinking water utility in the northeastern United States complained, "Competent management is lacking. [We] need to find leaders who focus on water quality and infrastructure rather than politics and money." An industry consultant from the southeastern United States voiced a desire (shared by many) for "good science to support or oppose regulatory requirements."

Technology, energy, and wastewater. Comments on technology focused on specific areas in which improved or more cost-effective technology was needed, in particular ultraviolet disinfection, membranes, and, in coastal regions, desalination. The major thrust of many responses was a hope to use technology that delivers water of increasingly higher quality while avoiding undesirable collateral issues—all at a low relative price. Among the few comments on energy were specific mentions of the soaring cost of energy needed to produce and distribute water. Wastewater issues remained the category least mentioned.

Regional results. The water industry across North America shares many challenges, but some of the issues cited demonstrated strong regional skews. The 2006 SOTI survey results were broken down according to AWWA's key regions:

• *Region 1:* Atlantic Canada, Quebec, New England, Connecticut, New Jersey, New York, and Pennsylvania

• *Region 2:* Chesapeake, Florida, Georgia, Kentucky, North Carolina, South Carolina, Tennessee, Virginia, West Virginia, and Puerto Rico

• *Region 3:* Ontario, Indiana, Illinois, Michigan, Minnesota, Ohio, and Wisconsin

• *Region 4:* Alabama, Arkansas, Kansas, Louisiana, Missouri, Mississippi, Oklahoma, Texas, and Mexico

• *Region 5:* Iowa, Montana, Nebraska, North Dakota, and South Dakota

• *Region 6:* British Columbia, Yukon Territory, Western Canada, Alaska, Idaho, Oregon, and Washington

• *Region 7:* Arizona, California, Colorado, Idaho, Hawaii, New Mexico, Nevada, Utah, and Wyoming (Idaho is shared by regions 6 and 7)

Regional differences were reflected in how respondents ranked critical issues in the near term (Table 1), long term (Table 2), and adequacy of how issues were addressed (Table 3).

Infrastructure, regulatory factors, and business factors—three of the top four critical issue areas—

TABLE 1   Near-term issues by region

Near-term Issues	Regional Response Rate*							Average
	1	2	3	4	5	6	7	
Regulatory factors	98	102	95	103	122	83	109	100
Source water supply	80	91	97	118	85	157	140	100
Business factors	101	107	110	82	99	81	92	100
Infrastructure	105	103	98	69	109	108	98	100
Workforce	83	107	93	97	97	131	139	100
Water treatment	98	104	82	97	118	123	115	100
Security	137	119	92	59	51	38	65	100
Macro factors	81	96	109	109	22	144	144	100
Drinking water quality	71	84	127	107	169	171	83	100
Consumers	125	77	116	56	29	109	85	100
Energy	91	114	129	79	–		164	100
Leadership	153	105	54	97	–	39	104	100
Technology	111	80	132	66	–	79	112	100
Wastewater	76	72		331	269		228	100

*Regions are defined on page 71. Values reflect indexes indicating the regional response rate (frequency of respondents mentioning that particular issue) compared with the response rate for the entire survey. For example, an index of 100 indicates response rates were identical in the region and the survey as a whole. An index of 75 indicates the response rate was 25% lower in the region versus the entire survey, whereas an index of 150 indicates the response rate was 50% higher in the region versus the entire survey.

Numbers in dark grey indicate issues more frequently mentioned by at least 25% (index of 125 or higher), pointing to hot spots or areas in which a particular issue is of greater concern. Numbers in light grey indicate regions in which a particular issue was less frequently mentioned by at least 25% (index of 75 or lower).

TABLE 2   Long-term issues by region

Long-term Issues	Regional Response Rate*							Average
	1	2	3	4	5	6	7	
Regulatory factors	105	94	96	106	108	78	106	100
Source water supply	88	92	95	129	86	130	127	100
Business factors	105	104	110	97	82	75	86	100
Infrastructure	107	100	93	78	107	103	103	100
Workforce	101	117	96	77	71	81	118	100
Water treatment	107	69	102	93	87	166	93	100
Security	118	123	103	59	91	51	68	100
Macro factors	96	112	77	69	16	164	158	100
Drinking water quality	87	92	121	72	83	149	108	100
Consumers	131	66	100	89	83	126	59	100
Energy	58	144	126	101	82	63	139	100
Leadership	92	121	57	116	37	139	170	100
Technology	109	58	105	105	81	146	105	100
Wastewater	114	97	112	100	162	124		100

*Regions are defined on page 71. Values reflect indexes indicating the regional response rate (frequency of respondents mentioning that particular issue) compared with the response rate for the entire survey. For example, an index of 100 indicates response rates were identical in the region and the survey as a whole. An index of 75 indicates the response rate was 25% lower in the region versus the entire survey, whereas an index of 150 indicates the response rate was 50% higher in the region versus the entire survey.

Numbers in dark grey indicate issues more frequently mentioned by at least 25% (index of 125 or higher), pointing to hot spots or areas in which a particular issue is of greater concern. Numbers in light grey indicate regions in which a particular issue was less frequently mentioned by at least 25% (index of 75 or lower).

TABLE 3   Inadequately addressed issues by region

Inadequately Addressed Issues	Regional Response Rate*							Average
	1	2	3	4	5	6	7	
Regulatory factors	89	89	101	124	122	100	122	100
Source water supply	86	91	92	130	87	126	139	100
Business factors	98	106	96	105	113	110	92	100
Infrastructure	118	86	97	88	72	89	97	100
Workforce	108	100	106	94	83	70	94	100
Water treatment	89	114	97	36	115	219	86	100
Security	123	78	113	126	22	67	75	100
Macro factors	96	79	87	54	44	165	185	100
Drinking water quality	138	32	117	101	52	157	44	100
Consumers	86	143	117	60	109	83	82	100
Energy	92	146	75	100	163	125	63	100
Leadership	115	101	80	57	71	121	121	100
Technology	103	97	128	51	41	156	70	100
Wastewater	141	100	178					100

*Regions are defined on page 71. Values reflect indexes indicating the regional response rate (frequency of respondents mentioning that particular issue) compared with the response rate for the entire survey. For example, an index of 100 indicates response rates were identical in the region and the survey as a whole. An index of 75 indicates the response rate was 25% lower in the region versus the entire survey, whereas an index of 150 indicates the response rate was 50% higher in the region versus the entire survey.

Numbers in dark grey indicate issues more frequently mentioned by at least 25% (index of 125 or higher), pointing to hot spots or areas in which a particular issue is of greater concern. Numbers in light grey indicate regions in which a particular issue was less frequently mentioned by at least 25% (index of 75 or lower).

drew similar levels of concern from region to region. However, rankings of other key issues demonstrated a regional bias. For example, source water supply and quality were of more concern in regions 4, 6, and 7—generally western or southwestern areas where arid climates, population growth, and drought tend to intersect. Regions 6 and 7 tended to have a concentration of longer-term and inadequately addressed issues, with region 6 indexing particularly high in terms of inadequately addressed issues. Respondents in region 1 voiced more concern about consumer issues in both the near and long term, but they indicated these issues are well addressed.

UTILITIES SPEAK OUT ON SPECIFIC ISSUES DEMANDING THEIR RESOURCES

As in previous years, many of the questions on the 2006 SOTI survey were geared specifically toward utilities.

Top three issues for utilities. The main section of the SOTI survey asked respondents to indicate critical issues facing the industry, providing an outward look at the industry as a whole. For the utility section of the survey, questions were designed to elicit an inward focus on those particular issues that most occupy the time and resources of individual utilities. Respondents were asked to review a list of potential challenges and rank the top three areas of concern at their facilities.

The critical issues occupying individual utilities tended to mirror those of the industry as a whole. Infrastructure (repair and replacement) was the top focus area and the issue most in need of significantly more attention. Compliance with new regulations was the number two focus area on the list, with almost 29% of utility respondents indicating that regulatory compliance required more attention. Adequacy of water supply and financial issues (e.g., capital needs, water rates, and cost of compliance with new regulations) were ranked third on the list. In terms of areas needing more attention, workforce concerns (i.e., attracting and hiring skilled workers) followed infrastructure, and funding capital needs was third.

Utility capital spending. Utilities were asked to project capital spending for 2006 and 2007 in five major categories: new source water supply, distribution system expansion, compliance with regulatory requirements, compliance with security requirements, and replacement and/or rehabilitation of existing infrastructure not related to changes in regulation (Table 4).

TABLE 4	Utility capital spending	
Spending Category	Projected Percent of 2007 Spending	Percent Increase/Decrease From 2006 Spending
Replacing/upgrading infrastructure	34	+11
Expanding distribution system	24	+16
Developing new source water supply	21	+42
Meeting regulatory requirements	16	−9
Meeting security requirements	4	+16
Total	100	+14

TABLE 5	Utilities' perceived barriers to rate increases
Barrier	Percentage of Respondents
Politics/politicians/councils/boards	26
Public acceptance/understanding of costs	24
Low or fixed income/aging customers	14
Dissatisfied customers	10
Current cost/price of rates	9
Bad economy	5
Increases in all utility rates	4
Higher taxes on end users	3

TABLE 6	Most common customer complaints received by utilities
Category	Percentage of Respondents
Rates	62
Water quality	36
Water quantity	13
Low pressure/flow	12
Taste	12
Water safety/source protection	11
Chemicals (including fluoride, chlorine)	7
Brown/off-color water	7
Odor	6
Maintenance/improvements	6
Reliability of supply	4
Customer service	4
Meter/billing issues	4
Hard water	3
Leaks/breaks	2

Overall, total capital spending is projected to rise 14% in 2007 over 2006, slowing somewhat from the 20% projected increase in 2006 over 2005. Of the total capital spending, 34% is slated for replacing or upgrading existing infrastructure in 2007, an 11% increase over 2006. Utilities indicated that 24% of their projected 2007 spending will be used for distribution system expansion and 21% to fund new source water supply. Of the areas surveyed, only capital spending to meet regulatory requirements is projected to decline.

Rate matters and customer complaints. The survey asked utilities to rate the factors that they perceived as the primary barriers to rate increases (Table 5). Politics and public understanding and/or acceptance were the most frequently cited factors at 26% and 24%, respectively. Utilities recognized that public resistance to rate increases was exacerbated by the fact that many end users are on fixed incomes.

Utility respondents were also asked to list the most common customer complaints they received (Table 6). Rate feedback dominated the list with 62% of utilities reporting complaints in this area, followed by water quality complaints at 36%. The next tier down included operational issues such as quantity (with 13% of utilities reporting these complaints), pressure/flow (12%), and taste (12%).

AWWA USES SOTI DATA TO PLOT A COURSE TO THE FUTURE

When AWWA embarked on the first SOTI survey in 2004, the organization hoped to provide its members with useful information on where the profession was headed and what roadblocks might be encountered along the way. Now, with three years of statistically significant data collected, the SOTI survey is gaining momentum and value as a planning tool. The 2007 survey will mark the next frame in the developing motion picture that will help the water community analyze where it has been, where it is, and where it is headed. By identifying developing trends early and focusing attention on inadequately addressed issues, AWWA is helping the industry define its agenda and plot a course for future growth and success.

ABOUT THE AUTHORS

Jon Runge is director of communications, marketing, and customer service for AWWA. John Mann is founder and president of Mann Consulting Inc., a marketing and research consulting firm based in Colorado.

This page intentionally blank.

Water: Resource or Commodity?

BY JOHN (WOODY) WODRASKA
(*JAWWA* May 2006)

Should water be viewed as a resource to be allocated solely for public benefit? Or should it be considered a commodity, owned by property rights, that can be bought, sold, and traded?

Well-respected experts and interests stand firmly in both the resource and commodity categories—and both camps have valid arguments. Water is essential to life, and public sources of water must be protected. At the same time, a market model offers an efficient means of allocating this vital, diminishing resource.

Rather than placing water under a heading and considering the problem solved, the water industry needs to engage in a comprehensive debate. The following questions should be addressed:

• What water resource management approach will best serve society in the twenty-first century?

• Which new technologies should be embraced?

• Which old concepts should be revisited?

Of course, the answers will not be cut and dried. They will, however, certainly point to a management approach that is somewhere between the extremes of resource and commodity.

A hybrid model called managed stewardship could serve as a jumping-off point for the debate. This new position unites a respect for water with options for fair, efficient delivery to urban, industrial, and agricultural users.

MODELS ABOUND

The standard water industry assumption is that a water shortage—a period with less than average rainfall—occurs in 7 out of 20 years, or about once every 3 years. This rule of thumb holds true for most parts of the United States. Added to the natural reality of periodic water shortages are growing urban demands and calls for water to support environmental mandates.

Given this situation, questions of water allocation have grown more contentious throughout the country. In fact, the marked differences between the eastern and western outlooks on water management have faded, and states on both sides of the Mississippi have now accepted that water is a precious and limited resource.

Water managers nationwide have wondered aloud if water allocation is more a "commoditized" property right or a protected public resource to be managed by government. Numerous water allocation models exist in the United States. Examples include riparian, common law, administrative, and appropriative systems, with hybrids combining several of these approaches.

COURTS ARE LARGELY HANDS-OFF

In the United States today, the overriding trend is moving away from viewing water as a property right and toward protecting the public interest through a regulated or administrative approach. In a famous ruling, the California Supreme Court kicked off this resource point-of-view trend by ruling that Mono Lake was for the public's benefit and use (*National Audubon Society v. Superior Court*). A recent New Mexico Supreme Court ruling (*Turner* v. *Basset*) upheld the state's system of administering water rights for the public's best interest.

For the most part, however, the courts prefer to avoid dealing with the complexities of water issues. Their only actions in the future might be to establish a custodian, as in the case of the recent Arizona-versus-California water debate, in which the US secretary of the interior was appointed water master.

This leaves states with the responsibility for legislating water allocation policies—policies that will attempt to balance urban, rural, and environmental needs while protecting the resource. With such a broad goal, not all stakeholders will be happy with every policy decision. States realize, however, that the time has arrived for tough decisions. They know they must set water allocation policies before water issues become more critical.

If the balance lies somewhere between the extremes of the resource and commodity models, states (and the water industry) must take a close look at both sides of the scale. Manifesting a viable water management approach will require the integration of many components.

GEORGIA PROVIDES RIPARIAN EXAMPLE

The state of Georgia has attempted to balance urban and rural water needs by adopting its regulated riparian water allocation policy. In doing so, Georgia has defined water as a public resource subject to regulation. Water use permits are issued by county. Although Georgia has moved away from supporting traditional riparian rights, the state's water use permits do not take away the riparian landowners' rights to use water.

Can this system work? Although it represents a noble attempt to balance water's characteristics as both resource and commodity, the term "regulated riparian" is an ambiguous one. At best—in the world of water allocation—ambiguity does nothing to ease stakeholders' concerns. At worst, ambiguous terms can lead to unexpected and lasting outcomes.

Take, for example, another area of water policy negotiations. During a water dispute with Native Americans, an agreement was drawn stating that the water could be used for "tribal customs." Water policy planners took tribal customs to mean hunting or fishing, which are uses with little environmental effect. The tribe's intent, however, was to construct a casino—a tribal custom from the Native Americans' viewpoint but one that had untold effects for water policy planners.

SPECIAL CONSIDERATION HELPS ALLEVIATE AMBIGUITIES

A means may exist to alleviate the ambiguity between preserving the water rights of good, long-term stewards while still making clear that water is a public resource. Under this line of thinking, the legal term *usufruct* may come to have an expanded meaning in the world of water rights.

Usufruct describes a provision in which one party owns a resource and another is given the "use and fruits" of the resource—as long as the resource is returned in good condition. Another term for usufruct is common law life use. A state could establish usufruct relationships with long-term riparian property owners who show a good record of water use. These agreements would only require review by the state if the water use changed or if evidence emerged showing the once good condition of the resource was not being maintained.

A usufruct relationship is not the same as that of trustee or steward—both of which are terms for a paid overseer of someone else's resource. Instead, usufruct relationships recognize the property owner and offer benefits from use of the resource for a period of time. This model would improve states' relationships with rural water interests and also make it clear that states own the water resources.

MANAGEMENT BY WATERSHEDS AVOIDS UNFORSEEN CONSEQUENCES

Another potential shortcoming arises when a state issues permits via counties, as Georgia does. At first glance, this administrative model appears logical, even practical. However, water resources such as rivers and lakes may be part of several counties. Tracking water use to be sure the resource is properly allocated in such cases would require compiling and updating information from multiple counties, a task sure to generate bureaucratic headaches, especially in states with a large number of geographically small counties.

From an environmental standpoint, water use permits issued through a county rarely take topographic features such as ridges and natural waterways into account. Thus water use in one county could yield unforeseen, potentially damaging consequences at a location within the same watershed but in a different county.

All the shortcomings of a county-by-county approach are solved when water resources are managed within natural watershed and basin boundaries. Under this method, water allocations and environmental considerations can be addressed more easily and accurately.

CONSERVATION IS KEY

Conservation is a key part of any water management program, whether instituted by states that are managing the public resources or by water purveyors. Although conservation appears to be improving, further cutbacks in water use are a tough sell in the United States.

In developing countries, either the cost of water is high compared with a typical income or the time needed to fetch water from the local stream, well, or water pump is significant. Conservation is a part of life in these countries, not a philosophy. For US customers, however, water bills are a small budget item, and water is as close as the nearest faucet.

Thriving plants and grasses along highways and in parks may belie a water shortage. Citizens may not realize the large volume of water that is needed to maintain the everyday scenery. For instance, the irrigation required to transform subtropical South Florida landscape into a lush tropical paradise can top 500 gpd for each person, compared with the nationwide average water use of 80–100 gpd per person.

Attitudes about water use must change. Across-the-board conservation will be a key component of any water supply management policy in the near future.

SUPPLY AND DEMAND ARE IN CONSTANT FLUX

State regulators cannot anticipate meteorological conditions 20 or 30 years into the future. Therefore the responsibility for another component to water management for the twenty-first century falls on the water supply industry. This group can help alleviate stress during temporary water shortages by finding ways to efficiently shift water supplies.

Investing in practical water storage solutions is one approach. In the past, communities invested a great deal of time and resources into flood prevention. These flood-control systems channeled water away from development, yet they often did not take water storage for future use into account. Today, the water industry is looking at ways to hold on to rainwater for longer periods of time before it is lost to the tides and oceans. In essence, this is the plan for South Florida and the Everglades as well as California's Bay-Delta ecosystem.

These approaches have an added benefit beyond water storage. The lesson learned in both the Florida Everglades and the Bay-Delta is that the environment needs a portion of the freshwater supply if it is to be sustained. Urban water interests around the country have an opportunity to seize the high ground in the inevitable fight that looms over future water allocations. Everyone wants a healthy environment and a good economy, and urban water interests are better positioned to create the collaborative network and alliances necessary for that outcome. Developing such relationships will likely prove to be the skill set most in demand when it comes to solving our water problems in the future.

TECHNOLOGICAL SOLUTIONS INCREASE SUPPLY

Aquifer storage and recovery (ASR), once the topic of sparsely attended water seminars, is now a commonly used tool. ASR helps to balance a region's dry, average, and wet years by storing water underground and then bringing it to the surface when needed. In many cases, depending on scale, ASR is more cost-effective than building aboveground reservoirs.

Desalination can also enhance water supplies—but probably not those supplies intended for agricultural use. Agriculture, to remain viable, must be supplied with lower-cost water. Desalination provides ultrapure water at a higher cost, which could be used to meet manufacturing and possibly urban demands. The technology has become competitively priced for the treatment of brackish water. In addition, the treatment of ocean water is on the horizon as a feasible component of major urban water supply systems—on both the West and East Coasts of the United States.

REDUCING AGRICULTURAL USE COULD HELP

Most water allocation estimates for the United States show that 70%–80% of the country's water supply goes to agriculture while 20%–30% supports urban interests. A means of changing that ratio slightly—to agriculture receiving between 65% and 75% of the water supply during a water shortage—could help a region get through the worst phases of a low-water period.

A solution may also be found in water supply option programs, which are part of an approach that stands more squarely at the commodity end of the scale. As has been demonstrated in practice in Southern California, urban dollars can be effectively used to encourage agriculture conservation. The most vocal criticism of such methods usually comes from the growth management community, which seeks to make water the primary tool for limiting growth.

AN ANSWER IS CLOSE

Is water more of a commodity or a public resource? Everyone in the United States pays for water. At the same time, water is a precious public resource to be managed creatively and carefully.

With diminishing water resources, the practice of applying simple labels has come to an end. An industrywide dialogue is now needed to determine the best means to serve society's water requirements in the most efficient ways possible.

One model for achieving this goal is the managed stewardship approach. Such a plan is characterized by
• a regulatory administrative system that weighs demand against long-term public interest;
• a clearly defined policy that manages water use on a watershed basis rather than by county;
• preservation of traditional riparian rights for long-term, trustworthy water stewards;
• conservation;
• a balance of urban and rural needs based on creative allocation solutions; and
• the support of legislation and established funding.

The last points—legislation and a consistent funding mechanism—are key to upholding a statewide policy and maintaining its operation, especially in the critical years ahead. Possible funding mechanisms include an ad valorum tax converted to property tax and user payment systems.

Water management programs may not fully satisfy every stakeholder. However, in a time of impending water shortages, states, citizens, and water purveyors require the best water allocation policy that can be created. Let the debate begin.

ABOUT THE AUTHOR

John (Woody) Wodraska is a nationally known water resource expert and the national director of water resources for the consulting firm PBS&J, West Palm Beach, FL. During his 30 years in the water industry, Wodraska has worked with sensitive water management issues as both executive director of the South Florida Water Management District and general manager and chief executive officer of the Metropolitan Water District of Southern California.

This page intentionally blank.

About the Author

Steve Maxwell formed Tech*KNOWLEDGE*y Strategic Group (TSG) in 1992 to provide strategic consulting and advisory services to the environmental industry. The firm offers investment banking services, as well strategic and management consulting assistance to a broad array of public and private entities in the environmental and water industries. Mr. Maxwell is involved with initiating and managing successful merger and acquisition transactions, developing more practical strategic planning processes, and helping to implement stronger management organizations for many of the leading companies in the environmental industry. Over the years, Mr. Maxwell has worked with many of the major industry leaders in the water and wastewater treatment business, and he has represented both buyers and sellers in approximately fifty successful merger and acquisition transactions.

Mr. Maxwell has twenty-five years of experience in the environmental and water industries. Prior to founding his current company, he served as Manager of Strategic Planning for Union Pacific Corporation, and Executive Vice President of Recra Environmental, an environmental testing and consulting firm. Mr. Maxwell is the founder and editor of *The Environmental Benchmarker & Strategist,* the industry's most comprehensive source of competitive and financial data. He is also the author of numerous reports on the environmental business and water commerce issues, and is active across the industry, serving on various editorial and advisory boards. Mr. Maxwell holds Masters Degrees in both Public Policy and in Geology from Harvard University, and is a Phi Beta Kappa graduate of Earlham College. He enjoys long distance cycling, photography, hiking and other outdoor activities. He lives with his wife Susan in Boulder, Colorado, and can be reached at (303) 442-4800 or maxwell@tech-strategy.com.

This page intentionally blank.

This page intentionally blank.

This page intentionally blank.